PROJECT **FEASIBILITY**

TOOLS FOR UNCOVERING POINTS OF VULNERABILITY

T0330551

Industrial Innovation Series

Series Editor

Adedeji B. Badiru

Air Force Institute of Technology (AFIT) – Dayton, Ohio

PUBLISHED TITLES

Carbon Footprint Analysis: Concepts, Methods, Implementation, and Case Studies,
  Matthew John Franchetti & Defne Apul

Cellular Manufacturing: Mitigating Risk and Uncertainty, *John X. Wang*

Communication for Continuous Improvement Projects, *Tina Agustiady*

Computational Economic Analysis for Engineering and Industry, *Adedeji B. Badiru &*
  Olufemi A. Omitaomu

Conveyors: Applications, Selection, and Integration, *Patrick M. McGuire*

Culture and Trust in Technology-Driven Organizations, *Frances Alston*

Design for Profitability: Guidelines to Cost Effectively Management the Development Process
  of Complex Products, *Salah Ahmed Mohamed Elmoselhy*

Global Engineering: Design, Decision Making, and Communication, *Carlos Acosta, V. Jorge Leon,*
  Charles Conrad, & Cesar O. Malave

Global Manufacturing Technology Transfer: Africa–USA Strategies, Adaptations, and Management,
  Adedeji B. Badiru

Guide to Environment Safety and Health Management: Developing, Implementing, and
  Maintaining a Continuous Improvement Program, *Frances Alston & Emily J. Millikin*

Handbook of Construction Management: Scope, Schedule, and Cost Control,
  Abdul Razzak Rumane

Handbook of Emergency Response: A Human Factors and Systems Engineering Approach,
  Adedeji B. Badiru & LeeAnn Racz

Handbook of Industrial Engineering Equations, Formulas, and Calculations, *Adedeji B. Badiru &*
  Olufemi A. Omitaomu

Handbook of Industrial and Systems Engineering, Second Edition, *Adedeji B. Badiru*

Handbook of Military Industrial Engineering, *Adedeji B. Badiru & Marlin U. Thomas*

Industrial Control Systems: Mathematical and Statistical Models and Techniques,
  Adedeji B. Badiru, Oye Ibidapo-Obe, & Babatunde J. Ayeni

Industrial Project Management: Concepts, Tools, and Techniques, *Adedeji B. Badiru,*
  Abidemi Badiru, & Adetokunboh Badiru

Inventory Management: Non-Classical Views, *Mohamad Y. Jaber*

Kansei Engineering—2-volume set
- Innovations of Kansei Engineering, *Mitsuo Nagamachi & Anitawati Mohd Lokman*
- Kansei/Affective Engineering, *Mitsuo Nagamachi*

Kansei Innovation: Practical Design Applications for Product and Service Development,
  Mitsuo Nagamachi & Anitawati Mohd Lokman

Knowledge Discovery from Sensor Data, *Auroop R. Ganguly, João Gama, Olufemi A. Omitaomu,*
  Mohamed Medhat Gaber, & Ranga Raju Vatsavai

Learning Curves: Theory, Models, and Applications, *Mohamad Y. Jaber*

Managing Projects as Investments: Earned Value to Business Value, *Stephen A. Devaux*

Modern Construction: Lean Project Delivery and Integrated Practices, *Lincoln Harding Forbes &*
  Syed M. Ahmed

PUBLISHED TITLES

Moving from Project Management to Project Leadership: A Practical Guide to Leading Groups, *R. Camper Bull*

Project Feasibility: Tools for Uncovering Points of Vulnerability, *Olivier Mesly*

Project Management: Systems, Principles, and Applications, *Adedeji B. Badiru*

Project Management for the Oil and Gas Industry: A World System Approach, *Adedeji B. Badiru & Samuel O. Osisanya*

Project Management for Research: A Guide for Graduate Students, *Adedeji B. Badiru, Christina Rusnock, & Vhance V. Valencia*

Project Management Simplified: A Step-by-Step Process, *Barbara Karten*

Quality Management in Construction Projects, *Abdul Razzak Rumane*

Quality Tools for Managing Construction Projects, *Abdul Razzak Rumane*

A Six Sigma Approach to Sustainability: Continual Improvement for Social Responsibility, *Holly A. Duckworth & Andrea Hoffmeier*

Social Responsibility: Failure Mode Effects and Analysis, *Holly Alison Duckworth & Rosemond Ann Moore*

Statistical Techniques for Project Control, *Adedeji B. Badiru & Tina Agustiady*

STEP Project Management: Guide for Science, Technology, and Engineering Projects, *Adedeji B. Badiru*

Sustainability: Utilizing Lean Six Sigma Techniques, *Tina Agustiady & Adedeji B. Badiru*

Systems Thinking: Coping with 21st Century Problems, *John Turner Boardman & Brian J. Sauser*

Techonomics: The Theory of Industrial Evolution, *H. Lee Martin*

Total Productive Maintenance: Strategies and Implementation Guide, *Tina Agustiady & Elizabeth A. Cudney*

Total Project Control: A Practitioner's Guide to Managing Projects as Investments, Second Edition, *Stephen A. Devaux*

Triple C Model of Project Management: Communication, Cooperation, Coordination, *Adedeji B. Badiru*

PROJECT **FEASIBILITY**

TOOLS FOR UNCOVERING POINTS OF VULNERABILITY

OLIVIER MESLY

CRC Press
Taylor & Francis Group
Boca Raton London New York

CRC Press is an imprint of the
Taylor & Francis Group, an **informa** business

Photos and drawings are by the artist Darloz[1] except where indicated.

CRC Press
Taylor & Francis Group
6000 Broken Sound Parkway NW, Suite 300
Boca Raton, FL 33487-2742

First issued in paperback 2021

ISBN-13: 978-0-367-78247-4 (pbk)
ISBN-13: 978-1-4987-5791-1 (hbk)

Library of Congress Cataloging-in-Publication Data

Names: Mesly, Olivier, author.
Title: Project feasibility : tools for uncovering points of vulnerability / Olivier Mesly.
Description: Boca Raton, FL : CRC Press, 2017. | Series: Industrial innovation series | Includes bibliographical references.
Identifiers: LCCN 2016030831| ISBN 9781498757911 (hardback : alk. paper) | ISBN 9781315295251 (ebook)
Subjects: LCSH: Project management. | Strategic planning.
Classification: LCC HD69.P75 M48 2017 | DDC 658.4/04--dc23
LC record available at https://lccn.loc.gov/2016030831

Visit the Taylor & Francis Web site at
http://www.taylorandfrancis.com

and the CRC Press Web site at
http://www.crcpress.com

Dedication

To my mom and dad, who passed away 15 years ago.

Contents

Abbreviations and acronyms

AB	Abilities
ACDI	Agence canadienne de développement international
AFF	Affinity
AIDA	Attention, Interest, Desire, Action
ANS	Autonomic nervous system
AOI	Area of interest
A-R	Resistant, defensive hostile trait, profile, or personality The "no" people
A-V	Avoidant personality (or the "I gotta go" people)
A-X	Anxious personality, also known as the "maybe" people
BDC	Business development center
BV	Benevolence
CAID	Canadian Agency for International Aid (ACDI in French)
CAPM	Capital Asset Pricing Model
CD	Consideration
CFIA	Canadian Food Inspection Agency
CG	Controlled group, when referring to scientific research
CNS	Central nervous system
CPM	Critical Path Method
CS	Contingency strategy
CSI	Construction Specifications Institute
DDM	Requests for changes or *"demandes de modifications"*
DP	Defensive position (prey position) or "lonely sheep position"
DS	Dominant strategy
DSM-IV (or V)	Diagnosis and Statistical Manual of Mental Disorders
EI	Exchange of information
FAST	Functional analysis system technique
FBI	Federal Bureau of Investigation
FEW	Family-wise error
FL	Flexibility
FMEA	Failure Mode and Effect Analysis
Four Ps	Plans, Processes, People, and Power

FP	Forces of Production in general
FP$_c$	Forces of Production under control or the "Fits"
FP$_{nc}$	Forces of Production that are not under control or the "Unfits"
FTA	Fault Tree Analysis
GM	General Motors
***g*-rate**	Rate analogous to an interest rate as applied to a project stage
***g*-spread**	Misstep measuring factor
HCA	Higher cortisol levels after the experiment
HGA	Hypothalamic–pituitary–gonad axis
HPA	Hypothalamic–pituitary–adrenaline axis
ICAO	International Civil Aviation Organization
IN	Integrity
IOC	International Olympic Committee
IP	(Instrumentally) hostile position, predator, or "hungry tiger position"
k	Constant, 1.3
KFF	Key Failure Factor (or Key Failure Fundamentals)
KPPP	Known predator–prey position
KSF	Key Success Factors (or Key Success Fundamentals)
LCA	Lower cortisol level after the experiment
MBTI	Myers–Briggs personality test
MOS	Montréal Olympic Stadium
MP	Means of production
MSDS	Material safety data sheet
NCC	National Capital Commission (Ottawa, Canada)
NHL	National Hockey League
NO	Nitric oxide
OBS	Organization Breakdown Structure
OG	Olympic Games
OPA	Optimal Path Analysis
OPM3	Organizational Project Management Maturity Model
P & L	Profit and Loss statement
PAG	Periaqueductal gray (brain region)
PBS	Product Breakdown Structure
PFT	Functional and technological program (*"Programme fonctionnel et technologique"*)
PIGS	Project implementation groups (!)
PIP	Profile of Instrumentally hostile individuals
PMBOK	Project Management Body of Knowledge
PMI	Project Management Institute
PMP	Project management professional
POE	Point of Equilibrium

POV	Point of vulnerability
POW²	Product, Organization, Work
PPP	Private–Public Partnerships
PR	Joint problem resolution
PRO	Pessimistic, Realistic, Optimistic scenarios
PWP	Work psychodynamics (model)
QMA	Québec Multifunctional Amphitheatre
$|R$	Essential and nonessential resources
R	Essential resources
RIO	*Régie des installations olympiques* ("Olympic Games Steering Board")
R_n	Nonessential resources
SCC	Standards Council of Canada
SKU	Stock-Keeping Unit
SOCAN	Société canadienne des auteurs, compositeurs et éditeurs de musique (music, writer, and editor association)
SS	Short strategy
ST	Summative Triangle
SVOR	Strengths, vulnerabilities, opportunity, and risks
SWOT	Strengths, weaknesses, opportunities, and threats
$|T$	Work and Technology
T	Work (*travail*)
TCI	Triple constraints of information
TG	Trained group
T_h	Technology, knowledge
TSO	The Stationary Office
UG	Uncontrolled group, when referring to scientific research
UPPP	Unknown predator–prey position
UTG	Untrained group
WBS	Work Breakdown Structure
WIP	Work in Progress

Preface

I believe this book to be like no other on the subject of feasibility studies. First, there is scant literature on the subject of project feasibility and even fewer writings on the concept of "points of vulnerability" (POVs). Yet, if I take a second to think about it, points of vulnerability permeate the evolution of mankind and of nature at large. Nature should have equipped man with a third eye, not on the forehead, but on the back of the skull—perhaps Caesar, given his dramatic death, would have been grateful for having been born with such an asset, although it is not clear whether Cleopatra would have found it attractive or not! A third eye would ensure that a man could see who is stalking him or at least guess who is closing in on him, perhaps with threatening behavior (e.g., Brutus). Or else, nature would have equipped man with a rabbit's eyes or the owl's capacity to make an exorcist-like twist with his head so that he could see all around and not be exposed to surprise attacks. Regardless of the prowess that man would be capable of, there would still be points of vulnerability: man would not run fast enough from danger (e.g., predators) or else he would not tolerate sunlight. Whatever form man would adopt, he would have points of vulnerability. This makes for predator–prey dynamics: predators (dominants), just like chess players bet on their opponent's vulnerabilities and force these vulnerabilities to emerge. POVs are costly: in chess, when taken advantage of by the opponent, they lead to defeat.

Put in the context of U.S. politics, recall what Colin Powell said: "(...) we will try to take advantage of whatever weaknesses exist there and play to our strengths and not act against their strengths."[3] Put in project management terminology, there are strengths (forces, which guarantee dominance) and weaknesses (which can be allayed by way of contingency planning) in any project, and a fair amount of risks that are not too distant. We can view POVs as performance gaps: they are costly, use time, and monopolize team members when they should be concentrating on their normal tasks, including ensuring quality outputs. A recent report found that 74% of the 1662 executives surveyed considered that success was achieved in a transformation initiative when both strengths and performance gaps had been assessed.[4] Certainly, identifying POVs implies some expenditures; but the costs of conforming to the highest standards (e.g., observations, studies, surveys, and training) must be weighed against the costs of nonconforming (dead inventories, lost opportunities, rework, scrap, warranty surcharge, to cite only those). Managing POVs means improved productivity, no less.

I will talk a lot in this book about Dominant and Contingency strategies. In short, a **Dominant strategy** (DS) is one that allows a project manager to complete all mandatory sets of tasks, to drive the organization to the level that has been predetermined, and to deliver the final product within the specific constraints of time, costs, and quality.[5] An example of an attempt to reach a Dominant strategy drawn from U.S. politics is the tug of war between the president, Congress, the Supreme Court, and the House of Representatives.

A **Contingency strategy** (CS) is one whereby the project leader is getting prepared to face risks and vulnerabilities in the worst of scenarios. PMBOK 5 (p. 119) refers to "preventive actions (as) an intentional activity that ensures the future performance of the project work is aligned with the project management plan (…)." A **Short strategy** develops when the manager is caught by surprise and loses control; he feels that he is outsmarted by uncontrollable human behaviors or naturally occurring events. Short strategies have a good side; failing everything else (Dominant or Contingency strategies), at least the manager can fall back on them to cope with unusual, threatening events. In cost analyses related to project management, a Dominant position is equivalent to a normal cost structure, a Contingency position corresponds to Contingency reserves, and a Short strategy to management reserves.

You may also think of this as a football game. The Dominant strategy is composed of all the meticulously planned and executed step-by-step approaches toward the goal: scoring deep into the opponent's part of the field. The Contingency strategy is put in place when the quarterback is under pressure to throw the football; failing that, he may actually get hurt by a hard hit from the rushing adversarial team and lose ground. The Short strategy is resorted to when the opponent has managed to catch your football and starts running, from your perspective, like a chicken (or a game perhaps) with its head cut off deep into your zone. Obviously, you'd rather have the chicken (or the game) on your plate than in your game! Throughout the game, team spirit is of the essence.

Dominant and Contingency strategies are what the project leaders are after, as they must assess, audit, and forecast numerous activities and tasks in order to bring a project to satisfactory completion.[6]

The literature on project management likes to delve into concepts such as mature organization and agile management. While it is fine to be agile (especially in the IT industry), to take only this example, from a POV angle, it is much more efficient to be robust. Being **robust** means being able to deal with points of vulnerability, whether this entails being agile at times or else standing firm in front of adversity.

A project leader I interviewed has explained vulnerability, as follows:

> When my father started this company at the age of 50, he faced quite a high level of risk, which we now call the "1-1-1." He had one client, a single product (algae) and only one market. He soon realized he had to diversify in order to mitigate the risks (and vulnerabilities) and in order to be less dependent. The fact that our initial product was a commodity made the small company vulnerable: it was subject to market forces, which decided on the price of the algae. We developed value-added products which entitled us to justify our prices, thus reducing our internal vulnerability.[7]

As the reader can judge, I have strong opinions about project management, and I'll endeavor to prove them in this book with numerous research results and analyses. The reader may find that I challenge some of the preset ideas or myths—such as project managers being free of biases, or the economy at large being an open system—but I'll try to provide support for my points of view.[8]

Secondly, while most books on projects, and project feasibility in particular, address the aspects of finance, technology (engineering), marketing, or human resource management, the present book adopts a very wide view of projects. I dig into modeling, mathematics, human behavioral sciences, neuropsychology, and many other fields. I see projects as multifaceted objects that must necessarily be examined with intense and diversified scrutiny. Thus, I will refer extensively to research of my own, including in neuropsychology, to substantiate some of my arguments.[9]

How the human brain functions has been just as important to understand as is the means by which human behavior forms or how a bridge linking two cities is built. I acknowledge that human behavior is at the heart of projects, and that it is a fine-tuned mechanic that can be examined with some kind of an engineering approach. It is very simple: a tiger that is hungry will hunt and kill; it is programmed to act this way. A newborn baby naturally looks for his/her mother's milk. There are a large number of behaviors that can be predicted or at least partly anticipated because they obey a certain mechanical scheme, which is anchored in the brain (e.g., in the hypothalamus). Hence, from my perspective, preparing a project feasibility study includes understanding how teams work from a *human* perspective (not merely from a human resource management perspective), how decisions are made with inherent biases that can at times compromise a project, and how plans and processes are conceived in order to build, say, a brand new commercial center.

Feasibility studies are not new: in the late 1950s, Khrouchtchev's son, Sergei, asked his chief commander in East Germany to do a feasibility study regarding the closing of the frontier between East and West Germany (the German wall was eventually put up on August 13, 1962, and lasted 40 years). Everybody knows about the consequences on people's lives and international politics following the erection of the wall.

I invoke the fact that organizations nowadays are no longer open systems; rather, they are **closed dynamic systems**.[10] First, they must protect their intellectual property[11]; secondly, the world has now completed a full loop during which all geographical areas have been explored and are open for business—what happens in one country affects what happens in many more[12]; finally, the Internet has made the world instantly reachable, so that each market player can retaliate against the other in a matter of seconds,[13] one way or the other, for example, by running a denigrating viral video. We will see how this concept influences the way managers do their jobs; in short, companies are no longer walking in an open field, they are belted by regulations, markets, and capacities. They are in the midst of a bounded project's reality, better yet, of bounded projects' realities. Nowadays, all business is a series of projects. Thus, this book extends beyond the scope of projects and may provide, I would hope, some useful tips on daily operations that can be regarded as a continuum of projects. In fact, the last case in Chapter 7—"BB's highs and lows"—should convince the reader that the knowledge transferred in this book can be readily used to analyze business operations and projects after they have been completed. I do not pretend, however, to act as management guru; my sole objective in this regard is to provide advice that may or may not be deemed useful, hopefully without sounding pretentious.

To assist me in my various demonstrations, I resort to a fair number of real-case scenarios (some of which have been slightly modified to conceal the identity of the market players when this was demanded of me),[14] both within the various chapters that this book contains and in the "case studies" at the end of each chapter.

Overall, humans have vulnerabilities inasmuch as machines and equipment have limits; both can fail, break, or not deliver as expected. As mentioned, I am fortunate enough to have had the chance to study a slew of groups participating in different projects during the last seven years, and to have been able to distribute questionnaires and conduct interviews that have allowed me to make general statements, and to develop a working model accordingly. Analyzing my database, I have been able to pinpoint some behavioral components that I believe are intrinsic to projects, and through interviews with project managers I have been able to substantiate my findings qualitatively. Statistics in project management are, indeed, a tool that can be used, including with respect to planning, estimating, and monitoring deviations.[15]

This book talks about important psychological phenomena, and especially about behaviors that steer projects in one way or another. The project is assumed to be well planned and to proceed smoothly toward success. However, some team members experience a certain level of apprehension; they may display, for example, some level of fear that the project will not be brought to term, think that their manager is not out there for their best interests, or hold the belief that some of their colleagues play obnoxious games behind their backs. For a project to be a change process, management has to prevent team members' resistance; to put it differently, members must comply with the schedule of activities and calendar of tasks, do all that they can to limit the costs and losses, and adhere to the preset norms of quality. These managers juggle between external risks and internal vulnerabilities and try to maximize the utility of their actions and decisions. In particular, they must keep the project under control, which means that they have to have the proper management skills, especially in consideration of potential or real difficulties with some employees, who resist or perform poorly, voluntarily or not.

The work culture is based on trust. Team members also want to be treated fairly. When there is trust and a sense of fairness, there is no reason not to cooperate, and this in turn encourages commitment. A team spirit is formed; everyone assumes his/her role and respects the plan, much like members of a football team. As the project evolves, management and team members get more and more satisfied with their progress; yet, they must remain vigilant at all times to avoid slip-ups. Overall, team members must learn to work with each other and find a point of equilibrium where everyone can perform to the level that is expected of them. At times, however, overly ambitious and optimistic individuals get in the way, or else panic kicks in and a negative mood prevails because the project looks less and less likely to see the light of day. Occasionally, a manager or a team member even becomes hostile. Some team members seem to drag their feet, others find any possible excuse not to do their job. Does it sound familiar? Sure it does, we have all lived through this; chances are that we experience this every week during work, be it project driven or not.[16]

The last argument about the virtues of the present book is that it provides what I believe to be a unique view of project feasibility while attempting to contrast the viewpoints of various experts and sources of reference. In fact, my references delve into three languages in which I publish and teach: English, French, and Spanish.[17] I believe that this widens the analytical perspective for the readers' benefit. I resorted to a research method called "data percolation" by which information is retrieved from five independent sources: expert opinions, literature, simulation, qualitative, and quantitative studies. I feel that this approach brings in a rich and pragmatic view of the subject. Everyone has seen one of those classic movies in which a car moves forward while its wheels move backward due to a stroboscopic effect. We do not want to be tricked by the effects of the reality of projects. When it comes down to feasibility analysis, it is not the best-fitted project that matters, but the one with the best grip on its points of vulnerability.

Many projects fail for having been erroneously evaluated, if it were only based on misjudging team members and project managers interpersonal competencies. The *Project Management Body of Knowledge*[18] (PMBOK), which is a standard book in the field of project management, does not deal in depth with the sometimes dreadful dynamics that bind stakeholders together. Field and academic experts point to the fact that divergent interests on the part of stakeholders are a major risk (for external stakeholders) and vulnerability (for internal stakeholders).[19] Some authors have also noted that, too often, it is technical competence that prevails over the interpersonal kind.[20,21] Yet, vulnerabilities (created by humans), if mishandled, soon enough lead to crises (and potentially heavy losses), if not to chaos altogether. The goal of this book is to offer the reader—business owners, project managers, project promoters, and students at the master's and doctoral levels—tools to identify POVs and ways to deal with them.

We cannot blame computers or machines for errors. We must instead turn to the human factor for the real answer. In the movie *2001 A Space Odyssey*, the space shuttle computer makes an error that the crew notices; crew members decide to act on it with dire repercussions as the self-conscious computer realizes its useful life is about to end. That computer was programmed by humans. Humans make mistakes and shape POVs. This is true at large and truer in a context of time pressure and stress as is the case when having to complete a project on time.

The effort of capturing what can go wrong in any particular project is worthwhile, even though it can be painstakingly demanding (and may appear depressing ... but it is actually fun!). Project management is big business. Global construction business alone has been valued at US$8.5 trillion as of 2015.[22] Between 2012 and 2022, it is estimated that 78,200 new construction managers will be needed in the United States alone.[23] Still in 2015, U.S. residential and nonresidential construction will amount to a whopping US$612 billion. The value of U.S. residential projects grabs a portion of US$354 billion of that amount, growing steadily since the posttraumatic effects of the 2008 predatory mortgage crisis. Since 1947, the cumulative number of projects supported by the World Bank amounts to 12,372, which have been sprawled across over 173 countries.[24] In 2014, private investments in infrastructure projects totaled over US$50 billion worldwide. Montréal's new Champlain Bridge is just one of several major infrastructure projects in the planning stages across Canada that could help boost the fortunes of the nation's engineering and construction firms. In 2015, a consortium led by Canadian engineering giant SNC-Lavalin Group Inc. was awarded a contract worth approximately C\$3 billion to build and manage this new bridge, a necessary piece of infrastructure that connects Montréal to its south shore and the roads accessing New York. Elsewhere in Canada, projects include the Eglinton Crosstown light rail transit in Toronto (approx. C\$5 billion) and a hydroelectric generating station in British Columbia (approx. C\$9 billion).

A project expert[25] whom I interviewed provided this example:

> I recall this particular project whereby a new building was to be constructed across the street from an existing building, which had become too small to accommodate the ever growing staff. It was of prime importance to verify that the staff in either building could communicate efficiently between them as the core of the business was about IT services to the community.
>
> Two of the four stories of the new building were to be rented, and two were to be occupied on September 1, 2009 by the staff. The plans were ready by August 30, 2008 and construction was to end by August 31, 2009. Moving from the old to the new building would have to be completed by September 30, 2009 in order to minimize the impact on the existing clientele.
>
> Two project teams were formed. The first one looked after the physical layout of each office in the new building. It was headed by a 20-year project management veteran. The second team was responsible for the technological infrastructures.
>
> By mid-August of 2009, 2 weeks prior to moving, everything looked in marching order when it was realized that, while all the equipment had just been installed and cabled inside the new building, there had not been a plan to connect the various pieces of IT equipment between the two buildings! This would affect the accounting department (and with it the pay to the employees) among other work teams. No one had noticed that there had not been a provision to dig the necessary channels underneath the building to install the IT cables. While there

were still 2 weeks left prior to the project's completion, correcting the situation would not be easy. Underground channels are subject to various laws and regulations set by governmental authorities, so that obtaining the necessary permits takes time; it also involves various outside suppliers on which the teams had no control, of course.

A short-term solution was adopted whereby a wireless system was installed. It is not until 2010 that the project was actually completed, with its proper IT cabling system finally in place. Said one engineer: "We dealt with this predicament temporarily, but still, how could well-trained experts manage to miss such an important element of the project, and for so long? This baffles me. The project surely was feasible; it wasn't rocket science after all. Nevertheless, its design contained a big hole. Perhaps it wasn't well defined in the first place." So no project is perfect, I suppose.

There is money to be made … but there is money that may be lost, in Canada, in the United States, and anywhere else in the world. Assume conservatively that projects present a 50% chance of suffering from POVs and that each POV costs, say, 0.01% to the project. On a US$ 3 billion project, this represents what could be the salary of one middle- or top-level manager who would have been able to prevent these very POVs from sticking out like a sore thumb. I suppose that it is worth sifting through POVs before they become a problem, and I speculate that not 50% of projects but, in fact, 100% present POVs. I will discuss this further in the general introduction of this book.

In short, all of the aforementioned projects contain points of vulnerability. POVs are no mystery.[26] Despite the best efforts made at planning projects, soon enough, shortfalls appear (e.g., in administration, human resources, or technical procedures; in building team spirit; in the capacity for managers to influence and impose authority or credibility upon staff, clients, and the general public; in change procedures; in endeavors requiring coordination; in relationships between stakeholders; in getting support from parent organizations; and in progress/status reports), or else excesses develop (e.g., in bureaucracy, complexity, insecurity, and structuring).[27] Shortages and excesses plague projects as inputs are transformed into outputs, as management and team members (Forces of Production, FP)[28] work with Means of Production (MP) to produce results, that is, deliverables.

Despite the fact that everyone knows that a substantial portion of projects experience considerable difficulties and fall short on promises, it seems that the concept of project feasibility remains largely underdeveloped. Yet, feasibility studies serve a number of goals, including evaluating opportunities for employment creation, forecasting market trends and innovation, giving a new life to once obsolete resources, improving standards such as security standards, performing financial and economic assessments, supporting government strategic business or community development (including antirecession measures), upgrading infrastructures, and so forth.[29]

POVs represent costs: they demand that meticulous planning be set in place and that monitoring be emphasized for the entire length of the project. When POVs develop and start affecting the project, they call for corrective actions. In the worst-case scenario, they negatively affect the calendar of activities, the costs, and/or the quality,[30] or a combination of these three core aspects of the project. Not taking into account POVs is tantamount to poor loss management. In fact, I propose that if PMBOK wanted to keep the current logic it uses for defining domains of knowledge, POVs should be added as one of them.

I do not pretend to compete with the PMBOK but merely wish to add to current knowledge. In fact, PMBOK 5 (p. 130) refers to such tools as Failure Mode and Effect Analysis (FMEA) and Fault Tree Analysis (FTA). However, I feel project feasibility analysts can be better equipped than they currently are. I do not contend either that what I present is the absolute truth—although it'd be great to hold such power!—or that I am inventing anything revolutionary. I humbly present some of my findings, hypotheses, and ideas and hope they can be useful.

My objective is somewhat demanding because I look at the dark side of things, with respect to Plans,[31] Processes, People,[32] and authority (Power). It's like being a forensic expert for a "beast" that is human, machine, and process based all at once, and to constantly probe the bad side of things. Maybe, at times, it's like playing detective Eliot Ness or the socio-psychiatrist Carl Jung. Fortunately, each project being unique, there is something new to discover around every corner.

I believe that developing tools and measurements that can enhance the evaluation of projects has its merits, and that this effort should include how humans interact strategically from conception to completion of a project.[33] Indeed, PMBOK[34] recognizes the importance of interpersonal competencies and offers a rough view on it.[35] As part of what it calls general knowledge in project management, it refers to interpersonal and managerial competencies, knowledge of the environment, norms, and rules.

In my opinion, a feasibility study must preclude the emergence of potential problems, refine its scope of analysis until all aspects of a project are deemed measurable, and help determine, in a credible way, whether a project should go ahead, be hunkered down, or else abandoned. In other words, the feasibility expert wants to identify ahead of time where the project might fail; that is, he/she wants to uncover its POVs, including those pertaining to team members and persons of authority (Power). A good feasibility study is also able to define a project succinctly; it uses proper modeling to unveil a project's components and their interactions, and it establishes scales and measuring indices.

Because this book walks on a different path than most books on project management, it cannot be used to pass Project Management Institute (PMI) certification (e.g., Project Management Professional [PMP]). However, it is my hope that the knowledge I provide will be deemed, one day, an essential part of project management and perhaps of contemporary management.

I provide some cases that are based on real business events and companies. In some of them, I have modified names and locations to preserve confidentiality as requested by the companies and people I have met. I have included some elements from the PMBOK so that they can serve as exercises for the theory that is taught in this book. This includes the project charter, the domains of knowledge, and the groups of processes.

Finally, please note that I favor a nonacademic tone and the use of humor at times. I believe that a few words in jest as well as expressions such as "hungry tigers", "lonely sheep", the "I gotta go" people, and the "no" people, as well as "bad apples", can serve as mnemonic devices for a subject that is, occasionally, a bit raw. In my opinion, active sentences and a writing style that offers a source of motivation to continue reading the book are important features of any text. So there you have it: a project is at times a bunch of hungry tigers and lonely sheep, including "maybe," "no," and "I gotta go," people, and bad apples that must be managed to ensure success!

What we have learned about POVs: Preface

POVs …

1. Make for predator–prey dynamics
2. Are costly
3. Are tantamount to poor loss management
4. Are shaped by humans

Author

Olivier Mesly[36] **(MBA, DBA)** is a business analyst who also teaches project feasibility and international project management in Canada (in both English and French), Latin America (in Spanish), and Asia (in English) at university level. He is a member of the Laboratory of Psychoendoneurocrinological Analyses of Stress and Health (LAPS) located in Canada.

Over the years, Professor Mesly has been responsible for launching new products in national and international markets and has investigated over 40 different teams assigned to a wide variety of projects. He acts as a consultant for various organizations, such as local government-sponsored development agencies as well as independent businesses. Professor Mesly completed a postdoctoral fellowship at *Hautes études commerciales* (HEC) *de Montréal*. He graduated as a doctor of business administration in a record time of two and a half years. He holds an MBA in international marketing from Guelph University, a BA in Japanese studies, with distinction, as well as a diploma in public relations from McGill University.

He has written several books, various case studies, and more than 30 peer-reviewed scientific articles (in English, French, and Spanish), a brief summary of which is contained in the present book.

Acknowledgments

I express my gratitude to Jean Rochette, chief project manager of the City of Québec's 400-million-dollar Multifunctional Amphitheatre, as well as to Clément Demers of the *Quartier international de Montréal* (a project that received the *Project of the Year Award* given by the Project Management Institute in 2005), and to Michel Dubuc. My appreciation goes as well to the many businessmen from the Nova Scotia community, especially along Bay Ste-Marie in Nova Scotia—Jean-Paul Deveau, André LeBlanc and his team, Graham Oakley, Brian Saulnier, David Saulnier, Gilles Thériault—and to my students from all over the world. A special thanks to Mount-Allison University, where I teach, and which has been named the best Canadian undergraduate university 17 out of 23 times by the influential *Mclean's* magazine, for its financial support in some of my research.

Thank you also to the many contributors who have generously provided pictures and/or who have agreed to discuss some of the topics covered in this book: Daniel Bourgoin, Daoust Lestage Inc., Éric Déry, André Girard, Stéphan Poulin, Pierre Martin Tardif, and Gaston Thérrien. Grégory Huchet has generously assisted me with the mathematics found in this book. Thank you also to my colleagues from various universities who have helped me put my ideas together: Julien Bousquet (*L'Université du Québec à Chicoutimi*), Michel Noël (University of Sherbrooke), Hélène Plamondon and Andra Smith (University of Ottawa), Stéphane Potvin (University of Montréal), and François-Éric Racicot (University of Ottawa). I extend a special word of appreciation for those who have supported me during the year I spent writing this book: Maria Arruda, Olivier Beaudoin and his team, Alain Belliveau, Stéphane Bouchard, Sophie Boyer, Jean Lepage, Michel Paiement, Karinka Tremblay, Marthe Saint-Laurent, Rick Thériault, and many others, including Me Daniel Johnson, former Prime Minister of Québec, who provided me with sound advice. I have been lucky to get to know all of these people.

I wish to mention that some portions of this book are inspired from my book on the same topic published in French by the *Presses internationales Polytechnique* in May 2015. This publisher worked diligently to bring the French book to the market, provided me with incredible support, and dealt very diplomatically with the fact that I suffered, at the time of the final editing of the book, from my third concussion in four years, which brought me many headaches. It is thanks to them, in large measure, that I came to be offered an opportunity to write a new book on project feasibility, in English this time, for the reputable publisher at Taylor & Francis. As if it were not enough, I also suffered another sporting accident during the writing of the present book, resulting in yet another concussion. I now qualify to be a professional hockey or football player, which will help my bank account!

Seven case studies

1. Case study Chapter 1: Maine East Pharmacy (MEP)
2. Case study Chapter 2: Recycl'Art
3. Case study Chapter 3: AF Thériault (AFT) and the Hammerhead military target boats
4. Case study Chapter 4: Sea Crest Fisheries (SCF)
5. Case study Chapter 5: Africa versus Haiti
6. Case study Chapter 6: The MID: The best managed project in the world
7. Case study Chapter 7: BB's highs and lows

Notes

1. I use [...] as an equivalent to (...), again to make the text clearer when needed (especially when presenting formulae within a sentence to maintain clarity).
2. I specify where I have translated from French to English.
3. I fully write out acronym with acronym in parentheses, and then use only the acronym thereafter.
4. I set some acronyms to be in the plural form unless indicated otherwise, so I don't add an s to them, again, to ease the reading. An example would be key success factors (KSF).
5. Names of people and places refer to actual people and places, except where indicated.
6. Generally, I identify a construct by starting its name with a capital letter (e.g., Trust). When a capital letter is not in use, the term is not deemed to be a construct but simply a normal word.
7. When a new concept is introduced, I put it in bold.
8. I refer to some tables, graphs, and expressions on a limited number of occasions to facilitate the reading, so that the reader would not have to go back and forth looking for information.
9. At times, I do not name the axes on the graphs, but rather include the information in the text; this is because some graphs are actually multidimensional. The reader can then use the name of the axes that are relevant depending on which curves in the graph are being considered. This multidimensionality is feasible because all curves revolve around a constant, named the k constant.
10. The curves are drawn in a simplified, stylized manner to ease the reading.
11. The questionnaire presented in Appendix 5.2 cannot be used without the author's written permission.
12. The data set and questionnaire on psychological constructs can be made available to the reader upon request.
13. None of the remarks or critical commentary are meant to be derogatory; rather, they must be taken constructively as a commendable effort to increase knowledge.
14. I have made the best efforts to substantiate my hypotheses and to express my findings in the most accurate manner.

Expectations

The present book offers tools for conducting prefeasibility and feasibility studies, ranging from identifying risks and points of vulnerability, establishing procedures for ensuring success, and recognizing the true potential of given proposed projects, to structuring a feasibility analysis in order to make effective decisions. Risks, as any project manager knows, cannot be ignored. For example, in the case of the Québec Multifunctional Amphitheatre project (QMA), which I discuss in the introduction, a special reserve for risks was initially set at 2 million dollars (C$),[37] with an additional 8 million dollars also put aside for construction contingencies (these numbers eventually changed). Risks may even require protection against rare occurrences: for example, the QMA was designed to sustain soil elevation pressures and seismic tremors. Risks and POVs are, however, two different concepts. This book clarifies this fact.

Feasibility haunts every project: in the QMA case, for example, two concepts were debated with respect to the choice of the main roof. In the end, it was decided by the City of Québec that a steel structure was the logical and realistically feasible choice (by contrast, the roof component turned into a nightmare for the Montréal Olympic Stadium—the MOS—also discussed in the introduction).

This book introduces a number of new concepts, including the four Ps of projects (Plans, Processes, People, and Power), the g-spread concept, the notion of POVs, the prefeasibility index, the six core competencies model, the six laws of project feasibility, the six Ps of strategic management thinking, and the k constant.

QMA upon completion.

I expect that the knowledge I share in this book will help project managers in reducing catastrophic cost miscalculations, counterproductive tensions among stakeholders, missteps in the conception stage, as well as organizational blindness and delays that could have been prevented. I also hope that company leaders will endorse the view that, nowadays, with the constant flow of innovation made necessary to compete on the international stage, there is no choice but to engage in projects and to do so wisely. More precisely, I theorize that the economy has changed and continues to change: we are entering into a closed dynamic system whereby business is nothing more and nothing less than a series of projects, confined by the limits of calendars, costs, and norms of quality. The free operation of economic laws without restrictions and regulations has led in the last centuries, particularly in the United States, to the birth of mammoth organizations, thanks to a large supply of cheap labor (immigrants) and the availability of raw material (wheat, cotton, iron, etc.). However, poor and unsafe working conditions as well as the competition between humans and machines, all eventually nurtured the grueling rise of the labor movement, which then turned into a mammoth machine in itself, especially with the 1933 passage in the United States of the National Industrial Recovery Act followed by the Wagner Act, which established the National Labor Relations Board. Similarly, the Sherman Act, aimed at curbing abuses in commercial practices, took effect at the end of the nineteenth century (1890).[38]

There is no way around it: the economy lives within frontiers and expands beyond them. More and more, the economy is moving from an open, static environment to a global closed dynamic system. Indeed, projects can serve as an antirecession measure fostered by governments (e.g., highway construction).

Finally, I hope the reader will appreciate this book for its theoretical content as well as for the many concrete examples it provides, in a written style that I anticipate the reader will find concise and pleasant at the same time.

Endnotes

1. See www.darlozart.com.
2. Nothing to do with "Prisoner of War" (POW).
3. U.S. Department of State (2001).
4. PM Network, February 2016, 30(2), p. 17.
5. In my view, quality is the result of the proper workings of the 4 Ps—Plan, Process, People, and Power—which I will discuss further in this book.
6. We will see that a Dominant strategy mixed with a Contingency strategy is equivalent, from a psychological point of view, to stability, or put differently, to a proper balance between Hostile and Defensive positions.
7. Deveau, J.P. Interviewed March 1, 2016.
8. I include a minor critique of the famed PMBOK and related material.
9. Mesly (2010, 2011, 2013, 2014, 2015a–c).

10. I view projects as closed dynamic systems and in this respect, we oppose PMBOK 5's (p. 48) statement to the opposite effect: "Projects exist within an organization and do not operate as a closed system." Indeed, projects are self-sustained entities that have all the characteristics of closed dynamic systems.
11. Apple is a case in point.
12. The 2006–2008 U.S. predatory mortgage crisis, which spread all over the world, serves as a perfect example.
13. We will see that the capacity to retaliate is an important factor in creating market chaos.
14. There are seven full-length cases in this book.
15. vom Brocke and Lippe (2015).
16. I call this "work psychodynamics."
17. Not to count the fact that I have been trained in Japanese studies and management and have worked for a large Japanese trading house (*sogo shosha*).
18. PMBOK, 5th edition (2013).
19. See the theory related to stakeholders' interests (for example, Donaldson and Preston, 1995; Flak and Dertz, 2005; Okunoye et al., 2008).
20. Kloppenborg and Opfer (2002).
21. Generally speaking, the project management literature identifies two categories of competencies: technical and relational.
22. PM Network magazine (June 2015).
23. PM Network magazine (June 2015, p. 17)
24. See: http://www.worldbank.org/projects. Accessed August 5, 2015.
25. Tardif, P.M. Interviewed April 2016.
26. There is a hint on POVs in the PMBOK 5th edition: (p. 57): "The Monitoring and Controlling Group also involves: Controlling changes and recommending corrective or preventive action in anticipation of possible problems."
27. See Cleland and King (1988).
28. In general, the terms FP, FP_c, FP_{nc}, and the term MP are meant to be in the plural form.
29. Shen et al. (2010).
30. The notion of quality has been made popular by such experts as Philip Crosby (1989), W. Edwards Deming (n.d.), and Joseph Juran (n.d.). Many quality-related concepts have been widely adopted around the world, such as six-sigma, total quality management, the failure model, and effects analysis. In the present book, I somehow group these concepts under the topic of POVs.
31. Sometimes referred to as "organizational strategies" in PMBOK 5 (p. 7).
32. PMBOK 5 (p. 27) mentions that "Organizational process assets are the plans, processes, policies, procedures, and knowledge bases specific to and used by the performing organization." People *per se* are regretfully excluded. In reality, it is by People that these elements are made use of.
33. Berg and Karlsen (2007).
34. PMBOK (3rd Edition, pp. 12, 13, 17, 18).
35. The newest edition includes two domains of knowledge that pertain to People: the one of the management of human resources and the one on the management of stakeholders.
36. See www.oliviermesly.com
37. As of beginning of 2016, the exchange rate was roughly 1 US$ = C$1.40.
38. On modern predatory pricing, see Besanko et al. (2011).

General introduction

So it is, said the Wise Eagle, that the cure to uncertainty is transparency, and the cure to chaos is control.

Darloz

Feasibility studies that can be accessed through various sources, such as websites, and companies' documentation show limited scope in analyzing projects that are put forth with regard to vulnerability. Yet, the notion of "vulnerability" is at the center of any project. Some authors see project vulnerability as a sensitivity to negative events that entail a difficulty in coping with them,[1] thus reducing a project's value. Others define vulnerability as the incapacity of an organization to deal with the threats it is exposed to.[2] Other definitions include the following:

> The characteristics and circumstances of a community, system or asset that make it susceptible to the damaging effects of a hazard. Comment: There are many aspects of vulnerability, arising from various physical, social, economic, and environmental factors.[3]

Examples are given thereafter, among which are (Plan and Processes) poor design and the construction of buildings; (People) lack of public awareness; (Power) deficient contingency measures, the inadequate management of assets, little concern for environmental impacts, and tainted recognition of the risks. Vulnerability is subject to the social context and time.

A model has also been offered[4] whereby vulnerability is linked to exposure to hazards, which trigger a disquieting response and which have some impacts. It has been explained that "vulnerability is the degree to which a system, subsystem, or system component is likely to experience harm due to exposure to a hazard, either a perturbation or stress/stressor."[5] I pretend that vulnerability does not relate exclusively to the effects of hazards, but also to any source that can have a damaging effect, such as voluntary deceitful human interventions.

Essentially, we prefer to see vulnerability as a "potential for failure," with this potential being augmented as the various negative forces that affect the project (hazards, human threats, etc.) become more vigorous.

I.1   *Definition of vulnerability*

> *Vulnerability*: condition experienced by any of the four Ps (Plan, Processes, People, Power) of a project that make it susceptible to failure.

I derive the definition of points of vulnerability from this explanation, as follows:

I.2   *Definition of points of vulnerability*

> *Points of Vulnerability (POVs)*: temporal and physical points along the various stages of a project that impede on the calendar, costs, and/or quality of the project as it faces adverse conditions (negative forces), whether these conditions are under human control or not.

From this, I derive the first law of feasibility (there will be six in total), as follows:

I.3   *First law of project feasibility: Law of positive and negative forces*

> A project is not feasible if the positive forces (which play in favor of the project, which maintain a

functional g-spread[6]) are smaller than the negative
forces (which play against the project, such as risks[7]).

Thus, a project[8] is not feasible if [F– > F+]. When positive forces (F+) are
coupled with opportunities, it provides a path for success; otherwise, may-
hem is likely in sight. If not managed, POVs weaken managers' Dominant
and Contingency strategies and fuel the likelihood of them resorting to
Short strategies. A Dominant strategy is one whereby the project manager
has more forces going in his/her favor (positive forces F+) than against
him/her (negative forces F–). A feasibility analyst would normally want
to ensure that he/she examines the negative forces, whether in the form of
incapacities or so-called threats that plague a project.

Table I.1 provides just a few examples of feasibility studies retrieved
from the World Bank.[9]

Table I.1 Examples of feasibility studies

Name of document	Year	Main method used
Feasibility study: Microwork for the Palestinian territories (2013). Report no: acs3685[10]	2013	SWOT Competitive analysis
Jamaica—Weather insurance for the coffee sector feasibility study (2011). Report no: 75653[11]	2011	Risk assessment Vulnerability study
Urban accessibility/mobility index feasibility stage report (2010). Report no: 69933[12]	2010	Market study
Guyana—Agricultural insurance component: prefeasibility study report (2010). Report no: 75652[13]	2010	Risk assessment Market analysis
Nepal—Agricultural insurance feasibility study(2009). Report no: 46521[14]	2009	Risk assessment Market assessment

I found many documents among a vast number of studies (including those from the World Bank)[15] in the public and private sectors that
addressed the issue of vulnerability—that of crops, not of the processes
or humans in the case shown in Table I.1. The concept has been used in
feasibility studies.

One project leader sees vulnerability as follows:

Before commencing a project, we at AF Thériault
pose ourselves four questions:

1. Do we have the necessary technology?
2. Do we have the labor skills?
3. Are we underestimating the challenges? and
4. Are we overestimating our capabilities?

Any one of these questions point to vulnerabilities and we make a continuous effort to root out these vulnerabilities as far in advance as possible. When well assessed from this perspective by using a proactive approach, our projects benefit from better projections (planning), improved cost management if not elimination of costly issues altogether and reduction of problems down the road. Correcting vulnerabilities early on allows us to remove potential hurdles along the critical path. We also learned that when people feel more in control, they naturally try to improve, which for us, translates in highest quality effort and products. When people do their best because they feel respected and protected from vulnerabilities, they tend to do their best. We have come to realize that every one of our team members is capable of doing quality work beyond what they ever imagined.[16]

Roughly speaking, about 30% of projects succeed (meaning that they evolve exactly according to the initial plan),[17] 45% experience difficulties, and 25% derail.[18] These numbers meet my estimates based on various sources, including consulting firms such as KPMG (merger of Peat Marwick International and Klynveld Main Goerdeler). Some authors mention that a third of software projects experience excess costs ranging from 150% to 200% and time overruns of 200% to 300% compared with the original schedule.[19] There are notorious examples, which everyone is familiar with, such as that of the Denver International Airport's automated baggage system that crashed on its opening day of operation. Cost overruns are no light matter: the U.S. 1983 Nunn–McCurdy Act requires the Department of Defense to report to Congress when they occur in certain projects. Average cost overruns are higher in Asia and Europe, followed by North America and to a lesser extent South America, Africa, and Oceania, perhaps reflecting the fact that projects are more complex in more advanced nations (as exemplified by the costs of projects).[20] It has been stated that project overruns are common,[21] and that problems arise from overcomplexity. Poor planning (Plans), lack of training (People), and poor management (Power) are also to blame.

Some authors are convinced that overruns are due to two causes[22]: (1) forces that are external to the project, such as community or political intervention, customer order changes, government changes, inflation, weather, and so forth; and (2) strategic miscalculations (these are internal forces). Of note, however, is that not all changes are detrimental; some can have a positive impact on a project. In the case of the QMA, as an example, a change in the surface of the exterior landing dock from concrete to partial concrete cut the project cost by C\$216,000 while maintaining the integrity of the infrastructure. Epoxy was used for the terrazzo of the main passageway and granite for the main hall, saving C\$2,012,500.

From my perspective, this means that at least part of overruns can be avoided by adopting a Dominant strategy during the planning stage. The other important argument here is that there are external and internal forces, and positive and negative forces that foster or sink a project. If I were to put this in a Strengths, Weaknesses, Opportunities, and Threats (SWOT) table, I would obtain the simple frame of analysis shown in Table I.2.

Table I.2 SWOT and SVOR

Positive forces	Strengths	Opportunities
Negative forces	Vulnerabilities (weaknesses)	Risks (threats)

The word "vulnerabilities" depicts with a greater sense of urgency the presence of weaknesses and implies a trust factor that I will methodically examine in this book (in Chapters 5 and 6 in particular). One can say that projects stand the best chance of success when strengths respond to the opportunities given that risks and vulnerabilities are kept under control, or better yet, are under control and at a minimum level of activity. In the case of the QMA, interestingly enough, there is a provision in the budget document for so-called risks and opportunities; this goes to show that opportunities exist not only in the conception/vision stage, but also during the entire project (e.g., an opportunity to improve on a design or to reduce costs). Based on my experience and after interviewing feasibility analysts, I feel indeed that the probability of the success of a project is related to strengths and opportunities, given set risks and vulnerabilities. For the same amount of risks and vulnerabilities, the project with more strengths that respond to a better opportunity is likely to have more value to a project manager than a project without such strengths and opportunities.[23] This seems to make intuitive sense. On the other hand, given a set of strengths and opportunities, a project with more vulnerabilities and facing more risks is more likely to fall flat than a project with fewer risks and vulnerabilities. This, again, seems to match the natural flow of things.

As for strengths, some examples have been mentioned by various authors, such as an economically rational project with cost benefit returns, appropriate management, and a safe and ethical social and political climate.[24] A fifth factor has also been added, that of the favorable corporate strategic environment.

One can impute delays, which are a sign of vulnerability, on one or more of the four P components (Plans, Processes, People, and Power) of projects.[25] Delays may be due to poor planning (Plans). Examples stem from the design and conception being riddled with errors, ineffective penalties (where applicable) weighing on the system, legal disputes poisoning the atmosphere, the original schedule being too tight, or else external conditions (e.g., weather) altering the plan. Delays may also be incurred due to poor processes (Processes): changes or last-minute modifications (changes are regularly cited as one critical cause of delays), material supply difficulties, payment issues, production errors, reworking, or else accidents. Projects may suffer delays due to stakeholders' actions (People) or a lack of proactive actions in regard to conflicts, cultural differences, inexperience, poor communication, regulators' intervention, or else a shortage of skills. Finally, delays make for blight projects due to poor management (Power): disagreements between managers and promoters, misalignment in the line of authority, slow decision-making, supervisory inefficiencies, or else corruption.

When one looks at who is blaming who or what between consultants, contractors, equipment, external factors, promoters, teams, or plans, by far the vast majority of the blame is thrown at others, not on the machines, plans, or processes. Contractors impute their woes first to promoters then to consultants. Consultants consider that their woes are due to labor, promoters, and contractors. For all stakeholders combined, those held accountable for delays are first promoters, then contractors, then consultants, and finally labor.[26]

In short, the human factor is what comes out as the number one cause of delays. This points to the importance of understanding human behavior and of treating it like a mechanism (a Swiss watch of sorts) that can be decomposed into various parts—something I will do in the section dedicated to People (Chapters 5 and 6).

I.4   Three examples

Let us take three examples to support my argumentation on the four Ps (Plans, Processes, People, and Power): the 1976 MOS, the 2015 QMA briefly discussed in the preface, and the 2015 Mervel[27] farm project.

I.4.1   *Montréal Olympic Stadium (MOS)*

On December 4, 1969, Montréal's mayor, Mr. Jean Drapeau, officially proposed his city as a contender for the XXIst Olympic Summer Games. The International Olympic Committee (IOC) was convinced that Montréal was the best choice to host the Games, to be held from July 17 to August 1, 1976. Montréal, the financial heart of the province of Québec (Canada) was the largest French-speaking city in the world after Paris, of course. In the early 1970s, the province became the victim of bombings and kidnappings orchestrated by a small political faction in an attempt to found an independent state. The tragic sequence of violence produced zero heroes. The city and the province probably needed to restore their image and would work hard to impress the world with the upcoming Games.

The Montréal Olympic Games took place 40 years ago, in the summer of 1976; yet to this day, one of the iconic structures of the Olympic complex remains problematic. The stadium's roof has never been completed according to the original specifications set by the main architect, Roger Taillebert. All of the Olympic complex's structures, including the stadium (which was supposed to cost C$300 million), far outstretched their original budget, culminating in a cost of C$1.5 billion (Figure I.1).

Figure I.1 The Montréal Olympic Stadium (MOS).

In April 1972, the international press was invited to the unveiling of the first model of the future Olympic stadium, which looks like a gigantic shell, composed of 12,000 prefabricated components. A major dual technical challenge was an integral part of the design: a 574 foot high inclined tower capable of supporting a retractable roof. Construction began a year later. Unfortunately, between May and October 1975, some trade workers went on strike. The City of Montréal was suddenly facing failure and, even worse, embarrassment in the face of the entire planet.

The Québec government replaced the City of Montréal as the key manager of the project with the *Régie des installations olympiques* (RIO), or Olympic Games Steering Board. The RIO was responsible for the entire complex, including the Vélodrome, the Aquatics Center, and the Olympic Village where the athletes were to live for two weeks. Changes from the initial plan had to be adopted in a rush in a desperate effort to complete most of the work in time for the opening day. It soon became obvious that the stadium's tower would simply not be completed on time. Furthermore, unanticipated challenges complicated matters. The foundation rock was found to be unstable in many areas. To make matters worse, U.S. steel producers bumped up their selling prices from US$200 to US$900 per ton in 6 months, and then to an astonishing US$1200 per ton!

In short, two Forces of Production jeopardized the project: trade people and steel dealers (they certainly didn't fit in the plan set for the project!).

One week prior to the opening of the Games, the stadium construction was formally declared complete (despite the roof being a portion of what it was supposed to be). On July 16, 1976, one day before the grand opening, the laying of the turf at the Olympic Park was finalized. The next day, 12,000 athletes from five continents paraded before 76,000 spectators and half a billion TV viewers.

I.4.2 *Québec Multifunctional Amphitheatre (QMA)*

As of 2014, the Québec arena was the 100th largest project in Canada[28,29] and has been a vibrant example of a successful venture.

Very few articles or books on project feasibility discuss the success of a particular project. I thought it would be most important in this book to do so; first because it feels good to see that projects can, indeed, be completed on time, within budget, and with all the forces working together—strategic planning, processes, people, and line of authority—to produce the expected quality.

Second, this will allow us to shed a positive light that may help us to understand how dealing with points of vulnerabilities (POVs) can lead to great outcomes. I now review briefly each of the four Ps of this project (Plans, Processes, People, and Power).

I.4.2.1 *Plan*

The QMA was conceived out of a cultural and sport-related necessity. The existing arena—the Pepsi Centre[30]—was built in 1949; it currently has a seating capacity of more than 15,000 people. It was approaching the end of its useful life when it was decided to build a new structure. The development of Québec City,[31] which is the capital of the Québec province (in Canada), and which hosts a population of half a million people, required that a new building be constructed to accommodate entertainment and sporting events.

The idea of a multifunctional amphitheater replacing the aging (Colisée) Pepsi Centre started long before the QMA was finally built. City planners saw an opportunity to provide the population with a state-of-the-art building. Several versions of the plan were proposed. A company was hired by the city as the project manager for all of the trades. Some of the key contractors are listed in Table I.3.

Table I.3 Some of QMA key contractors

Work	Contractor
Architecture and engineering	SAGP
Audit (energy simulation)	Groupe Conseil Technosim Inc.
Construction manager	Pomerleau
Geotechnical and environmental study	Labo SM
Needs analysis	SAGP
Project manager	WSP
Selection of the project director	ENAP
Traffic study	Dessau
Qualitative control	LVM Inc.
Costs for all professional services	C$45,000,000

On top of these services, of course, a number of other services were required, ranging from security to insurance.

I.4.2.2   Processes
Several processes took place even before the construction actually began. Among those were: audits and meetings with the public, the approval by the city and the various partners in order to go ahead with the project (based on a feasibility study), the design of the structure and competition for awarding the various contracts needed to complete the project, technical choices with respect to the exact location of the structure, and so forth. A massive public demonstration took place on October 2, 2010, in favor of the project.

Table I.4 shows the main participants in the project (each work awarded to a separate company).

Of course, a whole range of other suppliers provided products ranging from office furniture to woodwork.

I.4.2.3   Power
The project was headed by the City of Québec, with the mayor, Mr. Régis Labeaume, having the final say. An important financier helped with the funding—"Groupe Québécor."

Figure I.2 shows the line of authority along with the flow of money for the entire project.

Table I.4 Some of the contracts allocated
to the QMA project

Work	Costs (in 0,000 C$ as of 2015)
Steel structure	50,000
Outside frame	24,000
Interior design/work	21,000
Electricity	20,000
Ventilation	18,000
Plumbing	15,000
Excavation	13,000
Slabs	10,000
Framework/casing	9,000
Landscaping	8,000
Phone system	5,000
Concrete	3,000
Elevators	3,000
Lightning rod	200

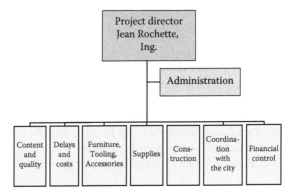

Figure I.2 Line of authority for the entire QMA project.

I.4.3   Mervel Farm project

I now turn to a smaller scenario, that of the Mervel Farm.[32] My goal here is to show that whether the project is a multibillion-dollar (e.g., MOS) or a few-thousand-dollar venture, the same principles apply. Some of the principles I will use are anchored in existing project management theory, as we shall see. Others are rather innovative, and this book is all about that—an innovative way of analyzing projects—because common project feasibility analyses often fall short of many essential steps that may explain why experiences such as the Montréal Olympic Stadium are repeated may times over at different levels of drama and in different countries.

In 2015, three entrepreneurs approached a local business center located in Pierreville, a small village squeezed between two large cities amounting to more than a million people and located across the river from Ottawa, the national capital of Canada. They wanted to develop a one-time summer project whereby the decaying Mervel Farm would be transformed into an "agritouristic" event that would include the participation of local farmers and artisans. The Mervel Farm is a historic site sitting in the heart of Pierreville. It consists of an 8.5-acre farmland that is owned by the National Capital Commission (NCC) and an old farmhouse adjacent to a barn. The land has not been put to agricultural use in decades. The house and barn have received basic maintenance, but little upgrades (e.g., to its electrical system) have been implemented. The site has been virtually abandoned. The idea of the three entrepreneurs was to capitalize on an emerging trend: that of "agritourism." Thus, they wished to receive funding from the local business center—an office that gets government funding aimed at encouraging local business ventures with the goal of creating jobs and promoting the area. Additionally, they had managed to convince the National Capital Commission that it would be worth attempting to revitalize this emblem of local history, at least for one summer. Should the project be successful, there would be an opportunity to turn the idea into a year-long operation, with winter activities that would include ice rinks, sledding, Christmas treasure hunts, and so forth.

The promoters estimated that they would need roughly C$150,000 in cash flow; also, C$50,000 would be spent to buy equipment and C$3,000 to acquire furniture. Funds would expectedly be invested by the promoters (C$20,000), the local business center (C$50,000), and the local tourism agency (C$100,000). Banks would loan C$50,000. The National Capital Commission would offer the site at no cost for the first year, providing that the promoters prioritize the proper maintenance of the land and buildings. The project would start at the end of May 2016 and end on the Labor Day weekend, for a total duration of approximately 3.5 months.

Research done by the promoters indicated that 46% of similar tourist-oriented agribusinesses received approximately 2500 visitors a year for average revenues of C$250,000. Very few of these businesses are located in town for obvious reasons. It was estimated that roughly 3% of the 5 million tourists in the greater area of Pierreville-Ottawa favored agritourism; in reality, only 2% visit the Museum of Agriculture located across the river, in Ottawa.

The City of Pierreville and the local business center see the concept as an additional opportunity to attract tourists, most of whom flock to downtown Ottawa, where the national Parliament buildings are located or venture in the nearby large Gatineau Park with its attractive outdoor sporting options (cycling, kayaking, swimming, etc.).

The Mervel Farm.

The core operating hypothesis was that an army of volunteers would come and work the land. These would be local people who cherish the idea of local farm products and teenagers who would want to gain some experience in a work environment. Tourists would show up and pick up, say, strawberries, and pay for the bucket at a premium of 20% over average local store retail prices. The cost of space for local producers and artisans offering their products out of temporary kiosks would be of C$15.00 per square foot.

There would be a full-time agronomist, a temporary site architect who would redesign the site given its new mission, one full-time project manager, and a small team of employees assigned to the various functions that the project entails (cleaning the site, coordinating the parking flow, guiding tours, etc.). The entrepreneurs would also hire a part-time marketing manager who would promote the site in all venues susceptible to attracting tourists.

The organizational structure would be that of a cooperative, a system that is very popular in Québec, whether in the agribusiness or in financial activities. Local farmers and crafts people who would set up shop around the farmhouse in the summer would have to become members of the cooperative, at a nominal fee of C$50.00 per year.

Some areas of farmland that would not be used for farming crops (e.g., pumpkins and strawberries) or for growing flowers would be rented out to other interested parties, such as small businesses offering donkey- or horseback-riding activities or else groups necessitating an outdoor space for special events (e.g., a company picnic).

Should the project not be found viable once the project is tested over the summer, there would be an opportunity to rent out the farming land to local citizens who would want to grow their own farm products on a dedicated parcel of land.

The Mervel Farm project was expected to have various impacts: it would attract tourists and generate an influx of money in the local economy (estimated at C$0.5 million/year). As it is now, the Mervel Farm is surrounded by houses owned by citizens much like a golf course in the middle of a town is bordered by houses. There is no parking site dedicated to accommodate a sudden influx of tourists, so that a new parking area would have to be built.

Once the project was completed, the project managers would write a report that they would submit to the funders (the City of Pierreville and the National Capital Commission).

I gave this example because it is easy to get a feel for what could go wrong. There can be no feasibility study worth its name without a full appreciation of the project's DNA. In fact, one of the major causes of project failure is the lack of understanding of the project's goal and structure. Being careless in evaluating all of the components that define a project is turning a blind eye on it, thus giving POVS an opportunity to cause havoc as the project evolves. This can spell considerable financial trouble down the road. POVs can invade inputs, transformation, and outputs.

Throughout this book, I will use a number of keywords and core concepts. The keywords of process modeling include "inputs," "transformation," and "outputs" (Figure I.3).

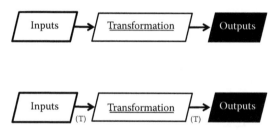

Figure I.3 A transformational/change process. (Note: (T) is time or calendar of events/work tasks.)

Figure I.3 is very simple yet it is at the core of the concept of any project. From it, I derive a number of submodels that create the canvas upon which feasibility studies can be commenced.

As put forth by a project leader: "We deliver our products in 80 countries. Each market is unique. Each market is a project in itself. The key is the capacity to handle change. A project is simply a change process."[33]

In the Mervel Farm project, inputs include all of the equipment that will need to be brought to the site, such as temporary sanitary facilities. Transformation refers to the changing farmland that has been abandoned for decades into a safe and friendly venue where families can come, walk, shop, and play. Outputs include waste blithely thrown away by visitors, and, from an intangible point of view, the positive city reputation that the event will engender if things go well.

One of the core models fostered in this book is, indeed, that of inputs–transformation–outputs. This is not news, of course. What is an addition to project management theory and especially to feasibility analysis is the four Ps.

I.5   Four Ps

Since the 1960s, marketing theory has adopted the four Ps system, which stands for "product, price, promotion, and place (distribution)." In project feasibility, four Ps are also used: Plans, Processes, People, and Power[34,35]

I suppose that the four Ps normally form positive forces driving the project (I know that there are also negative forces). They are most active during the transformation process, at which point I classify[36] them as Plans', Processes', People', and Power'.

I thus complement the input–transformation–output model by specifying what "inputs" mean, as shown in Figure I.4.

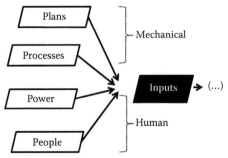

Figure I.4 The four Ps.

In my basic model, inputs are actually composed of four key elements: "Plans" and "Processes" deal with the internal mechanics of projects; "People" and "Power" represent its human aspect (the psychodynamics or an organization's internal cultural forces). The Plans and Processes, People, and Power, form the four Ps of project management, and are *sine qua non* elements that the analyst must dutifully take into account when doing a feasibility study. In the Mervel Farm project, roughly speaking, the Plan is the vision of a tourist venue for an old farmhouse; Processes are, for example, the means used to direct the traffic of cars and visitors; People are the tourists, the staff, and the army of volunteers; and Power is the site manager and the promoters.

The *sine qua non* condition is partly set forth by posing the question: "Can there be a project without a plan?" The answer is "No"; therefore, the plan is likely a *sine qua non* condition. In the present model, I assume there is no colinearity between the four Ps so that they don't interact with each other *before* the project actually enters into the transformation phase. The absence of colinearity is proved by the question posed at the input stage, prior to the phase of transformation: "Can the plan be established without team members actually transforming inputs into outputs?" The answer is "Yes." "Can the plan be established without all transformation processes being carved in stone?" "Yes." "Can the plan be set out without having found the right project managers to take on the challenge?" "Yes." Then, the same types of questions are posed for the other variables. For example, "Can the processes take place in the planning stage without committing specific people?" and so forth. Therefore, a *sine qua non* condition is set when (1) the variable must be present to define the project's inputs and (2) each variable is independent.[37] Of course, as the project moves from being an idea to being a concrete endeavor, colinearity builds up because people will handle certain processes, will need some lines of authority (Power), and will produce deliverables according to plan.

The four Ps will work together during the transformation stage, but for now, the plan is decided by the promoters (Power) independently of the staff (People) that will be hired later on, and the processes are being set but not put in motion just yet. Ultimately, a project is composed of inputs, a transformation process, and outputs, given a calendar of tasks and activities (time).

Time is entered as a constraint[38] because time is what links inputs to transformation and ultimately to outputs. For the four Ps set as inputs, prior to being brought together through the process of transformation, we have

$$\text{Inputs} = \text{Plans} + \text{Process} + \text{People} + \text{Power}$$

Assume that each P has its one beta (β) expressing the contribution of the respective P to the inputs; we then have[39]

$$\text{Inputs} = \beta_1 \text{ Plans} + \beta_2 \text{ Processes} + \beta_3 \text{ People} + \beta_4 \text{ Power} \qquad \text{(I.1)}$$

However, during the transformation stage, the four Ps are brought together and interact intensely with one another. In other words, the characteristics of each P affect the other Ps: as an example, what People do has an influence on the line of authority (Power), which could potentially trigger a change in the project plan or on some of its processes. Indeed, changes compared with the initial plan are one of the major complaints that project staff use when trying to explain their shortcomings.

In the case of the QMA for example, about 2 months prior to its completion, there had been no less than 1100 DDM (requests for changes or *"demandes de modifications"*), for a total value of approximately 17 million dollars, with each and every one having to be reviewed and approved or else dismissed by the chief project manager. On a 400 million dollar project, this represents a change request of roughly 4% (17/400); it turns out that 4% is also a usual measure for waste in a production line. For example, on a printing press, it is generally estimated that 4% of ink and paper loss is due to the warming up of the machines. When I look at the cost of changes related to construction and to equipment, I find that it is an average of 6% (for a budget of 30 million dollars involving 46 requests for change [DDM]). From this perspective and if I expand my thinking a bit (setting each change as a POV), I can estimate that there are a minimum of 4%–7% of POVs in projects in general. On a chessboard, even before starting the game, one POV exists among the 16 pieces of the same color (6%): the pawn on the F2 square is technically less protected compared with any of the other pieces on the board. This measure is by no means scientific, but serves to highlight the fact that there is a real possibility of POVs existing within any project and that, if they do, they are not insignificant.[40] If 100% of projects have POVs and knowing that POVs account for 4% of their content, it seems fair to say that the POVs must be dealt with.

Simply identifying the four Ps in the preliminary analysis of the project is useful (indeed, this is already a more complete view of the project compared with what is done customarily), but it is far from reflecting the manifold aspects of their interactions once the project gets into its transformation stage. In other words, there is plenty of room for points of vulnerability to be avowed *after* the project has been given the go-ahead; that is, once the project enters into the transformation stage. The project manager may be caught off guard because, on paper, everything seems to be in marching order. The more we can do to understand the four Ps that are tied to a particular project ahead of time, the more equipped we are to avoid unpleasant surprises. To account for this possibility, I add an error term (ε) to my project definition formula, but instead of having a positive sign in front of the error term, I use a negative sign. The reason is that in theory, the sum of the four Ps should deliver a perfect project, but points of vulnerability will come and reduce the potential for perfection. We all know that humans make errors (except for our wives!) and that processes don't always go according to plan, that the plans themselves can be faulty, and that power struggles can spoil a work atmosphere. We thus express the sum of all points of vulnerabilities as they materialize with their devastating effects by ε, which is, as we shall see later (see Equation 4.9), a measure of chaos (and more particularly, of deviance to norms of quality).We have

$$\text{Inputs} = \beta_1 \text{ Plans} + \beta_2 \text{ Processes} + \beta_3 \text{ People} + \beta_4 \text{ Power} - \varepsilon \qquad (I.2)$$

In terms of a "Plan" for the Mervel Farm project case, one must assume that the weather will be favorable, but nothing is less certain. What if it is a wet summer? Will tourists venture in an outdoor open field to buy regional products they can probably find one way or the other in a close, tidy, and warm environment such as a local specialty store? That is, for sure, a POV (no proper shelter) versus a given risk (that of bad weather).

In terms of Processes, the Mervel Farm concept relies heavily on the willingness of local craftspeople and farmers to adhere to its cooperative: an obligatory step in order to participate in the project. What if these people do not think the setup is worth their time and money? This is a risk.

In this same project, "People" refer to all stakeholders, including clients, fund providers, managers, staff, the media, and so forth. PMBOK would have it that the following groups would be present: initiators (Plans), clients (People), start-up group (Process), process group (Process), execution group (Process), surveillance group (Power), and control group (Power). The start-up group includes the three entrepreneurs that came up with the idea as well as the City of Pierreville and the National Capital Commission. To the process group, I add the project manager, the agronomist that will necessarily have to be hired, as well as disparate sources of financing such as banks, and a head coordinator of the crew of volunteers (as the project relies heavily on the participation of local volunteers). For all intents and purposes, I characterize these People' as inputs that participate in the chain of transformation, which then produce outputs that will be delivered to the end customers.

The execution group includes people specifically hired to do the work on the site. The surveillance group is simply the project manager and some of his key staff. The same applies to the control group, except that the National Capital Commission and the City of Pierreville will be sending their own inspectors to ascertain compliance with the farm's and city's strict rules of operation.

A typical POV when it comes to People, whom I refer to as Forces of Production (FP) when dealing with a transformation process, is the existence of so-called uncontrolled Forces of Production (FP_{nc}, or the "Unfits"), who are people acting in a detrimental fashion during the transformation stage.[41]

Forces of Production (who are People acting favorably during the transformation phase) are assumed to be under control (FP_c): a project manager directs them, evaluates them, and pays them. He thus has control over them. Technically, if one such Force of Production (team member) falls below a preset performance requirement, an action can be taken to rectify the situation, by way of docking bonuses, training, reprimand, or layoff. However, there are Forces of Production that remain amply uncontrolled or uncontrollable (FC_{nc}): these individuals or groups of individuals have their own agendas. An excellent example is the strike by the construction/trade individuals that plagued the construction of the Olympic Stadium in Montréal in 1976 a few months prior to its grand opening. Management had been "painted in a corner" so to speak: over a billion dollars had been invested to show the entire world what Montréal was made of. Evidently, there was a risk and a point of vulnerability (not having an internal specialized trade force) because a large workforce (apart from trade people) was uncontrolled. In the case of the Mervel Farm, there is also a certain number of possible Unfits (FP_{nc})—the army of volunteers whom the project promoters assume will rush to adhere to the ideology of the project ("a more natural way of life within a city"). However, what if this army of teenagers decides to not show up on the next workday for whatever reason (the teenagers went to bed too late, the nice weather commanded a picnic at the beach with their boy/girlfriends, etc.)? This is an ominous point of vulnerability (taking for granted that there are employees of this sort—albeit unpaid—available to the promoters).

A project is a cluster of people that is bound by a calendar of tasks and activities, costs, and norms of quality; that has specific objectives to meet; and that deals with problems and challenges. For example, a construction project such as the QMA includes agents such as architects, builders, experts responsible for establishing work schedules, subcontractors, and tradespeople (carpenters, electricians, etc.).

There are two broad categories of people: external (e.g., banks), to whom risks can be attributed; and internal (often busy performing daily operations), to whom vulnerabilities can be attributed. In the case of the QMA, external people are trade people hired by the construction manager, Pomerleau. Internal people are the crew headed by Mr. Jean Rochette, from the City of Québec. We can also divide People according to whether they are producers, consumers (such as the Olympic athletes and visitors to the Montréal Games), regulators (such as government or private professional associations), or so-called bad apples (who consist of unethical pressure groups, illegal workers, etc.—they act outside of the social norms and cause collateral damage). Unfits (uncontrolled Forces of Production) are one category of people acting up. All external uncontrolled people represent a risk, and may negatively affect projects by empowering points of vulnerability; all internal uncontrolled people represent a POV. We can represent People, before they become Forces of Production by way of the transformation process, as shown in Figure I.5.

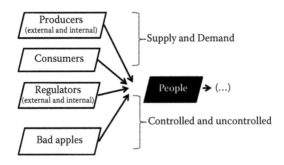

Figure I.5 A brief overview of People.

Some authors have separated people depending on whether they are customers (consumers), the parent organizations (producers), project teams (producers), or the public (consumers).[42] Neither they nor PMBOK[43] mention bad apples, but any experienced project manager will recognize their presence and devastating role without difficulty. By definition, internal bad apples are a POV. External bad apples were especially active in the construction industry in Québec—the 1974–1975 Cliche Commission exposed the fact that the mafia integrated many layers of construction projects with rippling effects in terms of costs and delays.

Power is the line of authority (both formal and informal) that sees that the project comes to a successful end. It includes the project managers and the various jurisdictional authorities who make sure codes and rules are respected. It is intimately linked to decision-making, a core topic of this book (see Chapter 7).

An example of Power is found in the QMA. Any change to the content of the project had to be formally authorized by the project director and scrupulously entered into a register.

As previously put forth, POVs may exist within each one of the four Ps. With respect to Power, there is a POV readily identifiable in the Mervel Farm project: the fact that the City of Pierreville (of provincial jurisdiction), which is an integral part of the project, must cooperate with the National Capital Commission (of federal jurisdiction). Often such political setups lead to bureaucratic or political upheaval.

Project promoters (acting as producers) design projects and find ways to fund them. Project leaders are more into day-to-day operations: they plan project activities, establish costs and schedules, lead and motivate teams, and implement a certain project culture. They measure risks, plan for the long term, and share their vision. Project managers, on the other hand, execute—they manage activities, costs, resources (including human resources), and schedules. They are task-oriented, they conform and administer and tend to avoid risk. They understand the importance of people in the organization, of competencies, culture, leadership, and performance.[44] Their approach to risk is what differentiates promoters from leaders from managers. Project leaders should be able to foresee POVs and project managers should be able to implement solutions that minimize them in earnest.

Judging from the Mervel Farm project, one can see that on paper all seems quite promising: the Plan is set and approved by expert committees, People who will work at the farm are assumed to be hardcore believers in the value of such an undertaking, Processes have been well laid out, and the promoters are enthusiastic and competent (Power). However, once all four elements (the four Ps) are put together, the scenario may not be as rosy. A group of volunteers (People) may not be willing to abide by the processes set in place. Intentions and reality differ. A project is by definition just that: intentions (hence the term "project").

So far, I have demonstrated that a project is a transformation process that receives inputs and that generates outputs, such as deliverables, thereafter. These inputs consist of the four Ps (Plans, Processes, People and Power) which interact during the transformation stage (Plans', Processes', People', and Power'). All four Ps are susceptible to POVs for whatever type of project, be it a construction endeavor or a touristy park of sorts.

I.6   *Projects and vulnerabilities*

Between 1990 and 2007, the interest in the concept of "vulnerability" in scientific literature has increased sevenfold.[45] Back in 1983, some authors[46] postulated that vulnerability had two sides: an external one consisting of risks and an internal one whereby the subject of risk lacks the means to cope without suffering from a damaging loss. Some scholars invoke the idea that vulnerability is at the center of the value-creation process.[47] Other authors present the reduction of vulnerability as an element of risk management.[48] The following formula has been proposed: risks are a mix of hazard, exposure, and vulnerability.[49]

Some of these authors rightfully assume a mathematical link between risk and vulnerability. This makes intuitive sense. Suppose a person is a champion swimmer. Going out and swimming in a turbid river with its load of rapids and hazards (rocks, turbulence, etc.) is indeed a gutsy venture. However, because he is an excellent swimmer, he does not feel vulnerable and will take on the challenge. On the other hand, a person who has never learned how to swim will judge that a still pond that is no deeper than 12 inches (approx. 30 cm) is a great source of danger. The perception of danger (threat) is tied to one's own perception of one's vulnerability.

I prefer to separate risk and vulnerability and not to consider that they are two sides of the same coin, they certainly are interconnected. My vision of projects is thus to differentiate between risks and vulnerabilities. Risks are actually external and vulnerabilities are internal, just as opportunities are external and strengths are assumed to be internal. In Table I.5, we have what we called the SVOR (strengths, vulnerabilities, opportunities, and risks) replacing the SWOT (the SVOR will be included in the six analytical Ps that I will discuss in Chapter 7 under the simple heading "POVs").

Table I.5 SVOR revisited

	Internal	External
Positive forces	Strengths	Opportunities
Negative forces	Vulnerabilities (weaknesses)	Risks (threats)

It is possible to establish degrees of gravity with respect to POVs. In Table I.6,[50] I connect high and low vulnerability with the four Ps.[51]

Table I.6 Low and high vulnerability contexts

Low-vulnerability (internal) context	High-vulnerability (internal) context	4 Ps
Appropriate resource commitment	Poor resource commitment	Plan
Carefully planned changes, solid results	Rapid changes, quick results	
Realistic expectations	Unrealistic expectations	
Set responsibilities	Ambiguous roles	
Operational changes	Strategic changes	Processes
Shared vision	Conflicting perceptions	People
Supportive top management	Uncommitted top management	Power

As can be seen, POVs can affect any one of the four Ps. This is important. In the literature on project management, SWOT analyses rarely encompass all of the major components of the project—meaning the four Ps. However, forgetting that each of the four Ps may contain a POV is diminishing the chances of success. We know that colinearity builds up between the four Ps as the project moves from the conception to the transformation phase; this means that Plans, Processes, People, and Power become more and more entangled and interdependent as the project is being materialized. The net result is that forgetting POVs on one or more of the four Ps at the project's initiation stage is really inviting problems to gain in momentum—the bad apple will contaminate the rest of the tree. Brutus will act on his hidden agenda.

Let's revert to the example of the Mervel Farm. Suppose for a moment that the army of volunteers that is assumed to be happily working at the farm has a sudden change of heart and decides that it wants better working conditions. The plan is altered; processes such as welcoming visitors are shaken, of course, and the few managers at the helm start to worry.

The QMA provides another example of how POVs can be regarded. The crew addressed potential problems with what it called "main concerns" (*Principales préoccupations*), which were concerns worth keeping a close eye on. In the main logbook, these were classified as follows: initial level, present level, and critical level, with each level having a status: critical, high, medium, low, and resolved. Concerns are included in Table I.7 as examples.

As can be seen, POVs in the case of the QMA evolved around four elements: Plans, Processes, People, and Power. These in turn are mandatorily tied to three constraints—those related to time (the calendar), those

Table I.7 Concerns as expressions of POVs in the QMA case

Plan	Calendar	Fear of delayed completion of lot plans and quotations
		Some lots operate solely on their critical paths, with no operating margin
	Costs	High levels of changes requested in some areas indicate poor planning prior to tenders
		Three years prior to completion, a 3-million-dollar excess is forecasted
	People'	Incomplete study of the vertical circulation of users within the structure
		Technical documents are clouded and may lead to claims by various parties
	Plan'	Spreads between original plans and actual PFT (*Programme fonctionnel et technique*)
Processes	Calendar	Potential delays in the approval of lot allocation
	People'	Incapacity to handle large volumes of tenders
	Power'	Possible delays in issuing certificates of conformity by legal authorities

related to costs, and those related to norms of quality—that interacted with the four Ps themselves once they were in the transformation phase (which are then referred to as Plans', Processes', People', and Power'). As explained by a project manager,[52] "**Constraints** are parameters that deplete the number of solutions that can be implemented to solve a problem." We can illustrate this classification of POVs as shown in Table I.8.

Table I.8 POVs and the constraints

POVs related to	... Given ...	Constraint
$\begin{pmatrix} \text{Plans} \\ \text{Processes} \\ \text{People} \\ \text{Power} \end{pmatrix}$	$\mid$	$\begin{pmatrix} \text{Time (calendar)} \\ \text{Costs} \\ \text{Norms of quality} \end{pmatrix}$

Of note, risks and POVs exist in both national and international projects: they are without frontiers. Let's call them "Problems without borders!" Factors deemed to harbor an international project's risks and vulnerabilities include inappropriate localization (Plans), insufficient funding (Plans), dubious contractor's business behaviors (People), a lack of desirability in the host country and poor public support (People), strenuous relationship with the government (Power), and unfair contractual conditions (Power), among many others.

Even though risks are at times said to be the result of an encounter between threats and vulnerabilities,[53] I prefer to refer to the term "Apprehension" instead of using the term "threat" and to separate this concept from risks. **Apprehension** is an internalized reaction to a "perceived threat," which is a well-documented psychological phenomenon that affects Dominant, Contingency, and Short strategies in different ways. It is an internal phenomenon. Assessing threats (and feeling apprehensive) is a useful self-protective mechanism[54]; it is an assessment upon which Dominant, Contingency or Short strategies are drafted.

The concept of vulnerability can apply either to a system or to persons.[55] There could be a large threat and a sense of invulnerability and hence, no risks would be deemed to exist. Therefore, I choose to allocate the word "threat" (under the form of "apprehension") to a psychological phenomenon that affects stakeholders (People). The coming of a snowstorm does not actually threaten the pouring of concrete to set the foundation of a new building. It's hard to imagine that inanimate objects or events actually threaten anyone, including a house builder. Rather, they represent a risk. Therefore, I keep the expression of "threat" where it belongs: in the psychological field, and I relate it to Apprehension, and I use risks instead of threats. This is why I do not use the SWOT system, but rather the SVOR system. Being circumspect is sensing a possible threat or danger from the outside world; this in turn leads to reduced trust. Apprehension is a cognitive state resulting from an appraisal of the outside world, which can take at times a mere fraction of a second, whereas distrust is the emotional state that takes time to build (yet less time than to build trust[56]). When one is apprehensive, one anticipates to a certain degree that one can get hurt in vulnerable weak spots. We will see how this plays out in the context of a project where people and teams interact, often in the midst of great uncertainty. In sum, Apprehension can be viewed as the cognitive equivalent of distrust, which is considered to be an emotional state. However, beware! With Apprehension comes a sense of vulnerability; this is not necessarily true with distrust. One can distrust another person without feeling vulnerable. In fact, he will probably distance himself and feel untouchable.

The term "vulnerability" has been given different meanings by a slew of authors, ranging from adaptability, capacity, exposure, potentiality, resilience, robustness, sensitivity, or even wound.[57] Some scholars propose a methodology to deal with vulnerabilities[58] as follows: There should be a management group assigned to identify vulnerabilities; once POVs are identified, they should be analyzed. Then, a response plan must be prepared (the model excludes the fact that it must be implemented). Finally, monitoring and control must take place with a provision for lessons learned.

The present book proposes a more thorough methodology (the six Ps of strategic management thinking—PRO, the four Ps, POW, POV, POE, and PWP), but at least we are comforted by the fact that attempts have been made to establish a procedure to address vulnerabilities, something that escapes the various versions of PMBOK. In the same vein, some authors envision to measure risks and vulnerabilities along a supply chain.[59] I provide some samples of measurements that could potentially be utilized to gauge the presence of POVs in Table I.9.[60]

Table I.9 Examples of measurements of potential POVs

For outputs items	For time	For costs	For norms of quality
Forms processed	Cycle time	Budget variances	Complaints
Items assembled	Equipment downtime	Contingency costs	Defects
Productivity	Late reporting	Cost by account	Error rates
Sales (e.g., ticket sales)	Overtime	Delay costs	Number of accidents
Units produced	Response time to complaints	Overhead costs	Rejects
Work backlog	Repair time	Penalties/fines	Rework
—	Work stoppage	Unit costs	Scrap
—	—	Variable costs	Waste

This is not to say that these items should be actual measurements of POVs, but simply that feasibility analysts and project managers are not short of potential metrics—these are incremental when having to meet norms of quality. In other words, there are no managerial reasons to dodge the reality of POVs.

Vulnerabilities, just like risks, nourish uncertainty; uncertainty intensifies the sense of looming external risks and vulnerability (and thus the level of apprehension). Uncertainty is forged by both external and internal forces, such as the novelty of particular equipment and materials (Processes), or by political changes (Power). Ambiguity is often fed by questionable planning (e.g., with respect to contracts, financing, material procurement, and scope changes)[61] and reckless individuals (e.g., people with poor competencies—unfits in their own right). Adding to the ambiguity is the existence of what I call **hidden truths**[62] or what is also known as hidden agendas when it comes down to people's behaviors—something we will study in detail in the sections on People (Chapters 5 and 6). In this respect, some authors refer to nonquantifiable factors, which include labor disputes, a lack of discipline, and the overlapping of project stages.[63]

The point I make is that POVs cannot be taken lightly; they exist and they thwart efforts to ensure that the calendar of activities, cost structure, and preset norms of quality are respected. They can be measured, and strategies can be implemented to deal with them. Dominant and Contingency strategies are available.

I highlight the fact that the human factor (People) is critical in POV assessments, even when they primarily concern equipment, machinery, or mechanical processes, for example. Finally, I emphasize the fact that POVs can be somewhat tricky, especially when they involve human behavior.

I.7   Need and opportunity

The term "opportunity" is the other important term in the SWOT or its equivalent, the SVOR (strengths, vulnerabilities, opportunity, and risks, which really have a role to play in the mental process of apprehension) analysis. Most books and articles on project management refer to the concept of "need," but this concept must be enlarged. Quite a few projects have commenced not because of needs *per se*, but by pure opportunism: a political party in power will grant a massive fund for a community project not based on the population's needs but on electoral motivations.

Hence, I prefer to resort to the term opportunity. I like to classify the opportunities into four categories: actual needs, desires, problems, or sources of discomfort.[64]   Let us explain. A customer's (the people who will use the deliverables) motivation can be a need: there is an urgent need to provide a Canadian aboriginal community with a proper water supply system and with electricity. What makes this example a need are the dire consequences of not responding to the situation the community is in by way of a project. A customer may have a desire—it has come to the attention of some promoters that people in Pierreville (Québec, Canada) liked the idea of a farm located right downtown. Citizens don't need a tourist-oriented farm; they can live without this farm producing crops of fruits and vegetables, or flowers for that matter, as they have for decades. However, it sounds desirable to revitalize tourism in the region, including agritourism. A customer may also face a particular problem: say the customer is a city with high levels of pollution. The problem may be addressed by developing better access to cyclists in order to reduce car usage. This is neither a need nor a desire, but certainly, it is a situation whereby a problem could potentially be solved or alleviated through a particular project. Finally, a source of motivation for a customer may be discomfort; say, for example, that a group of office employees has been working in less than adequate conditions in an older building. Some even get sick due to dust and mold. There is no need *per se* to build a new office, nor is there a desire or a fancy to get into a new building with pink walls, nor is there a problem *per se* since employees who will use the new building continue to work while it is being built. However, perhaps their productivity would be higher if they were relocated to a modern building. The bottom line is that there is a source of motivation for the client or potential client (need, desire, problem, or discomfort) which, from the project initiator's point of view, is an opportunity.

We somewhat encompass not only need, but also desire, problems, and discomfort as a source of motivation from a client's point of view and as an opportunity from the project promoter's point of view as what justifies, in the end, a project. This is not just pure semantics: the public will usually be much more critical if the project addresses a specific need (e.g., a sewer system) than if the project serves a fancy political agenda. Hence, feasibility analysts will look at POVs with a critical eye trained to recognize the nature of the opportunity the project is conceived upon.

Put simply, there is no requirement for a feasibility study if there is no opportunity, in whatever form it takes. A so-called needs analysis should really be an opportunity analysis.

I have referred in the previous paragraphs to the notion of a "mathematical link" between risk and vulnerability. I have conducted a number of studies on the subject and results are communicated throughout this book.[65] More specifically, I assume that there is a mathematical link between Total strengths and Opportunities, as shown in Figure I.6.

Figure I.6 Relationship between total strengths and opportunities.

The mathematical function is assumed to be:[66]

$$\text{Total strengths} = \frac{x}{\text{Opportunities}} \qquad (I.3)$$

This is the function of a rectangular hyperbola when x is a constant (actually, we only use a portion of the hyperbola). Large opportunities weaken total strengths because they entail more POVs. The model I present here posits that both strengths and opportunities are important, at the same time. A company in charge of realizing a project may possess substantial strengths related to technical expertise, for example, but if the opportunity does not exist, what is the point of spending money on the proposed project? Similarly, the opportunity may be vast, but if the project is flaccid (e.g., if it does not respond to a real need), how can one benefit from it? Perhaps the MOS is a good example in this regard: the opportunity to present Montréal to the world was outstanding ... but were there enough strengths within Québec society to take on the challenge?

Hence, Total strengths and Opportunities go hand in hand. A rectangular hyperbola has a particularity: the surface under any point along the curve is the same regardless of this point. In essence, a project manager will elect to be consistent with himself:[67] he will choose to adjust his strengths to the opportunities by augmenting one and reducing the consideration of the other, or vice versa. The numerator x is, in this case, a variable that belongs to each project; the larger the value of x, the more strengths are assumed in order to complete the proposed project given the opportunity. From this perspective, we assume that x refers to infrastructures, that is, anything that is already in place and that can provide a solid base for the project and which is pretty much a given (it is hard to change infrastructures—see the Africa vs. Haiti case, Chapter 5, for example). A company wishing to construct a well in a remote part of Africa on behalf of an international organization may be full of resources (strengths) and see a great opportunity to provide a community with water, but if the infrastructures are not in place (roads, electricity, etc.), the project will be that much more difficult to realize; the strengths will be of no use. For the Mervel Farm project as another example, the infrastructure that is particularly lacking is a large parking area.

For all intents and purposes, therefore, let's set our equation as follows:

$$\text{Total strengths} \,|\, \text{Time} = \frac{\text{Infrastructures}}{\text{Opportunities}} \tag{I.4}$$

Note that our model exists given time (in fact, we will see that time is just one element of a triple constraint). Infrastructures decay with time; thus, time must be part of the equation.

We can set a similar equation with respect to the relationship between risks (which are external) and vulnerabilities (which are internal), as shown in Figure I.7.

The mathematical relationship is expressed as follows:[68]

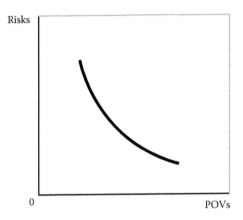

Figure I.7 Risks versus Vulnerability.

$$\text{Risks} \,|\, \text{Time} = \frac{k}{\text{Vulnerabilities}} \tag{I.5}$$

I provided the example of the swimmer in a previous paragraph. Placed in a project context, this equation means that risks are risks only with respect to vulnerabilities, and vice versa. The project manager cannot escape risks nor can it be assumed that there are no POVs among the four Ps of the project. Instead, the project manager must learn to find a balance between them both, a stand which allows him to remain equal to himself, that is, to respect the project plan, to properly use processes, people's efforts, and power. I discuss in more details the nature and value of k that I estimated based on my multiple studies in further chapters.

For now, what matters is that, overall, a project manager is best positioned when he balances each element of the SVOR (strengths, vulnerabilities, opportunity, and risks), given their mathematical interactions. We have Table I.10.

Table I.10 SVOR re-revisited

	Internal	Mathematical link	External
Positive forces	Total Strengths	Total Strengths | Constraints (Time) = (Infrastructures/ Opportunity)	Opportunities
Negative forces	Vulnerabilities	Risks | Constraints (Time)= (k/Vulnerability)	Risks

Of course, an ideal scenario is one whereby there are strengths matching opportunities (positive forces) and few risks or vulnerabilities (negative forces). A remark for the feasibility analyst to be made here is to note the importance of the state of the infrastructures where the project takes place.

It would be nice to know what the mathematical relationship is between strengths and vulnerabilities and between opportunities and risks. To help us with this challenge, I resort to two disciplines that seem disconnected from the present reality, but which will help us throughout this book to comprehend project feasibility.

In the game of chess, there are two basic strategies: an offensive strategy, and a defensive strategy consisting of two options: contingency retreat and short retreat (most players only refer to offensive and defensive strategies though). The offensive strategy is aggressive and seeks to take advantage of the opponent's vulnerabilities. A contingency option is elected when the player wants to secure his position and his pieces, anticipating that he may have some vulnerability (some POVs). For example, typically, after acquiring an opponent's piece, a player will not immediately deploy an additional attack. He/she will make sure his position is not subject to assaults given the new piece layout on the check board. He is adopting a Contingency strategy. A Short option occurs when the player is forced into a defensive position because he is under relentless pressure from his opponent. In chess, a defensive retreat strategy of this kind is mandatory after suffering the loss of an important piece, such as a bishop or a strategic pawn.

From a neurobiological point of view, there are also two main behavioral positions anchored in the brain—offensive and defensive—when experiencing apprehension (perceived threat). The lateral hypothalamus hosts the Instrumentally hostile aggression center often referred to in this book as simply "Hostile position"[69]—IP—or "predator dynamic"; the direct perceived threat in this case is the fact that the predator's needs will not be fulfilled if action is not taken[70]—the tiger is hungry. The medial hypothalamus is responsible for what is called "defensive aggression" (prey position or, as referred to in this book, "Defensive position"—DP, or else, figuratively, a "lonely sheep" position).[71] This is true for both humans and animals (e.g., in cats). If someone inadvertently walks on a cat's tail, the cat may react by attacking the leg of the offender. This reaction is not planned, but it helps the cat survive or else escape danger (defensive aggression). By contrast, a cat will methodically stalk and attack an innocent, naïve, and cute wandering mouse (!). This is predatory, instrumental (offensive) aggression. In cats, most particularly, the lateral and medial hypothalamus cannot operate at the same time: the cat acts in either an offensive or a defensive position, but absolutely no behavior takes place with part of one and part of the other. In humans, it seems the frontier is not as sharply divided—this remains to be further studied by neuroscientists.

What counts for us with the foregoing explanations is that we can assume that "strengths" in the SVOR model refers to total strengths, which would be the sum of a Dominant strategy, a Contingency strategy, and a Short strategy, as follows:

$$\text{Total strengths} = DS + CS + SS \qquad (I.6)$$

As previously touched on, **a Dominant strategy** (DS) contains strengths that allow the project manager to take a responsible lead on a project. A **Contingency strategy** (CS) encompasses strengths that allow the project manager to minimize the effect of risks and vulnerabilities to the project. A robust project is one that sees both DS and CS operate at the right time. The point here is that while total strengths are mathematically connected to opportunities, Short strategies—SS—(calculated as [Total Strengths - (DS + CS)]) are more directly connected to vulnerabilities.[72] A Short strategy is a strategy that reveals itself when a project manager is caught by surprise following the surge of an unsuspected POV in front of an unsuspected risk. If a project manager cannot develop Short strategies in the heat of the action, his vulnerabilities augment. Think of a gazelle caught by surprise by a hungry tiger; if the gazelle cannot find a way to deal with the threat (the risk of being eaten), it will make itself completely vulnerable.

People use Dominant, Contingency and Short strategies to deal with their vulnerabilities at large, not only with the unsuspected vulnerabilities. Thus, let's assume that there is a reverse relationship between "Total strengths" and "Vulnerabilities," giving us the following:

$$\text{Vulnerabilities} = \frac{1}{\text{Total strengths}} \qquad (I.7)$$

We know that Total Strengths is equal to Infrastructures over Opportunities (given time, in particular), so we have

$$\text{Vulnerabilities} = \frac{1}{\left(\dfrac{\text{Infrastructures}}{\text{Opportunities}}\right)} = \frac{\text{Opportunities}}{\text{Infrastructures}} \qquad (I.8)$$

We can now complete our SVOR table as shown in Table I.11.

I do not pretend this to be the Holy Grail of management, but it can serve as a guideline when trying to capture the relationships between the elements of the SVOR (more traditionally, of the SWOT). Assume for now that k expresses the ideal position of a project manager (an efficient position, by which he would want to remain stable over the entire course of the project). We will see that k is related to the sense of apprehension

Table I.11 SVOR and hypothesized mathematical links

	Internal	Mathematical link	External
Positive forces	Total Strengths	Total Strengths | Constraints = Infrastructures/ Opportunities	Opportunities
Mathematical link	Vulnerabilities | Constraints = 1/ Total Strengths	$k = 1.3$	Opportunity | Constraints = 1/ Risks
Negative forces	Vulnerabilities	Risks | Constraints = $(k/$ Vulnerability)	Risks

in Chapters 5 and 6 addressing People and to control and transparency in Chapter 7 dedicated to the study of Power. For now, what matters is that we have found a useful way to connect vulnerabilities (or more exactly POVs) to important managerial considerations such as strengths, risks, and opportunities.

Overall, the SVOR model seems much more complete than the traditional SWOT analysis. We know that the probability of the success of a project is related to the interaction between the strengths, opportunities, vulnerabilities, and risks. Hence, a feasibility expert should keep the SVOR method in mind when analyzing a project.

The lesson learned here is that it is advisable to link POVs to managerial considerations. Hence, in my opinion, evaluating POVs should be part of any project managerial action, for whatever type of project is undertaken.

I.8 Book layout

This introduction gives you a flavor of how this book deals with project management theory, and more particularly with project feasibility analysis, and more particularly still, with analyses of POVs. As mentioned, I will endeavor to tie in new concepts to existing concepts and terminology found in project management literature so that the reader does not feel overwhelmed by the flow of new information.

This book offers a range of tools that allow project evaluators to anticipate what can go wrong with the project proposal that is put in front of him. The theory I present here is in large part my own; in fact, I rely heavily on my own experiences, hypotheses, intuition, previous writings, and on researches I have conducted, especially with respect to neurobiological research that I completed in 2013, addressing the process of decision-making under uncertain circumstances. I would like to claim that my theoretical assumptions are well documented, but I will let the reader be the judge of the usefulness of the tools that I present. I invite the reader to refer to my previous works, some of which are listed in the reference section but not abundantly documented throughout the text, so that my name does not keep popping up, thus giving the reader the impression that I suffer from narcissism, which is certainly not the case!

The next sections are as follows: first, I discuss Plans (Chapters 1 through 3). I focus in particular on the definition of a project, and then offer an overview of prefeasibility and feasibility studies. The subsequent section focuses on the second *P*—processes (Chapter 4). I provide examples of real firms engaged in project and production processes and elucidate some elements of modeling in project feasibility analyses. In Chapters 5 and 6, I discuss People, beginning with an attempt at understanding key stakeholders; I examine how People interact with each other. A number of psychological studies performed in the course of multiple projects provide valuable data that shed light on the intricacies of human behavior in the context of project deployment. Chapter 7 is dedicated to Power; I resort to neurobiological research to show how the brain goes about making decisions (and decision errors) in the context of apprehension and opportunities, considering both project leaders and project funders as decision-makers. The conclusion outlines the various unique findings of this book. Each chapter ends with real cases pertaining to a food plant facility in North America, an artistic venture, and so forth. I believe these cases may help the reader to apply the knowledge I put forth to particular contexts. I also include some fun brainteasers to get those neurons firing after the reader has become exhausted by reading the present book!

As the reader can already judge, I refer throughout the book to various projects in which I have been participating either closely or remotely, some valued at many hundreds of millions of dollars and others brought to completion within a minimal budget. The Table of Contents lists the major sections of this book with their key topics.

I.9   Conclusion to the general introduction

I hope I have given the reader a fair picture of reality. In all, from the three examples we have discussed—the MOS, the QMA, and the Mervel Farm—it is easy to see, or at least to get a gut feeling, that something in the project went wrong, could have gone wrong, or will continue to deteriorate. This emphasizes the importance of fully comprehending a project's DNA before it takes place. This is a bit of a soothsayer work, but there are tools that we shall discover that will help us in this endeavor. The other important element that calls for attention in my three examples is the role of the stakeholders—it is nice to examine a project from a financial, marketing, or technical point of view, but in the end, it all comes down to people working together, and in the worst-case scenario, not wanting to work together because of some Machiavellian agendas.

I have defined some key concepts, examined and contrasted some expert opinions, made a brief effort at modeling, and provided concrete examples. Further sections of this book continue along these lines with a study of each of the four Ps, starting with Chapter 1 on Plans.

I.10   *What we have learned about POVs: General introduction*

POVs ...

1. That are not managed impair managers' Dominant and Contingency strategies
2. That are managed can help a manager to achieve success
3. Spell trouble down the road
4. Are present in any project at a minimum level of 4%–7%, in 100% of the projects
5. Exist within each one of the four Ps
6. Have degrees of gravity
7. Are sensitive to the triple constraints, but are mostly associated with costs
8. Exist in both national and international projects
9. Can be somewhat tricky
10. Are directly connected to risks

I.11   *Key managerial considerations: General introduction*

1. POVs cannot be ignored
2. POVs exist for each of the four Ps: Plans, Processes, People, and Power
3. Managers must strive for a Total strength position (Dominant + Contingency + Short strategies)
4. Risks cannot be assessed without assessing POVs at the same time

Endnotes

1. Le Moigne (1990).
2. Durand (2007).

3. http://www.unisdr.org/we/inform/terminology. Accessed August 13, 2015 (United Nations).
4. Turner et al. (2003, p. 8075).
5. Turner et al. (2003, p. 8074).
6. See Chapter 6 on processes.
7. Especially uncontrolled Forces of Production active in a poorly planned projects.
8. Please note that I use brackets [...] to isolate formulas from the text for ease of reading and understanding, and that when used for equations, again, this is meant to replace parentheses in order to ease reading and understanding.
9. http://documents.worldbank.org/; World Bank (2010)
10. By Paradi-Guilford (2013), The World Bank.
11. By Ministry of Agriculture and Fisheries (2011). The World Bank.
12. By The World Bank. Transport Anchor.
13. By The World Bank.
14. By Bagazonzya et al. (2009). The World Bank.
15. La Banque Mondiale (n.d.).
16. Oakley, G. Interviewed March 2nd, 2016.
17. See for example Assaf and Al-Hejji (2006) for a measure of success in the case of Saudi Arabia's projects.
18. See Dalal (2012).
19. See https://www.standishgroup.com; Doloi (2011).
20. See Mackenzie (2011).
21. Buchanan (1991).
22. Morris (1988).
23. In Chapter 7, we will see how the fear of missing out on an opportunity (greed) inflates a positivity bias (positive outlook or overoptimism) assigned to a project.
24. Morris (1988).
25. See, for example, Sauser et al. (2009).
26. Assaf and Al-Hejji (2006).
27. The case is real but the names have been changed.
28. See http://top100projects.ca/2015filters/?yr=2015. Accessed August 1, 2015.
29. See http://www.ville.Québec.qc.ca/en/webcameras/amphitheatre/amphi-theatre_ensemble.aspx. Accessed August 15, 2015.
30. See http://www.Québecregion.com/en/theatres-auditoriums/colisee-pepsi/. Accessed September 12, 2015.
31. See https://www.ville.Québec.qc.ca/en/. Accessed August 21, 2015.
32. Names have been changed.
33. Deveau, J.-P. Interviewed March 1, 2016.
34. Alternatively, "Propósito, Procesos, Personas, Poder" in Spanish and "Plan, Processus, Personnes, Pouvoir" in French.
35. Note: throughout this book, when I want to emphasize the fact that I am referring to a concept or construct, I use a capital letter at the beginning of its name.
36. Note the prime.
37. See Chapters 4 and 5 on modeling.
38. In fact, we will see two more constraints in Chapter 3.

39. Note: To facilitate the reading, I do not transform most of the text into symbols; many equations are simply meant to be a way of illustrating the text. Most of my equations are thus meant to be a stylized expression of the core concepts being discussed.
40. In statistics, a common *p*-value is of 0.05, or 95% level of confidence. It is the most common error allocation value used in general in marketing research, for example. This value is the proportion of false alarm that can be tolerated and that is related to a population of events or people.
41. I will be using a number of similar expressions and terms in this book and will proceed to explain them and link them to existing project management theory so that the reader can easily adopt them and understand their usefulness.
42. Meredith and Mantel (2009, p. 11).
43. For example, PMBOK 5 (pp. 31–36).
44. Mantel et al. (2011, p. 86).
45. See Vidal and Marle (2012).
46. Chambers (1983).
47. Bogataj and Bogataj (2007). See also Schneider (2008).
48. See Neuvel et al. (2015).
49. Blaikie et al. (1994).
50. Inspired by Buchanan (1991).
51. This, of course, is subject to the context of the project.
52. Tardif, P.M. Interviewed April 2016.
53. Birch and McEvoy (1992).
54. Frijda (1986); Lazarus (1991a,b); Levenson (1994); Ekman (1999); Keltner and Gross (1999); Kunzmann et al. (2014).
55. See Nienaber et al. (2015).
56. One moves more rapidly to distrust than to trust; hence, in this book, distrust is not considered in a pure sense or to be the exact opposite of trust. One may have zero trust toward someone without experiencing any distrust *per se*. This is why I introduce the term "negative trust" which is simply trust placed on flipped axis, as we shall see in Chapter 5.
57. See Deng et al. (2014); Kelly and Adger (2000).
58. Vidal and Marle (2012).
59. Bogataj and Bogataj (2007).
60. Inspired by *The project management scorecard* (Phillips et al. 2002, p. 144).
61. Khodakarami and Abdi (2014).
62. Hidden truths are what People know and refuse to admit, or don't know but that still exist and infect People's motives.
63. del Caño (1992).
64. PMBOK 5 lists opportunities (p. 10)—for need: market demand, business need, social need, customer request; for problems: environmental concern, technological advance (solved problem); and for sources of discomfort: legal requirement (improving life conditions). Also listed are (p. 69)— organizational needs and problems in relation to ecological impact.
65. All of my studies were performed in an academic context, meaning that they followed a strict research code and had to have the clearance of an ethics committee.

66. I did not have a chance to test this hypothesis but I use it as an assumption worth its while in my analyses.
67. Consistency with oneself is a core tenet of my model.
68. This equation revealed itself out of the analysis of my database, as we will see later in this book.
69. I feel constantly referring to "Instrumentally hostile" would make the text cumbersome. So I simplify this by using two different behavioral positions: "Hostile" and "Defensive".
70. For example, a tiger will adopt a predatory position when his life is threatened by hunger, which is an internal state.
71. See Chapter 6, and in particular, the section entitled "Hostile and Defensive behaviors" as well as Appendix 6.1.
72. Total strengths is the sum of Dominant, Contingency, and Short strategies; these are three strategies the manager can choose to face his challenges, hence the name "Total strength position."

chapter one

Plan—Project definition

So it is, said the Wise Eagle, that everything had been planned from the very beginning, even the unplanned.

Darloz

1.1 Introduction to Chapter 1

In the general introduction, I examined the notions of risk, vulnerability, and points of vulnerability; I introduced the constant k, and I expanded somewhat on the notion of opportunity among other themes. I have noted that the concept of points of vulnerability (POVs) is underdeveloped both in the scientific literature and in feasibility studies. I also decided to discuss feasibility studies and POVs by using the four Ps model. Put simply,

POVs are inherent in any one of the four Ps—Plans, Processes, People, and Power—when they combine during the transformation phase to become Plan', Processes', People', and Power' that sees where the initial project idea becomes a reality. A product will eventually be delivered to eager clients.

We now know that feasibility studies must focus on POVs, even if this involves playing devil's advocate when it would be much more pleasant to believe in the project with, at times, our eyes closed.

The present section, consisting of three chapters (1, 2, and 3), addresses the first of the four Ps: Plans. A Plan is about establishing a vision: proper modeling so that all facets of the project are laid out, with set measurements.

All projects worthy of this name have a plan. In the case of the Québec Multifunctional Amphitheatre (QMA), a master plan was designed and the main architectural working document, called the PFT (*Programme fonctionnel et technologique* or "Functional and Technological Program") was roughly composed of the following sections: levels, lots, sectors, special areas (e.g., corporate rooms, lounges, press rooms, and suites.), utility areas (e.g., hydro room), parking, and so forth. The main costs were listed as

1. Art (for art works displayed within the amphitheater—roughly C$1.5 million).

A different view of the QMA.

2. Construction: Construction manager's fees (the firm Pomerleau), construction work, equipment and material, legal costs, owners' fees (Québec City), professional services (including the firm Genivar for project management), and professional trades headed by the firm SAGP, and various other costs).
3. Reserves: For risks (C$21.1 million), contingencies related to conception and planning (roughly C$7.6 million), construction (C$10 million), inflation (C$12 million for the work, C$2 million for the furniture), as well as conformity and budget reserve.
4. The city's infrastructure.

The reserve funds are monetary provisions set aside to address unplanned and unfortunate events; including, when these are internally controlled, what could become POVs in that they express a Contingency Strategy (CS). In this case of a project that was four years in the making (from initial budgeting in 2011 to completion on September 1, 2015), it is fair to assume that it was susceptible to some construction mishaps. Contingencies account for approximately 10% of the total estimated cost of the project (C$39 million/C$400 million). This significant amount goes to show the importance of risks and POVs.

In the present chapter, I will discuss further project elements such as Dominant (DS) and Contingency (CS) strategies. I will start by defining what a project is and the different facets it can take. I will propose a way of modeling projects. Modeling is a useful tool for finding dark or wobbly spots. A proper model is much like an architectural plan; a faulty plan would prompt a manager's reaction because he/she would detect the possibilities of land mines, so to speak. As some authors have put it, the goal is to read the early warning signs of what could go wrong.[1] A project that is not adequately conceived will contain languid spots that the experienced manager will recognize.

I complete my modeling effort by examining what prefeasibility and feasibility studies are composed of, and will end with a conclusion.

1.2   Toward a definition of projects

The word "project" comes from the Latin words *projectum* and *prociere*, which mean to "throw something forward";[2] it is implied that something is put forth with some sense of anticipation.

According to one project leader that I interviewed:

> A project is a self-contained activity (or set of activities) that has three stages: a start, a production stage, and a completion, aimed at achieving a defined outcome. Generally, projects abide by a budget and a discrete timeline as well as by specific quality requirements.[3]

Different models have been proposed with respect to projects; in some, a project consists of benefits, hypotheses, and plans to list just these items. For some authors,[4] a project is composed of a program, goals and effects, and of some sense of rationale (reason for being). Various scholars[5] compare projects with systems that encompass cost challenges, manpower, priorities, schedules, technical know-how, as well as a spectrum of personalities that may be entangled in tugs-of-war. Projects have also been defined as a dynamic "set of interrelated units"[6] that share a common purpose. Indeed, during the transformation phase, the four Ps come together to produce forecasted outputs, including deliverables. Raising the question of the feasibility of a project is to ask whether it can hold up for all of its life cycle and achieve the desired results.

The newest version of the *Project Management Body of Knowledge* (PMBOK)[7] describes projects as follows: they are a temporary effort that aims to create a product or a service. They have a beginning and an end and can be short, medium, or long term. They are bounded by a set of parameters, which, if not respected, cause them to lose their reason for being. Typically, a project team dissolves once their project is completed.

Projects are temporary efforts by definition. Suppose the Mervel Farm project turns out to be a success and that the entrepreneurs, as well as the City of Pierreville and the National Capital Commission (NCC), decide to repeat the experience for as long as it is profitable over the subsequent years after the first year trial. It is no longer a project: it becomes an operation. Some authors indicate that projects have no fixed tools while operations do, that projects are unique while operations consist of repetitive activities,[8] and that a project is singular while operations can run different processes with different objectives at a time. It has been said that projects differ from operations in that innovative versus conservative solutions are preferred; furthermore, change is of the essence in projects as opposed to the incremental procedures of regular operations. Finally, projects develop in uncertain conditions, whereas operations assume stability.

Projects are "unique, temporary packets of efforts"[9] where cohesion among stakeholders is more important than drive, which was the old industrial paradigm. Some sources also cite time and uniqueness as defining elements of projects,[10] and add the notion of change (transformation), the multifunctional role of employees, and uncertainty.

According to one project association,[11] a project consists of a series of actions aligned according to a specific goal and a set mission, which has a beginning and an end. Some authors define a project as a singular, nonrepetitive activity, framed by a calendar, which is subject to uncertainty and which sees the union of distinct and complementary competencies in order to instill change or progress.[12] Field and academic experts argue that project management is about problem resolution, a process that is sometimes clustered in highly flexible horizontal or rigid vertical organizational structures. Other scholars present project management in terms of financial, human, and technical allocation effectuated in order to reach an objective by a particular deadline.[13] It has also been said that project management is about behavior, action, and knowledge.

It is relevant to delineate a project by what it is and what it is not—this palliates for the many imbroglios that could exist among stakeholders. The Mervel Farm project is a venue for urban families to experience some kind of farming lifestyle through activities and the acquisition of agriproducts. What it is not, by any means, is an amusement park. The core of the Mervel Farm project is in educating the local population, who, it is assumed, have a stake in the values of local environment-friendly products (produce and crafts); hence, two of its target customer groups are families and primary schools. It is not designed to be an entertainment site or a zoo: would-be customers would be disappointed if they didn't find paintball activities and wild animals or academics locked in a cage!

Properly defining and modeling a project is crucial. One of the most common causes of project failure that is consistently listed by experts, project management authorities, and marketing companies as well as academics having researched the field, is the fact that the mission of the project is poorly understood by the stakeholders, most particularly by the team members.

Light at the end of the tunnel.

While the PMBOK stipulates that a project generates products, services, or unique results, I wish to clarify that in reality, products and services are results by themselves. A project is actually a concrete effort aimed at producing a deliverable, which may be a product (such as an amphitheater), a service (such as an improved way of distributing the mail), or a process (such as a new way of making an electronic component or floor tiles). Some authors pretend that scientific research is a project, although this research is not done within a limited duration, it is periodic, and time (calendar) is of the essence in the definition of projects. Researchers have a schedule (a beginning and an end) to their projects. To summarize, the deliverables of a project are a product, service, process, or scientific research—nothing else. A project can conceivably deliver both a product and a service, the Mervel Farm being a case in point (a product would be a craft item and a service would be renting parts of the site to groups wanting to run social activities at that location).

Another definition has been provided: a project is a process in itself.[14] This process calls for the transformation of resources to produce a result in order to meet preset objectives, conditional to budget, human, material, and temporal constraints. The concept of the transformation of inputs into outputs within a framework of constraints is certainly adequate. As seen in the introduction, transformation (or change[15]) is at the heart of any project. I use parallelograms in Figure 1.1 as a code: they entail that I am referring to processes, which necessarily implies a set of constraints (we shall see that I use other geometrical forms, such as bubbles, for other kinds of modeling in Chapter 5). There are three types of constraints: **walls, ceilings**, and **floors**. To make matters clearer, I draw lines around the parallelograms where I believe that constraints prevail within the realm of projects (Figure 1.1).

In Figure 1.1, each step has a ceiling and a floor and there are two walls, one at the beginning of the process and one at the end of the process— these relate to the calendar (beginning and end), costs (ceilings), as well as norms of quality (floors). The ceiling is the maximum amount of money the promoter wants to dedicate to inputs, transformation, and outputs,

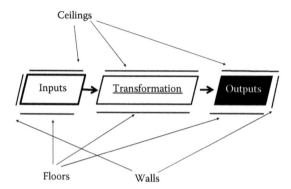

Figure 1.1 Project definition with walls, ceilings, and floors.

and the floor represents the minimum quality required to guarantee that the process is functional. The beginning walls indicate the mandatory start-up time for the system. The end wall is when, for example, the outputs are due (in the case of the Montréal Olympic games, the infrastructures had to be functional, even if not complete, a day before the opening of the Olympiads at the latest, that is, on July 16, 1976).

Concerning the QMA, the entry ceiling consisted of 40 million dollars' worth of equipment and material invested to start the construction project in a timely manner. The ceiling for the entire project was the monetary cap set at 400 million dollars. The end wall was the deadline for completion of September 1, 2015.

In the case of an operation, there is no absolute end wall because there is no deadline as there is with a project. Note that the fact that there is an entry wall only at the start of the diagram and an end wall only at the end of the diagram, as well as ceilings and floors on every step along the way (inputs, transformation, and outputs), is what makes this system a closed dynamic system. This will have important repercussions in my mathematical analysis when I treat the People aspect of a project and when evaluating the effect of the error term ε (chaos) seen in the general introduction and to be reviewed in Chapter 7 on Power. Put it this way: as a member of the project team, my strategic behavior is going to change if I think I'll never see a colleague again after I start arguing with him versus if I know that we are in it together for the long run. In the first case, I have an escape; in the second, I don't and that person may have an opportunity to retaliate against me in insidious ways, which could temper my bad character. My Dominant strategy is dramatically altered by the fact that the system is closed.[16]

We can use these entry and end walls, ceilings, and floors in an intuitive way. Say, for example, that we are planning a concert whereby all the Stradivarius violin musicians/owners of the world will perform Maurice

Ravel's famous bolero. The concert will only take place once—after all, regrouping all the Stradivarius violin musicians/owners in one location at one point in time is a feat, and a hazardous one at that. What if the stage collapses while the musicians perform? Let's assume, very conservatively, that the project necessitates an initial investment input of US$50,000 and a cash flow of US$25,000 to carry it until tickets start selling. The beginning ceiling is US$75,000 because the project will simply not materialize without this amount of money but will be deemed too expensive if more money is initially needed. Let's also assume that the maximum a bank is willing to loan for the hiring of the musicians and music director is US$35,000. One can proceed with the same kind of logic throughout the project model. One would obtain something like that shown in Figure 1.2 (I provide a simplified version).

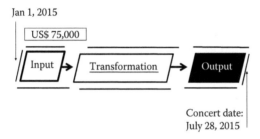

Figure 1.2 A music project with walls, ceilings, and floors.

The beginning floor would contain the norms of quality assigned to the project: only the authenticated Stradivarius violins would be eligible to participate. The point I am making is that if it is true that a project has a beginning and an end, is a transformation process, and is subject to some constraints (such as time), then we should be able to quantify this. Figure 1.2 represents an overview of the project that anyone can understand. That's a very good start because, as mentioned, one of the major causes of project failure is misalignment among the stakeholders.[17] A poorly defined project leads to a lame implementation of the plan.

Remember that inputs are structurally composed of four components, the four Ps, so that we could refine the model displayed in Figure 1.2 in more details if we wanted to, by showing the four Ps pointing at the input parallelogram.

Each project is unique.

Each project is unique and glaringly "contextualized"—the Montréal Olympic Stadium (MOS) is a prime example. It simply cannot be integrally copied so that each project has its own model with its own set of walls, ceilings, and floors.[18] Each project has positive and potentially negative impacts (e.g., promoting the City of Montréal worldwide while, in the process, accumulating a huge debt that 3.5 million Québec taxpayers had to reimburse over the course of 30 years).

If we rely solely on the PMBOK definition, the Montréal Olympic Stadium as seen by an individual back in 1970 would have appeared feasible: it had a beginning (the opening ceremony on July 17, 1976), and an end (the closing ceremony held on August 1, 1976), it was affordable, it was a unique product, and it would have a positive impact. However "bad apples" and inherent POVs such as the presence of "Unfits" (uncontrolled Forces of Production, FP_{nc}) eventually transformed the positive expectations into a painful reality.

We now have set a tentative definition of the concept of "projects" and have started to produce a model that exemplifies this definition. I will now attempt to improve it.

1.3   Completing the definition

In my opinion, some project definitions must be refined.[19] The model in Figure 1.3 presents the major aspects of the definition of a project from the output point of view; I will resort to this model extensively when discussing my feasibility analytical method.

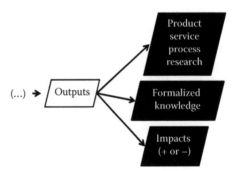

Figure 1.3 The output side of projects.

This model suggests that outputs are expressed by one or more of three items that are deemed to be tangibles (as opposed to intangibles).[20] Note that for large projects, all three components should be included in the definition of the project; for small projects, however, the requirement for the three outputs may not be obligatory or realistic (e.g., smaller projects may not produce a Book of Knowledge). We have (1) an innovative product, service, process, or research—also called the "deliverables"; (2) some official documentation (any large project is vastly documented—referred to as "Formalized Knowledge" or else at times "Corporate Knowledge Base"[21]); and (3) some impacts that can be positive and/or negative.

If one person tells us, in a game of charades, that he/she has seen something that is a product, we may think that they are referring to a "project". If they add that the something generated some formalized knowledge, we would be even more inclined to think that project is the miracle word. If they add at last that it generated some impact, then we will be 100% sure that they are talking about a project. The functional variables lead us toward finding the identity of the subject being discussed.

Each component is dependent on one another. The impacts are intimately linked to the product and knowledge is always knowledge of something, such as a product.

I will complete this model as I move along with my analysis. For now, the reader should probably get a feeling that it is not enough to simply define a project in tentative terms. The more precise we are with its definition, the better we will be able to tackle its POVs. Note that just as we quantified the input portion of the model and could have quantified its structural components (the four Ps), we could quantify the three different kinds of outputs displayed in Figure 1.3. As mentioned, it is also a good idea, from a very practical point of view, to spell out what the project is not and, more succinctly, what is not part of the project.[22] Indeed, my own experience with young entrepreneurs is that the majority of them do not fully comprehend the limits of their proposed business and expand beyond (often way beyond) its core definition, extrapolating into areas that are not at all related to the realm of the initial project.

1.4   Documentation

PMBOK[23] provides a roster of documents that are typically included in a full-scope project. They are gathered to serve the collective memory. I am inspired by PMBOK in classifying key documents as shown in Table 1.1.

To the PMBOK list, we must add documents that are relevant to People, such as those that address job description, marginal benefits, pay, and so forth. We find here the four Ps yet again. I assume that the four types of documents listed in Table 1.1 are not an essential part of an average project (for large projects, however, they should be). A small project could simply use a semiformal document addressing guidelines and have no set of highly technical standards. If we wanted to adopt the same kind of approach that we relied on so far in terms of modeling our findings, we would have Figure 1.4.

Table 1.1 Project documentation

Type of documentation		Four Ps
General documents[58]	Historical databases on management and finance	Plan
	Models (e.g., risk models)	
	Registers	
	The project charter	
	The project files	
Documents related to organizational control	Financial control procedures (e.g., journals of expenses and disbursements)	Power
	Problem management procedures	
	Quality control (e.g., audits)	Processes
Documents related to organizational guidelines	Communication instructions	Power
	Internal policies (e.g., health and safety policy)	
	Work instructions	
Documents related to organizational standards[59]	Including criteria set for evaluating tenders and for performance measurement	Processes (and Power)
Not listed in the PMBOK (e.g., a curriculum vitae)		People

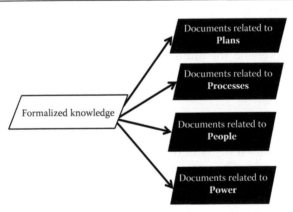

Figure 1.4 An average project's documentation with the four Ps.

The project feasibility analyst must support his/her decisions based on accurate documentation, which should include an investigation into POVs. I strongly recommend that the analyst finds existing measures and metrics or else that they formulate a set of measures so that all key items found in the various documents have a quantitative aspect. This is in line with his/her role to ascertain that norms of quality are respected.

Suppose the analyst believes that confidence (Trust) in a project team is important,[24] then he/she must find ways to measure this in advance. In the realm of a project, trust is established when group members recognize each other's skills and/or when they are seen as honest.[25] Often, because team members are brought together and work within a tight time line, they have no choice but to display so-called "swift trust"[26]—trust is a given from the get-go because there is no choice in the matter when the tasks must be accomplished. Measuring skills and honesty should be planned. Should the cooperation between stakeholders and the ability to quell problems jointly and listen to others be deemed important for the evolution of the project,[27] then the analyst would have to develop a scale or else use an existing one. It is up to the project evaluator to decide what should be measured (and become part of an official set of documents) and how to measure it. This may be important to secure the realization of the project. As an example, some authors have determined that trust, mutual respect, and stakeholder partnerships are features that lead to success. As stated by one of the project managers in an interview that I conducted:

> Through the course of my career, I have been called to oversee a number of international projects, one of which, as an example, was with the aluminum company Alcan in Arvida, Québec, Canada. The C$4.7 MM project was realized on time, below budget (at C$2.96 MM), and within the quality specifications set up front. What allowed us to successfully complete the projects can be attributed to a number of factors. We were inventive; we took great care in defining the project and in addressing environmental concerns right up front. We hired an extremely competent team, which was very much committed and which worked with flexibility and rigor. It is this combination of efforts and talents that drove us to the finish line.[28]

1.5   Impacts

Most projects (especially large ones) spawn positive and/or negative impacts. A tendency for promoters is to dispel negative impacts or to hide them altogether in order to get financing; this is a mistake, of course, because recognizing negative impacts is a way of improving the project before it starts. If the impacts, whether positive or negative, are found to be overwhelming, then the project should not go ahead. Some authors who discuss the case of the Montréal Olympic Games[29] mention impacts such as the upgrading of the transportation system, the redesign of some neighborhoods, and the reshaping of the city topography. The Montréal Olympic Games project called for improvements in these three areas. There are few projects worth this name that do not produce positive and/or negative impacts of any sort: a promoter who doesn't think so is probably hiding some POVs. Put it this way: you can never satisfy everyone. So someone, somewhere, will feel that they are the victim of the project, whether it's because they judge that there is excessive dust, crowding, noise, or that they face a hike in the cost of living or have to be displaced. They may express environmental concerns or displeasure with the layout, and so forth. Take Yale University, for example; the town of New Haven should have been thrilled with the expansion of this emblem of academic achievement, consistently ranked among the top five universities in the United States. Yet, some residents are not pleased with it: traffic has worsened, tax revenues are lost on what could have been residential lands, and neighborhoods were razed.

Not all impacts need to generate concerns, however. In the QMA case, a number of delays were deemed to have no impact on the project, as they were not critical and were easily corrected. Examples are shown in Table 1.2.

Table 1.2 QMA delays with inconsequential impacts (examples)

Physical area	Delays (*in days*)	Observation
Press room	18	No impact on end of project
Main hall	26	Stone steps cover—delays
Living room 2	41	Woodwork delays—no impact on end of project

While delays were experienced, the impacts were not significant; for all intents and purposes, the project would be (and was) completed on time.

So far, we have seen that a project can fit the emerging model shown in Figure 1.5.

We can read this flow of diagrams as follows: producers (e.g., project promoters), clients (existing or potential), regulators, and so-called bad apples interact as People. A plan is devised, Processes are conceived, and a line of authority is set (Power). The four Ps form the project's inputs, which are transformed and which then produce, over time, some outputs. Transformation implies change.

Of note, some last-minute modifications may actually be beneficial. For example, in the QMA project, the construction of a lot labeled "1-B 3.1" saw a cost reduction of some C$26,000. In fact, the final cost of the project was 30 million dollars below initial estimates, which goes to show that savings can be made along the way. This may be an indication that identifying POVs could actually help reduce costs, rather than merely preventing them from rising above budget. This is no scientific proof, but it

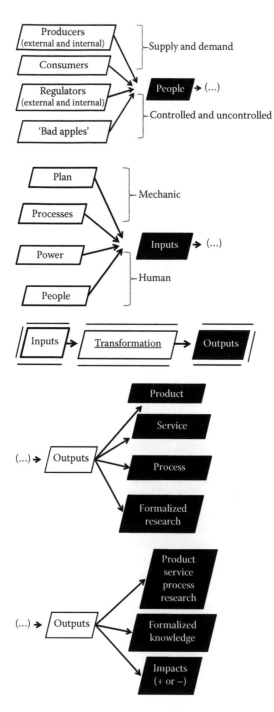

Figure 1.5 Toward a definition of an average project.

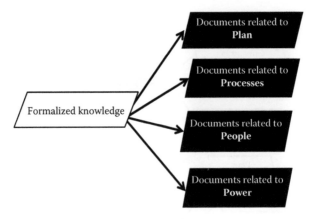

Figure 1.5 *(Continued)* Toward a definition of an average project.

remains a notion to keep in mind. I assume, therefore, that tackling POVs can present a dual advantage: the possibility of reducing costs (by way of finding inventive ways to rethink a particular process) and preventing the swelling of costs due to processing problems.

An examination of the reasons for changes made that deviate from the initial plan (i.e., vs. the PFT document) in the QMA case shows that the majority of them were related to functionality (more than 99%) as opposed to design (less than 1% in the case of the style of the floors for the artists' foyer), and that they were explained by a limited number of reasons, as shown in Table 1.3.

The outputs, according to my model, are expressed by a deliverable (e.g., a product), by some formal knowledge (e.g., Book of Knowledge), and by some impacts (positive and/or negative). In the case of the QMA, the end book (the final report) was prepared by the chief engineer of the company William Sale Partnership (WSP) Global, the project managing company. This book contains various topics in the following sequence: an executive summary, some comments with respect to the calendar of events, costs, risks, supply chain management, construction operations, management of changes, management of quality, main concerns (i.e., the identification of potential POVs), equipment, Key Performance Indicators (KPIs), critical path analysis, content analysis, and recommendations. The appendixes include a follow-up on tenders, environmental concerns (Leadership in Energy and Environmental Design [LEED] certification), and so forth.

As mentioned, small projects do not naturally lead to formalized knowledge or have an impact. Large projects, however, see the creation of some form of official knowledge and do have impacts on the way of life of the population they serve. A large project is all of this and no less! For each element (each parallelogram) and among each element (the arrows), there is a potential for POVs to sneak in.

Table 1.3 QMA reasons for changes[60]

Reason for change	Example (functionality)
Additional functionality	Addition of a drain on the main stage
	Larger area at the ticket sales desks
	A larger area near the elevators to facilitate the flow of people
	Relocalization of the kitchenette to make the artists' foyer more multifunctional
Conception improvement	Single door vs. double door in the athlete treatment room to facilitate the movement of medical stretchers
Durability (design)	Better doors in the first-aid unit (1070 mm[61] instead of 1220 mm)
Excess functionality	Doors 2.75 m in height where 2.34 m is the norm in the NHL (National Hockey League)
Omission in the PFT	Boot grids to remove snow
Respect of norms unaccounted for in the PFT (design)	Emergency lighting in the corporate rooms
	Epoxy floor in pantry room instead of sealed concrete floor
Safety	A 120 V plug in the men's toilets to prevent the use of long electrical cables during construction or maintenance, thus limiting the risk of accidents

A project has a certain number of characteristics, which we have touched on so far: it is unique (Plan), it spans over a predetermined period (Plan), it requires some level of innovation (Process), it faces challenges (Process), it is an answer to a need/an opportunity (People), and it involves stakeholders (People and Power). Should any of these features not be present, one must conclude that it is not a "project" in the pure sense of the term. Each one of these characteristics must be present; strictly speaking, a promoter who brings forward a plan to put new asphalt on his/her office's driveway is not dealing with a project in a way that would require project management knowledge. On the other hand, a project must be expressed by at least two of the following outputs: deliverables, formalized knowledge, and/or impacts. Essentially, this is what the feasibility expert must look for when evaluating projects. The expert will want to fill in Table 1.4 to double-check that he is indeed dealing with a project truly worth its name.

The rule dictates that each input characteristic and at least two outputs should be present; if so, the feasibility analyst can define the proposal as a project worth its name.

Table 1.4 Characteristics of a project

Characteristics		Present or not
Inputs (structural elements: all are *sine qua non* conditions)		
Plan	Is unique	
	Incurs some costs	
	Has a set time frame	
	Respects preset norms of quality	
Processes	Are innovative	
	Offer challenges	
People	Respond to an opportunity	
Power	Has a defined line of authority	
Outputs (at least two out of three must be identified)		
Provide deliverables		
Generate formalized knowledge		
Have impacts (+ and −)		

I arrive at the following definition of a project:

1.5.1   Definition of project

For the purpose of a feasibility analysis, a project is a concrete and orga-nized effort that leads to the realization of a unique and innovative deliverable, which can be a product, service or process, or even a science research initiative, which is conceived based on a perceived opportunity. The project has a beginning and an end, which can sometimes serve as new bedrock for a different project. It involves a plan, some processes, people and a line of authority; it contains inherent challenges and prob-lems. It is bound by a specific calendar, a cost structure and pre-set norms of quality. Finally, each project tends to generate some official documenta-tion as well as positive and potentially negative impacts.

1.6   Intangibles

Intangibles refer to elements of the project that cannot be measured and that are not necessarily an integral part of the project. PMBOK 5 (p. 51) men-tions that "examples of intangible elements include goodwill, brand recog-nition, public benefit, and trademarks." Negative intangibles include "loss of reputation, reduced moral staff, and loss of valuable corporate knowledge when many experienced technical and managerial staff are forced to leave the organization through (…) staff reduction."[30] A positive example is the strong bond that develops between stakeholders such as team members, as they embark on a project. Perhaps this friendship may be useful in another venture should the same people be brought together again. This will make the new project that much easier to get underway. Intangibles are immaterial. Some impacts are intangibles. For example, CAID—the Canadian Agency for International Development or *l'Agence canadienne de développement international* (ACDI) in French—developed an aid program for a rural community in Africa to build a small slaughterhouse.[31] The need arose from the fact that local village members slaughtered their cattle in an open field where cattle manure was scattered, in the full heat of the sun.

The project definition actually changed dramatically depending on the stakeholders. The engineers at CAID wanted to build a slaughterhouse that met Canadian regulations (at an astronomical cost), a feat impossible to accomplish given the lack of infrastructure[32] and experience from local builders. The local government was content with four walls and a roof (at a reasonable cost; yet unduly maintaining the danger of meat contamination). The villagers did not care so much because they had been slaughtering cattle in the open field for generations.

The slaughterhouse was eventually built with somewhat of a faint adhesion to the strict Canadian construction code. However, a problematic intangible impact emerged: the head of the village, seeing that other local communities wanted to either use the slaughterhouse or else the open field that had been cleaned of waste, saw the opportunity to invigorate his power and earn money by leasing the house and the field. This lack of cultural understanding and cohesiveness led to social problems and some animosity that had not been predicted by the Canadian agency.[33] Cohesion was certainly not achieved, far from it, regretfully.

Said one project leader in a different circumstance about the importance of cohesiveness:

> AF Thériault is probably the best kept secret in this 9000-people community (Clare). There is a high degree of pride in what we are doing and this, along with human bonds, carries over from one project to the next when team members are allocated to different projects. These transfers, which can be lateral or spread over time, provide a natural boost to the project right from the start. It reduces our project costs because our teams are readily cohesive.[34]

1.7 Classification of projects

Projects can take many forms, of course; China's Great Wall was a response to a need for security against northern tribes' invasions, the Concord airplane was seen as a business opportunity, the Egyptian pyramids were constructed as a tribute to the Gods and to guard the Pharaohs, the Mona Lisa painting was a project apparently based on a commission by a wealthy individual, the movie *2001 Space Odyssey* materialized out of a vision of the world to come, and the Sherman Antitrust Act was designed to protect consumers. Many more projects can be listed: the Callaway Nuclear Plant, the Long Island Lighting project, the Shoreham project, the Supersonic Transport program, the Trans-Alaska Pipeline System, and so forth. Our economy is all about projects.[35] According to some authors, projects are at the basis of our economic action given our increasingly complex and uncertain environment.[36]

Projects respond to a need, a desire, a problem, or a source of discomfort (overall, to an opportunity as seen from the point of view of the promoter), take different forms (e.g., work of art), and require some level of innovation.

Not all projects are born out of known needs. Some ingenious promoters actually design projects around so-called latent needs. These needs have not been formalized or universally recognized yet; they are dormant, or put differently, unconscious. For example, while Facebook did not exist for millions of years, its growth points to the fact that it initially addressed a latent need. In our fast-paced society, people found it more and more difficult to communicate *de visu*, yet they became more and more open to communication with strangers or with past, long-forgotten acquaintances. Often, it can be said that many projects are born out of the realization of a latent need (read: opportunity). The relevance of this fact is that with the need being latent, it is actually hard to ascertain—its full characteristics are unknown. Uncertainty[37] inhabits every project, hence the importance of pinpointing POVs.

Generally, besides monetary concerns,[38] the needs of projects fall within eight categories: communication, clothing, education, entertainment/sport, food/beverage/tobacco, health (and related industries), housing, and transportation. Money is actually a means to fulfill the needs associated with one of these eight categories of basic human life.

If we take the Québec Multifunctional Amphitheatre as an example, obviously, this is a structure that belongs to the sport and entertainment industry. But the inside fabric of it also takes into consideration specific needs that pertain to one of the other eight categories listed in the previous paragraph. In particular, of prime concern to the builders were the following: the availability of toilets; the control of sediment,[39] heat, and light; noise; rainfall and water infiltration; the quality of air, including ventilation, smoke and dust control (this was on the agenda when deciding on sealed concrete floors vs. decorative floors), both during construction[40] and once the project was completed; and finally, security (cameras, etc.).

In the general introduction, I discussed the notion of opportunity; this relates to one of these basic human need categories. Products that address these needs are paid for by a banking/financing system. Financial projects are, ultimately, designed to satisfy one of these eight basic needs. It is most useful for the feasibility analyst to recognize which need or needs is or are addressed by the project, because this sets the context in which this project matures. It also provides points for comparison with other (past) projects.

Projects are private, public, or a mixture of both (PPP[41]) and come in different sizes.

Some authors attribute the following engineering hours depending on project size:

- Small projects are 10,000–100,000 engineering hours
- Medium projects are 100,001–500,000 engineering hours
- Large projects are over 500,000 engineering hours

We have already established that larger projects tend to see deliverables, formalized knowledge, and impacts as defining conditions. Because these three elements have a high level of colinearity, they cannot be characterized as structural process elements; however, the feasibility analyst wants to keep in mind that they are, nevertheless, fundamental considerations in larger projects.

Whatever type or size, I prefer to adopt a system that classifies projects according to the four Ps (three Ps in reality because a Plan is not an option), as in Table 1.5.

There are soft and hard projects. In the first case, the deliverable is said to be intangible (e.g., a new television program), and in the second, to be tangible (e.g., a building).

A three-dimensional model that includes projects of all kinds (nuclear, petrochemical, roads, etc.), has been presented[42]; in it, projects are positioned according to market risk (x-axis), technical risk (y-axis), and social/institutional risk (z-axis). However, this is debatable: are all contextual risks included in this three-dimensional framework? Probably not. Does a society and do institutions represent risks in a pure sense? Unlikely.

1.8 Value

We can also define the project based on the concept of "value". A project is more than simply manufacturing a product, developing a service, designing a process, or conducting scientific research; it is generating **perceived value**.

Perceived value is the total value attributed to the project, including what it produces in terms of intangibles, as seen by the users/clients.[43] Figure 1.6 illustrates this modified project definition.

Intangibles, as demonstrated by the aforementioned African example, can turn out to be major POVs. To predict them, an analysis of the motivation (the opportunity[44]) behind the project from the point of view of each

Table 1.5 Project classification according to the four Ps (examples)

Type of P	Examples of project sectors
Process	Architecture/buildings
	Automation, automotive
	Financial services
	Manufacturing; new product development
	Mechanics: aerospace/defense, oil/gas/petrochemical/energy
	Pharmaceutical
	Process improvement
	Soft/hardware, including e-business and information systems, telecom
	Waterways and shipping, railways, and so on
People	Agriculture, fishing, and forestry
	Community development (e.g., international development)
	Communication/network
	Consumer products
	Environmental, governmental, health-care services/hospitals
	Event projects/media/movies/concerts/sports
	International aid
	Medical
	Retail
	Scientific[62]
	Utility industry
	Urban public transportation, and so on
Power	Aerospace/defense/military
	Business: acquisition/merger
	Central government administration
	Finance; banking
	General public administration (e.g., compulsory pension and unemployment insurance, education, industry, trade, ports, and transportation)
	Organizational restructuring
	Plant closing, opening, upgrading
	Public administration, law, and justice
	Some social services, and so on

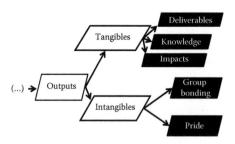

Figure 1.6 An evolving model with intangibles.

important stakeholder is mandatory.[45] In this example, the opportunity is based on a needs assessment by the CAID, on the desire of the local government (it is an opportunistic maneuver to gain the village's political support), and on virtually nothing for the local community (from its point of view). This chasm in the motivations (or put differently, in opportunities) is undoubtedly a POV, but one that starts poisoning the project *after* its completion. Indeed, divergent stakeholders' interests are a major risk and POV.

The role of the project feasibility analyst is to determine the value of the proposed project;[46] the value estimate cannot be valid without an appreciation of the POVs. From an analytical point of view, there are actually three kinds of value: **perceived value**, **added value**, and **residual value**. A large gap between them is also an indication of potential POVs.

Perceived value is in the eyes of the users of the project. The perceived value of the slaughterhouse evolved from being nothing but a disruption of a lifestyle for the local community chief to a source of wealth (as he eventually rented land to nearby village dwellers). It is calculated as follows:

$$\text{Perceived value} = \frac{\text{Perceived quality}}{\text{Costs}} = \frac{\text{Functionality} + \text{Design}}{\text{Costs}} \quad (1.1)$$

I have shown in the previous section (Table 1.3) that optimization of functionality and design (to a lesser extent) is indeed the core reason for changes in the QMA case. My various examples of research show that consumers systematically allocate their perceived value at the ratio of quality to cost, where quality is defined by the summation of functionality and design. Put simply, consumers will say: "This product provides excellent functionality (meets my needs) and has a beautiful design while being reasonably priced—it's a great buy."

The QMA from a different angle.

Recall that the four Ps are subject to norms of quality. Plans, Processes, People, and Power must strive to meet the standards set up front before the project starts; failing that, completion is likely hindered, costs bulge, and quality suffers. On the topic of standards, it has been said:

> When dealing with improving efficiency of project management practices within an organization, it is important to put some standards in place. As binding as it may seem to the project manager (each project being different from the previous ones) standards give a basis to build upon in terms of both

quality assurance and sharpened project monitoring and control.

It is difficult to evaluate the efficiency of a process if the project manager does not have a reference to compare it with. Standards help establishing ways to execute the project that the management processes. Preliminary schedules and budgets as well as a project charter presented in a standardized way do simplify the analysis and the decision process of the project manager.

Moreover, it may be time consuming to establish a new set of documents to run a project (project charter, change requests, issue register, etc.). Having some pre-defined standards avoids 're-inventing the wheel'. Sharing information in a multi-site organization is often a considerable challenge; standards serve as facilitators for communication, monitoring and reporting.[47]

Added value is the actual total cost of the project, computed by adding the cost of each material and service unit used along the transformation stages. It is constructed from an accounting point of view only (including interest charges), as follows:

$$\text{Added value} = \sum_{t=1}^{5} \text{Actual charges attributed to the project} \quad (1.2)$$

where:
$t=1$ is the starting point when expenses first occur
$t=5$ is the end point when the project is completed

Added value is not concerned with intangibles. Perceived value is most concerned with them. Thus, a large dichotomy between the two sets of values indicates the presence of POVs, because often intangibles host or else foster POVs.

There is a third type of value—the residual value. It is often forgotten, but is best exemplified by way of the car industry. What brought Toyota ahead of General Motors (GM) over the decades has been the fact that the reselling price of Toyota cars and trucks is higher, on a par basis, than that of GM cars and trucks. Toyota's vehicles do not always depreciate as much as GM vehicles so that over the long term, a consumer ends up paying less; he has fewer worries buying a Toyota versus a GM car. At least many consumers and car rating organizations think so. The same concept applies to projects: once the Montréal Olympic Stadium's main reason for being had ended, that is, once the Olympic Games were over, what could

be done with the infrastructures determined the residual value of the project. As it turned out, it has been somewhat of a nightmare since the City
of Montréal eventually lost its baseball team (the Expos), which played at
the stadium; furthermore, the facility has experienced many difficulties
due to the harsh Québec winters. Parts of the structure have fallen down.
Residual value is related to norms of quality as quality persists over time.

We can express the three types of values by way of the stylized diagram in Figure 1.7.

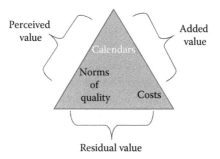

Figure 1.7 Three types of values.

A feasibility analyst spends some of their time looking at the lack of
harmonization between the three types of value in order to uncover POVs.

Projects address latent needs in the vast majority of cases. Part of
the reason why these needs are latent is that they did not have a way of
expressing themselves out in the open. Innovation is the key to awak-
ening latent needs. Thus, latent needs and innovation go hand in hand.
This can be exemplified in a natural context. Darwin, during his trip
around the world, observed a flower (named *Angraecum sesquipedalein*) in
Madagascar; its spur was so deep it would have required a bird with a
long beak to reach its nectar, but such a bird (or even such an insect) had
never been seen. Based on the assumption that the shape of the flower
served a purpose (was an innovative process), Darwin concluded that
there had to be a bird with a long enough beak to enjoy its nectar, and that
bird had a need to feed. This was eventually found to be the case.

A typical POV occurs when a latent need and the corresponding innovation don't match—there is no need to discuss whether the innovation is
too flimsy to fulfill the need. Thus, we can assume

$$POV \approx |\text{Perceived value} - \text{Added value}| \qquad (1.3)$$

Functionality is the capacity of an innovation to respond to a need. In
short, if a project does not offer the necessary functionality that it commands, then, most probably, there will be hurdles down the road: the Plan

will have to be revised. Processes are not going to prove efficient and/or efficacious, People will start arguing, and management will have to deal with all kinds of issues and headaches.

The functionality of the QMA is well established since it can adequately meet the need to host either sporting or entertainment events, one of those sporting events being the formerly latent need/desire to have an National Hockey League (NHL) franchise (it now is no longer latent as steps have actually been taken to try to get the franchise).

1.8.1 Errors and risks

Errors made during planning may or may not represent a potential for failure. Errors can even be beneficial. According to the Darwinian principle of evolution, species evolved by way of accidental genetic mutations that, by pure chance, have turned out to benefit the survival of individuals. Through sexual reproduction, changes are transmitted to the next generation, a process that may facilitate the making of more errors.

Errors may result from "acts of nature" or "acts of God" as is commonly known. Essentially, the feasibility analyst can ignore no POVs. They are generally classified according to their treatment,[48] just like for risks:

1. They have been resolved and documented
2. They are still active and pose a potential hazard
3. They deserve continuous monitoring

The use of tables such as Table 1.6[49] is common in project management—we saw that the QMA managers had such a table named "concerns" (*préoccupations*).

One way of classifying risks and POVs is by their levels of complexity, based on the assumption that complex projects present more risks (and imply more POVs) because there are more variables to control and possibly more uncontrollable variables.[50]

Table 1.6 Probability associated with POVs (or risks)

Probability of POVs (%)	Calendar	Costs	Four Ps
Very high 70–100			
High 50–69			
Average 30–49			
Low 10–29			
Very Low 0–9			

1.9   Innovation

Innovation is an intrinsic characteristic of projects. It necessitates walking off the normal paths set for doing things: this is what makes projects so exciting. As it has been once indicated when discussing research: "(…) any significant research requires that one tries out new paths and faces ambiguity to define new variables (…)."[51] Indeed, projects are spoiled with ambiguity and unknown variables that make them good targets for POVs to carve a niche in, mature, and take their toll on management and team members' efforts.

Patents and copyright certificates are indicators of the innovative aspects of projects. A truly innovative project must meet three criteria:

1. It must be unique
2. It must be useful
3. It should not consist of existing scattered elements that are put together without adding real value (functionality and design costs)

From this definition alone, one can guess that innovation is a diagonal process,[52] because it certainly means walking off the normal path, and venturing toward the unknown.

All innovations are based on an underlying concept; for example, ink, papers, parchment, pens, even sand or rocks can be used to convey the concept of written communication, and more precisely, take for granted the ability of receivers to read a given message. Similarly, airplanes, balloons, Delta planes, kites, and zeppelins all have "flight" as an underlying concept, or more precisely, the capacity to use air to travel. Most innovations emanate from a source inspired by nature. Velcro, for example, is based on the plant *Arctium lappa* L.

In essence, a project is innovative if it is expressed by all or most of the following functional elements (*F*): it has some kind of evolutionary goal; it has the capacity to take various forms; it may have various degrees of evolution; it implies that a compromise has been made (in a project, the compromise is between time, costs, and norms of quality); it has been inspired by a source such as Nature; it is a response to a problem (e.g., social or technological), need, desire, or discomfort; or finally, in the case of a project, it must have a certain utility (not all innovations are useful).

Some authors have presented a model that suggests that complexity and innovation define the two extremes of one single axis.[53] This implies that novel ideas cannot be complex, an observation that I question. What seems to occur in real life is that engineers try to balance functionality and design: a great design may not provide high functionality whereas great functionality comes at the expense of design. Products that succeed, such as the iPad, excel at mixing both contingencies through innovation.

From this perspective, innovation is the response to the merging of functionality and design, or, put differently, an expression of their matching levels. The best scenario is that 100% functionality is achieved while a top (ergonomic and hedonic) design is conceived. Since we know that functionality and design participate in the concept of value, we can affirm that a good innovation creates value given that costs are kept under control. Thus, when [Innovation | Costs] increases (↑), theoretically, the value of the project is boosted (↑). I can state this using my process modeling system, which I will explain in Chapter 4 on Processes (Figure 1.8).

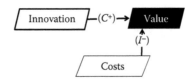

Figure 1.8 A process model linking innovation and value.

Noting that costs are a factor (a negative influence factor I^-), and taking into account that a great balance between functionality and design is usually achieved after much research (in fact, the complexity often comes from having to reconcile the demands of both functionality and design), finding the ultimate solution is a challenge. I propose that the horizontal axis of a model describing innovation be representative of costs and that the vertical axis be representative of functionality versus design. I thus temporarily present the following stylized model with respect to innovation (Figure 1.9).

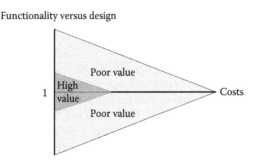

Figure 1.9 A stylized model for innovation.

Because this model includes all three components of the project value formula (functionality, design, and costs), it can be labeled "initial value proposition of a project". This proposition contains, *de facto*, the critical elements that a potential investor would want to see when a project is first presented to them (they'll generally worry about Processes—e.g., calendars of activities—People, and Power later on). It would precede the project charter, which once approved "formally initiates the project."[54]

Thus, the initial value proposition of a project is the first formal representation of a plan for a project. I would venture to say that all plans start with this. It is from this initial value proposition that the requirement for a prefeasibility study derives.[55]

1.10   Conclusion to Chapter 1

To the best of my knowledge, this chapter has done what few books or articles on project management and feasibility assessment have done. I have defined projects in a detailed, logical way so that a profound understanding of projects can be attained by the stakeholders. I have presented eight areas of human activities. I have introduced the notions of the three values—added, perceived, and residual—showing how they are linked together. I have recognized the importance of intangibles and of latent needs. I have established the notion of the initial value proposition. I have briefly presented the concept of closed dynamic systems, defined by their walls, ceilings, and floors. I have continued to articulate project knowledge around the concept of POVs.

This knowledge allows us to move forward and discuss prefeasibility studies, which we will do in Chapter 2.

1.11   What we have learned about POVs: Chapter 1

POVs ...

1. Are seldom discussed in scientific literature and in feasibility studies.
2. Express a CS, when internally controlled.
3. Are exemplified by such elements as Unfits (FP_{nc}).
4. Can be hidden (and often are indeed).
5. Can present a dual advantage, if tackled: the possibility of reducing costs (by way of finding inventive ways to rethink a particular process) and preventing the rise of costs due to processing problems.
6. Can sneak in along the process chain: inputs–transformation–outputs.
7. Can jeopardize a project *after* its completion.
8. Create uncertainty.
9. Are associated with ambiguity and unknown variables.
10. Are hosted or fostered by intangibles.
11. Require an analysis of the motivation (the opportunity) behind the project from the point of view of each important stakeholder.

12. Can be spotted by a large difference between perceived value, added value, and residual value.
13. Typically occur when a latent need and the corresponding innovation don't match.
14. Are associated with a Short strategy (when not tackled).

1.12   Key managerial considerations: Chapter 1

1. Ensure a full and accurate definition of the project
2. Explicitly define inputs and outputs
3. Identify the three values: perceived, added, and residual
4. Examine functionality, design, and costs

1.13   Case study Chapter 1: Maine East Pharmacy[56]

This pharmacy is located in a small town in the northeastern United States. It is a pharmacy that doesn't have the support of a larger group or chain, which could provide merchandising advice and sales force assistance. It is active on a number of Internet networks such as Facebook.

The owner has paid for market research over the years in an effort to boost his sales and to accelerate product turnover. His project is to redesign his store and customer interface to improve his business. Some of the key findings from past research are listed in Table Case 1.A.

Table Case 1.A MEP past business reports

Date	Past main recommendations
2013	Better display of products
	Better promotion of product lines
	Better signage
	Promotions during holidays
	Target nearby university students
2014	Better display of products
	Better outside signage
	Make store family friendly
	More promotion
	More use of colors
2015	Increase social network with Instagram and Twitter
	Encourage employee training
	Appeal to women

The store is located in a small town where the population is aging, with 21% being over 65 years old, and with the median age being 50. The percentage of children aged 0–14 is about 13. The average yearly income is US$20,000.

The store employs four pharmacists, five pharmacy assistants, one dietician, one part-time nurse, one part-time orthopedic insole specialist, as well as 15 employees assigned to various duties. The customers judge the staff to be pleasant.

Competition comes from two food retailer outlets located four miles away, a specialty store (five miles away), a local alcohol outlet (10 miles away), three drugstores (15, 24, and 40 miles away), and one Walmart (40 miles away).

Maine East Pharmacy (MEP) offers a large assortment of products and services. With respect to products, profit centers are as shown in Table Case 1.B.

Table Case 1.B MEP profit centers

Main category	Category	Sales 2015 First quarter (in US$0,000)	# SKU's[63]	Average sales US$/SKU
Health	First aid	980	3,600	272
	Baby	30	200	150
	Body care	22	2000	11
	Subtotal	1,032	5,800	178
Impulse	Tickets (Lotto)	30	30	1,000
	Smoking	9	50	180
	Chips/snacks	165	1,500	110
	Magazines	2	85	24
	Subtotal	206	1,665	124
Female	Hygiene and female	40	400	150
	Cosmetics	56	1,900	29
	Subtotal	116	2,300	50
Home	Durable home	59	1,200	49
	Cleaning home	20	500	40
	School	3	200	15
	Subtotal	82	1,900	104
Special occasions	Cards	55	200	275
	Christmas	17	1,500	11
	Others	72	1,700	42
	Subtotal	144	3,400	328

MEP also offers a number of services: ear piercing, foot care, passport photos, photocopies and lamination, prescription refills by phone or online, as well as an adjacent coffee and food bar (which has a door connecting to the pharmacy) and an adjacent gas station (owned by a different owner), both of which attract a lot of customers.

When a customer enters the store for the first time, they can immediately notice, consciously or not, that the floor plan is somewhat hectic and cumbersome. There are dead spots along/between shelves, the flow of products does not follow a standard pharmacy floor plan, there is an unused large room at the back, and there is an empty room leading to a storage space that is antique looking. Many products are not combined by type/category; products are located in antagonistic ways (HABA[57] next to pet foods, etc.). In short, the place is cluttered, offers mismatched items, has dead spots, and presents obstacles to the eager customers moving through the store.

There are two entry/exit points, but each one has major efficiency problems. The main entrance provides erroneous information: it promotes the adjacent coffee place, which in fact has its own door.

In the pharmacy retail business, the main hall that the customer faces once they walk in should normally be devoted to high-priced items (usually perfumes and the like), as the customer's wallet is full. In the case of the MEP, precious space is dedicated to low profit items such as inexpensive candies, which may convey the wrong image (Table Case 1.C).

The main exit provides a hindrance to purchasing. The items on sale cannot be purchased until *after* the client has paid for these other products and is leaving.

The customer is unlikely to step back and wait in line once more in order to buy a magazine or a phone card.

Entrance of the MEP.

Many products are put side by side when in fact they do not belong together, thus rebuffing customers (e.g., pet food beside ladies" products). Many aisles are cluttered: this creates dead spots, which disrupt the customer's line of thinking (and purchasing objectives) and confuse them (see Table Case 1.D).

An analysis of the computerized inventory system shows the following:

Exit of the MEP.

1. The coding system does not make it easy to group products by category or function, which renders in-depth analysis of data nearly impossible.

Table Case 1.C MEP's floor

Table Case 1.D MEP's areas of concern

The floor on the right side represents a potentially dangerous obstacle.	More cluttering that confuses the customer: he just does not know if this basket is someone's, or else if the products it contains are for sale. This is both a mental and physical obstacle.	The shoe and socks section is spread all over the store, which breaks the continuity of his shopping experience.

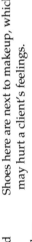

As another example, shoes are placed beside party balloons, a set which confuses the customer, who is then less inclined to buy them.

Shoes here are next to makeup, which may hurt a client's feelings.

Socks are placed about 10–15 steps away from the shoe section.

All clothing items are spread in opposite directions of the store: the customer is deprived of easy complementary product purchases.

(Continued)

Table Case 1.D (Continued) MEP's areas of concern

More cluttering and item misplacement (shoes).	Products are not being placed in a complementary manner (makeup with baby stuff).	Shoes in yet another section; this may create a sense of regret in the customer for not having seen the products before, when he debated the purchase of similar products in other aisles.

Revenue items are not being placed where they can be seen and accessed; in the world of the pharmacy, HABA receive a preferential display and treatment.

(Continued)

This display stops the natural flow on the planogram. A dead spot is created.

Here, nail clippers are adjacent to diapers: this sends the wrong message as to the actual use of the product.

Table Case 1.D (Continued) MEP's areas of concern

Food items next to shoe products and female hygiene products will push sales down. Specialty foods require a separate section in the store, in the general food area.	The specialty foods section beside the orthopedic section is contradictory and may even produce a negative image in the mind of the customer.	Another shoe section backing onto a window can be perceived as "threatening." This fact compounds with the many obstacles that clutter the floors, elbows,[64] and the planograms.
	 	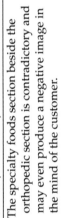 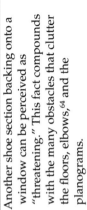

The orthopedic section is so cluttered it may represent a potential threat; customers may simply avoid it.

The sock displays face the shoes/chairs, but are short of all the socks on offer. This may make customers hesitant in buying.

Intimate products are placed on elbows; people need their privacy. This may turn customers off.

(Continued)

Table Case 1.D (Continued) MEP's areas of concern

More cluttering and dead spots.	More confusion, cluttering, obstacles, and dead spots. Here, the display prevents access to products.	People do not want to purchase behind people's back (people sitting on the chairs). The latter may not want to feel people behind their back.

Excess information may turn the consumer off.

More obstacles and mismatching.

Generally, signs are not easily legible.

(Continued)

Table Case 1.D (Continued) MEP's areas of concern

Tea is sold in the stationary section.

Customers may not want to deal with pictures that are potentially very personal in front of a waiting line.

The wipers do not belong in the food section.

The ATM requires privacy.

Possibility to build on a family concept.

Proper parking spots.

2. In some cases, the same kinds of products are entered in two different categories (e.g., "Glad zipper freezer" in "kitchen/bath/closet" and "Ziploc Freezer bag" in "household products."
3. Some products do not seem to be properly identified (e.g., "U by Kotex wipes" is entered as "paper products").
4. Cards, which are very profitable, must be better described (e.g., according to event instead of simply "cards, greetings") so cross-merchandising can be planned with special occasion items.
5. The coding is not systematic: sometimes the description mentions "lady" products, sometimes "women," sometimes "feminine."

This business offers many unique opportunities that can be easily transformed into profit centers. MEP would be advised to put itself in the eye of the customer and see how the store's current layout could cause apprehension for customers, who enter at an average rate of 30/h in this 1000 square foot area.

Overall, the observations of the flow of customers wandering into the store point to the fact that many sales are lost because the store (and the adjacent services—the café, the gas station) is cluttered; its layout and product placement impede the customer's ability to move around and make fast decisions. Products are not placed in a way that entices customers to open their wallets. More typically, expensive items are offered first, with impulse items being placed near the cashier, which in this case is not happening.

1.13.1   Questions related to Case 1: Maine East Pharmacy

1. What kind of feasibility study would this case entail? (environmental, etc.)
2. Discuss how a customer could experience some form of apprehension by walking into the store.
3. Prepare a short prefeasibility study and list all POVs, rank them, and group them where applicable (e.g., by level of potential lost sales).
4. Can perceived value be augmented by revamping the store layout?

Endnotes

1. Haji-Kazemi et al. (2015).
2. Dalal (2012, p. 2).
3. Oakley, G. Interviewed March 2, 2016.
4. Such as Dingle (1985).
5. Morris and Pinto (2004, p. 9).
6. O'Shaugnessy (1992, p. 10).
7. PMBOK 5th edition.
8. Cleland and King (1988).

9. Gilbreath, in Cleland and King (1998, pp. 3, 13).
10. For example, The Stationary Office (TSO) (2009, pp. 4–5).
11. Association francophone de management de projet (AFITEP): « Le projet est un ensemble d'actions à réaliser pour satisfaire un objectif défini, dans le cadre d'une mission précise, et pour la réalisation desquelles on a identifié non seulement un début, mais aussi une fin » (Afitep 2000, p. 3).
12. Asquin et al. (2005).
13. Austin and Luecke (2011).
14. O'Shaugnessy (1992, p. 2).
15. More than ever, we live in a world of changes—fast, intense, revolutionary.
16. In the economic system for the last 200 years, opportunities to retaliate were minimal; a producer could escape another producer's vengeful actions or a government's investigation by moving into a remote country, or by silencing consumers by way of gag lawsuits. This is becoming less and less true; greater accountability and accessibility put a fence around the economic system.
17. O'Shaugnessy (1992, p. 51).
18. The increasing capacity of firms to tailor their products to individual needs goes along these lines—Dell computers being a prime example.
19. For example that of the PMBOK and of O'Shaugnessy (1992).
20. In fact, by at least two of the three forms of outputs: deliverables, knowledge, and impacts.
21. This includes historical records and lessons learned—see Mulcahy (2013, p. 33).
22. See Mulcahy (2013, p. 173).
23. PMBOK 5th edition (2013, p. 122).
24. For example, Berg and Karlsen (2007, p. 8).
25. Russel and Stone (2002).
26. Meyerson et al. (1996).
27. Bstieler and Hemmert (2010, p. 485).
28. Girard, A. Interviewed October 2015.
29. Roult and Lefebvre (2012, p. 1).
30. Bourne (2015, p. 53).
31. Source: Therrien, G. Interviewed March 2015.
32. We have seen the importance of infrastructures in the introductory chapter.
33. Relate this to the African v. Haiti case toward the end of this book.
34. Oakley, G. Interviewed March 2, 2016.
35. See Corriveau (2007).
36. Courtot (1998).
37. I will discuss uncertainty in detail in Chapter 7 on Power.
38. Besides from the sex industry and the illicit drug industry or similar activities.
39. Another plan exists to that effect: *"Plan de gestion pour le contrôle de l'érosion et des sédiments"* (Plan for the control of erosion and sediments). The translation is by me as is the case for every translation in this book.
40. A full plan exists to that effect: *"Plan de gestion de la quality de l'air durant la construction"* (Plan for the control of air quality during construction). The translation is by the author as is the case for every translation in this book.
41. "Private–Public Partnership." These have been described as the new modern economic paradigm (see Cartlidge 2006; Bolz 2012, p. 11).

42. Miller and Lessard (2000).
43. PMBOK 5 (p. 15) refers to "business value" as the entire value of the business, which amounts to tangible and intangible assets; however, the link to projects appears unclear to us.
44. From my perspective, motivation and opportunity are two sides of the same coin. Motivation belongs to customers; opportunity to suppliers.
45. This will be discussed in more details in Chapter 7 on Power.
46. We will see that the higher the *g*-spread level, whether positively or negatively, the more the project value is compromised.
47. Bourgoin, D. Interviewed February 2016.
48. Morris and Pinto (2004, p. 37).
49. Adapted from Morris and Pinto (2004).
50. Complexity will be discussed in Chapter 7 on Power.
51. Parkhe (1993, p. 229).
52. We will discuss straight linear and diagonal processes in Chapter 4 on processes.
53. Sauser et al. (2009).
54. PMBOK 5 (p. 67).
55. There are many tools available to foster innovation such as conceptual maps, brainstorming, and the likes.
56. The name has been changed.
57. HABA: Health and Beauty Aid.
58. Other examples: forecasting reports, trend reports.
59. Examples: earned value reports, progress reports, status report, and variance reports.
60. Units are as follows: 24.5 mm = 1 inch, 0.3048 m = 1 foot.
61. The metric system is used in Québec and in Canada.
62. Of course, some projects can belong to more than one axis (more than one P).
63. SKU: Stock-keeping unit.
64. Elbows are located at the edge of aisles.

chapter two

Plan—Prefeasibility study

So it is, said the Wise Eagle, that nothing is impossible, as long as you know ahead of time what is feasible.

Darloz

2.1   Introduction to Chapter 2

The present chapter offers a methodology to quickly assess a project with as much information as possible; with an eagle eye, so to speak. The operational details are set aside for now; nevertheless, shedding light on the project from different angles is a way of uncovering obscure spots, of identifying points of vulnerability (POVs). As an example, one can look at a stone lying on the ground while the sun is throwing its rays on one side of it, thus generating a shadow on the opposite side. This is a dark spot, a potential POV. What we want to do is use as many sources of light as possible all around the object we are viewing—in the present example, a stone instead of a project—in order to eradicate the chances that dark spots, ominous troubles, or covert agendas could develop. In

Many angles.

fact, meteorology has it that the understanding of hurricanes, as an example, comes from simultaneous observations from different vantage points.

2.2   A definition of prefeasibility

I define a prefeasibility study as follows:

2.2.1   Definition of a prefeasibility study

> The prefeasibility study, which follows the initial value proposition of the project, offers a general view of the said project using various analytical frameworks that allow the feasibility expert to make a recommendation on the suitability of a feasibility study.

Five frameworks (suns throwing light) are deployed in a prefeasibility study. They are (1) the frame of definition (S and F process elements), (2) the contextual frame of risks, (3) the project's potentiality, (4) the project's parameters, and the (5) Key Success and Failure Factors (KSF and KFF).

I now discuss each framework in detail.

2.3   Frame of definition

This frame serves to confirm that the project has all the mandatory features that make it a project. Common sense dictates that if a project contains no technical challenges, has no elements of innovation, or else does not respond to an opportunity, there is no justification to conduct a prefeasibility study. A project's promoter should be able to convince the analyst that his/her project is unique, is contained within a limited time frame, is innovative, that it includes challenges, that it guarantees the leader's and the team's involvement, that it requires some level of investment, and, of course, that it responds to an opportunity (Figure 2.1).

Figure 2.1 Frame of definition.

In Table 2.1, I take the example of the Québec Multifunctional Amphitheatre (QMA) and check its characteristics.

Besides checking the characteristics of the project, the feasibility analyst will want to verify that all aspects of the project[1] are properly identified and defined. Again, let us use the QMA example, as shown in Table 2.2.

Finally, the analyst wants to determine the size and the nature (Plan) of the project, according to the remaining three Ps: the thrust of this effort may be Processes (e.g., a NASA undertaking), People (e.g., a slaughterhouse in Africa), or Power (e.g., an election) depending on the project, or a mixture of these Ps.

One of the entrances of the QMA.

2.3.1 Points of vulnerability and the definition frame

Through his/her analysis, the feasibility expert should try to identify the various POVs that may weigh on the project. Clues that he/she can use are given in Table 2.3.

All of these ways of looking at the project complement each other and guarantee, in theory, that the project is well planned and understood from a descriptive point of view. The analysis should not merely consist of listing characteristics or process elements; rather, the focus is on uncovering POVs.

Table 2.1 Assessment of some characteristics of the QMA project

Some characteristics of the project	QMA example
Responds to an opportunity	Replacement of the old Colisée stadium
Guaranteed stakeholder involvement	City of Québec
Is unique	Unique design
Innovative	Design
Is subject to a time constraint	Due date of September 1, 2015
Requires some investments	From multiple stakeholders
Contains challenges	For example, building during winter

Table 2.2 Structural and functional elements of the QMA project

Description element of project		Example from the QMA
Structural Elements		
Plan (discusses...)	Actual plan	Yes ("*Programme fonctionnel et technique*")
	Calendar or activities	Yes, with grand opening in September 2015
	Costs	Preset: C$400 million
	Norms of quality	Preset
Processes[46] (has a transformation phase)		For example, construction stages
People (involves...)	Users/customers	Citizens/sports teams/artists
	Producers	For example, the City of Québec
	Regulators	For example, Pomerleau
	Bad apples	None
Power (makes use of...)		Line of authority clearly defined
Functional Elements		
Generates deliverables		Entertainment and sports venue
Creates formalized knowledge		Yes; by the company WSP
Has some impacts (+ and −)		Boosts Québec's image

2.4   *Contextual frame of risks*

As I discussed in the general introduction, risks and vulnerabilities are two different concepts. Some authors see hazards as a source of threats[2] with these threats being capable of causing damage or, in terms of projects, of compromising the calendar of activities, the cost structure, or the norms of quality for the given project. Vulnerability is the sensitivity that one (or a project) has toward the threat.

Table 2.3 Definition frame and POVs

Assessment of	Example from the QMA
Dominant strategy	All key managers are working in the same office, but are there some external stakeholders that are not present?
Contingency strategy	Reserve funds, but are they sufficient?
Existing infrastructures	On existing stadium ground, but traffic problems?
Functionality	Entertainment and sport venues, but overly ambitious?
Design	—
Perceived value	Will be subject to complaints because of possible increases in taxes?
Added value	—
Residual value	—

Viewed from this angle, a risk is simply the probability that a threat can cause harm (cognitively speaking, can cause an error in the decision-making process).

The contextual frame of risks lists the major areas in which the potential for setbacks could occur. It may remind some readers of the PESTEL framework, but in fact, it is different: PESTEL refers to some "political, economic, social, technological, environmental, and legal" work frames. I use the wording "sociocultural" when referring to external risks, and the term **work psychodynamics** to describe the internal cultural and social interactions between stakeholders. As we shall see in Chapters 5 and 6 devoted to People, human interactions can be examined much like a mechanism with its own nuts and bolts. I illustrate the risks in Figure 2.2.

Figure 2.2 A project's contextual frame of risks.

In the case of the QMA, environmental concerns were treated by way of Leadership in Energy and Environmental Design (LEED) certification, obtained through credits allocated to various aspects of the project such as the efficiency of water management, the efficient use of materials, resources, the conservation of energy (avoidance of wastes), sustainability (e.g., durable materials), the presence of ecological sites, the quality of innovation and design, and the quality of the interior spaces (ventilation, etc.) during construction and during occupation (once the structure is complete).

A political risk assessment model has been presented[3] whereby the risk source (in the host country) leads to possible events (e.g., a government intervention), which then leads to a risk—seen as the interaction between a threat and vulnerability—which then has some consequences (such as cost overruns). In this model, the interaction between threat and vulnerability is stated. Assuming that threat is the probability that a risk can cause harm, the authors of the aforementioned model rightfully correlate risks and vulnerability, and rightfully point to potential grief (under the heading "consequences," which, in this context is a positive causal bond C^+).

I have separated risks into eight different contextual categories, which means that POVs may appear in any one of them. I represent the context of risks as influence variables ($I^\pm$) in Figure 2.3.

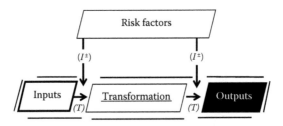

Figure 2.3 A project's risk contexts and the core model.

The reader will note that I consider the influence to be moderating ($I^\pm$)—thus, the arrow emerging from the process element called "contextual risks" points toward an arrow linking two process elements and not toward a single process element/parallelogram[4]. This means that for some stakeholders, risks are actually a source of motivation while for others they are a nuisance. Hence, the influence can be positive or negative ($I^\pm$). It is not so much risk that is the problem (there will always be risks) but the way it is managed.

2.4.1 Points of vulnerability and the frame of risks

We have seen that there is an intimate link between risks and vulnerabilities. Put simply, risks are external forces that affect a project while vulnerabilities are internal points within a project (within each of the four Ps) that weaken its chances of success. As previously discussed, my research shows their relationship,[5] which can be seen in Figure 2.4.

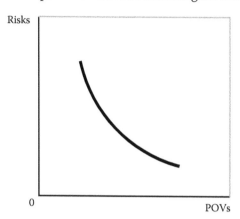

Figure 2.4 Risks and vulnerabilities.

The function is, as seen in the general introduction:[6]

$$\text{Risks} \mid \text{Constraints} = \frac{k}{\text{Vulnerabilities}} \tag{2.1}$$

where k = constant, valued at 1.3 in an ideal scenario.

I will examine this thoroughly in the section on work culture (psychodynamics) in Chapters 5 through 7. For now, all there is to know is that risks exist only where there are vulnerabilities—obviously, if an individual, for example, were not vulnerable to anything, no amount of risks would trouble them.

The feasibility analyst wishes to determine what risks can negatively affect the project [risks (I^-) → project].[7] In the case of the Mervel Farm example, see Table 2.4.

One way of determining risks is to list everything that can go wrong around a tantalizing project. Although this may seem like a negative approach, it is actually the most thorough method the analyst can use to perfect his/her analysis.

2.4.2 An example: A car oil filter

This case is inspired from a real venture; I have modified some elements in order to preserve confidentiality.

In the 1990s, three would-be East European inventors concocted a car engine oil recipient and filter. The promoters conceived a gilded plan that

Table 2.4 Contextual risks for the Mervel Farm project

Type of contextual risk	Four Ps	Example
Financial	Plan	Disagreements between the National Capital Commission (NCC) and the local town
Organizational		Difficult decision processes from NCC and the local town
Environmental	Process	Weather, pollution
Technological		None
Marketing	People	Local residents ranting about traffic
Sociocultural		None
Legal	Power	Issues of ownerships and rights over the cooperative
Political		NCC versus local governments

went about like this: They would sell two containers about 12 inches in diameter by 12 inches in height to each customer. Each customer would install the container in their car, connecting wires and pipes in various ways in order to reroute part of the engine oil into the container. This container would hold a certain amount of engine oil in its filter as the customer would drive his car. After about a month or so, the customer would remove the container and pack it in order to send it to a laboratory for analysis, at their cost (e.g., using a special delivery service). They would take the extra container and install it in their car to replace the first one. After a few weeks, the analysis done at the laboratory would provide the customer with information on the quality of their engine oil, including various minerals and dust levels. The promoters of this idea were convinced that their "invention" would entice customers to make better use of engine oil and motivate them to keep their engine clean, and thus making it more efficient, leading to savings and a better environment. They did not deem it necessary for the customer's insurance company to be notified of the change, nor the car manufacturer to be informed how such a such new element in the mechanics of the car would influence the riding experience.

When I give this example to engineering students, the following risks (all due to outside influences[8] and not to inherent vulnerabilities) are commonly listed: (1) financial—the cost of the patent, the lack of outside financing, and the requirement for a double purchase by the consumer with two units being required at a time; (2) marketing—competition (e.g., from companies such as Midas), a lack of interest from consumers, a lack of a real need/ opportunity, or a lack of understanding of the product by the consumer; (3) organizational—complex relationships between the car owner, the three promoters, and the external laboratory; (4) environmental—contamination in the external laboratory; (5) technical/technological—accidents, breaks,

erroneous use by customers, loss of the unit during transportation, product malfunction, poor or tainted laboratory analysis, or troubles at customs
(POV: poor capacity to obtain and defend patent); (6) sociocultural—bick-
ering with regulatory authorities (government, insurance companies, etc.);
(7) legal—possibility of the falsification of data, lawsuits consequent to
accidents (POVs: difficulty to offer warranty, disrespect of norms); and (8)
political—no risk identified.

These risks can be weighted in regard to presumed internal vulnerabilities: Plans—the project seems overly optimistic; Processes—the control of many processes is out of the hands of the promoters; People—key
staff (e.g., laboratory technicians) are outsourced (possible "bad apples");
and Power—the three promoters have little credibility in the domain.

A feasibility analyst is trying to set measurement scales so that their
analysis can be compared with other projects. Assume the analyst arrives
at the evaluation of the project given in Table 2.5.

Table 2.5 A quantitative measure of the contextual frame of risks

Type of contextual risk	Four Ps	Evaluation on a scale of 0 (no risks) to 10 (high risks)
Financial	Plan	9
Organizational		9
Environmental	Process	5
Technological		7
Marketing	People	10
Sociocultural		7
Legal	Power	9
Political		0
Average		7
Prefeasibility index		10 − 7 = 3 or 30%
Decision		No go

The **prefeasibility index** works as follows: strictly based on a risk
assessment, a value of 96%+ suggests that the project can proceed[9] ("go");
with a value ranging from 80%[10] to 95%, inclusive, the analyst would
impose strict conditions and would request more information about the
project; with a value ranging from 0% to 79%, a "no go" would be advisable.

Playing the devil's advocate role is what must be done when seeking
to unearth POVs. In the case of the oil recipient, the major concern is the
lack of demand for the product and the major POV is the lack of control by
the promoters. Remember that one of the key characteristics of a project
is the presence of a need (opportunity)—in this case, a consumer would
simply go to the garage every six months and change the oil as a routine
effort. There are no latent needs to speak of.

2.5   Frame of potentiality

This frame of potentiality is based on the *Project Management Body of Knowledge* (PMBOK) and is a way of ensuring that projects, big or small, are well articulated. A few items must be checked, according to PMBOK. I classify them using the four Ps system, as shown in Table 2.6.

Table 2.6 Frame of potentiality

List	Four Ps
Scope: Is the plan realistic?	Plan
Well-articulated plan: Is it so?	
Change processes: Are they adequate?	Processes
Measures of success[48]: Are they well established?	
Group of managers: Have the best managers been chosen given the circumstances?	Power
Monitoring system: Have the necessary safeguards been put in place to prevent derailment?	
Not addressed in the PMBOK (not needed yet)	People

At this point (prefeasibility) it is too early to measure potentiality in terms of People and in terms of efficiency/efficacy.

Various authors differ in their opinion on such an assessment. Some say, for example, that critical parameters relate to what is called "a market/ economic hypothesis", which includes, in their views, legislation, the market opportunities for the new investment, the natural environment, the social and political environment, and so forth.[11] A large vortex of candidate POVs is offered: an impoverished image for the company due to poor use of new technologies in the production process; the absence of cooperation and problems with personnel training; a lack of flexibility in new processes; the presence of dangerous, tiring, monotonous, and routine tasks for labor and so forth.

Certainly, such lists can be of value, but it seems that they can go on forever. The best approach, I feel, is to resort to a simple framework as per Table 2.6, especially in the prefeasibility assessment phase.

The PMBOK specifies that the questions listed in Table 2.6 must be posed throughout the project life cycle, not only at the preliminary stage. Indeed, they may act as a barometer; the more precise the answers are, the more useful the analytical frame is for the project manager that works day in and day out in the field.

As a brief overview, let's go back to the Mervel Farm project. If we go by our frame of definition, we must confirm that all four structural variables (the four Ps) have been defined. Is the plan well defined? Will the planned processes likely take place? Does the proposed project deal with stakeholders (People) and lines of authority (Power)? Moving on to the contextual frame of risks, have the promoters taken into account the risks in all of their forms (marketing—neighbors' complaints about noise or odors; environmental—e.g., farm pollution)? As for the frame of potentiality, how effective are the control/monitoring systems (for the loads of volunteers, for the flow of tourists, or for vehicle traffic)?

During the prefeasibility stage, the analyst's evaluation is quite subjective, but this effort is worth it nonetheless. Working through each framework at a time allows the feasibility analyst to get a tentative picture of the dark spots (POVs) dotting the project even though he has only taken a rough look at it. It may raise some flags without sounding the alarm; this is a way of avoiding catastrophe (chaos) down the road. As put forth by a project manager:[12] "A catastrophe is a situation which impacts exceed anything that was initially forecasted." Chaos is easy to see in the Mervel farm case: a massive arrival of foreign tourists showing up unannounced in various coaches, huge traffic jams build up downtown, annoyed neighbors and local businesses can't profit from it and file complaints. The tourists finally find their way to the farm as a thunderstorm kicks in. People seek refuge inside the farmhouse. Soon enough, there is a melee; not enough toilets; people walk involuntarily onto crops; they start whining in English, French, Spanish, Chinese, Urdu, and Esperanto(!); the army of volunteers panics, and it goes on and on.[13]

2.5.1   Points of vulnerability and the frame of potentiality

Table 2.7 focuses on some core questions recommended throughout the PMBOK. These relate to internal processes, and thus to potential vulnerabilities. It appears useful at this point to offer a wild guess at what would breed these vulnerabilities from outside of the projects, indicated in Table 2.7.

Hence, the frame of potentiality makes a lot of sense when it is accompanied by the contextual frame of risks.

Table 2.7 Risks and POVs in the frame of potentiality (examples)

Potential risks (↑)	Potential vulnerability (↑)	Four Ps
Community opposition	Unrealistic plan	Plan
Unstable economic environment	Poorly articulated plan	
No contracts signed with customers and suppliers	Diffuse measures of success	Processes
Poor impact on community	Inadequate change processes	
Social pressure to perform	Incompetent managers	Power
The presence of bad apples	Lack of safeguards to prevent derailment	

2.6   The parametric frame

We have very briefly seen that we could model processes with the use of parallelograms. However, I have not yet explained in detail how we go about specifying measurement scales, although we have sparingly touched on walls, ceilings, and floors. In fact, there are three ways of modeling, which are as follows: (1) process modeling, (2) psychodynamic modeling, and (3) parametric modeling. The first type uses parallelograms and arrows, the second bubbles and arrows (which we will discover in Chapters 5 and 6 on People), and the third one uses boxes and lines— the three lines representing walls, ceilings, and floors. In short, boxes and lines are a way of indicating that what is referred to in the model is a measurement. For example, a ceiling of US$25,000 is a measurement that says that the project requires US$25,000 in order to commence (below that level, the project has not commenced, hence the reason we call it a ceiling: it is the ceiling of the non-start-up of the project).

With respect to walls, ceilings, and floors, the basic assumption in a project feasibility analysis is that a project evolves in an ascending linear fashion (Figure 2.5).

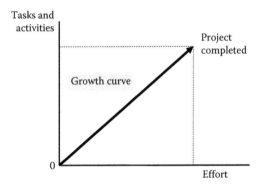

Figure 2.5 Basic model for the evolution of a project.

This assumption is simplistic (we set it with a slope of +1), yet it leads to important consequences. If one thinks about it, the assumption in classical economics is that buyers and sellers articulate the quantity of goods demanded and offered around price alone (when in fact functionality and design count as well). It is very simple and yet, from this simple model, a vast area of knowledge and activity has developed over centuries. Here, we simply state that as more efforts are put into the projects, the number of tasks and activities related to completion increase. In fact, we can also reverse the statement and conclude that as more tasks and activites are planned, more efforts will likely be needed. Here, we are in the planning stage; by reversing the graph, we would be in the mobilization and deployment stages.

Walls, ceilings, and floors transform the linear evolution of projects. The analyst can place them around the linear growth curve exhibited in Figure 2.5 and specify what they represent; conversely, he/she can examine a curve and break it down into stages by inferring that the presence of walls, ceilings, or floors is what transforms the straight curve (with a slope of 1) into a nonlinear curve. The scenarios given in Table 2.8 depict these concepts.

This way of seeing a project allows the feasibility analyst to determine which of the calendars (walls), costs (ceilings), and norms of quality (floors) are most important during the evolution of each curve. In the case of a negatively sloped curve (at −1), for example, only the ceilings (which push the curve down instead of up) and the floors are active, with no walls playing any "pushing" role. This means that costs (ceilings) and norms of quality (floors) rather than the calendars (walls) play a role *between* the beginning and the end of the project. In the case of a positive concave curve, a series of walls (calendar activities) push the linear ascending curve (with a slope of 1) to speed up (hence the concave shape). Eventually, ceilings (costs) start kicking in. In this latter case, therefore, it is calendar and costs that drive the project, not costs and norms of quality. S-curves, bell-shaped curves, and the like are simply combinations of these six basic patterns appearing in Table 2.8, which are also forged by the interplay between walls, ceilings, and floors. These forms represent the mathematical functions that appear regularly in project processes. For example, some authors propose that calendar (x) and accuracy of estimates (y) form a type CP (concave positive) curve; that calendar (x) and ability to control costs (y) form a type CN (concave negative) curve; and that calendar (x) and project risks (y) form a type CN (concave negative) curve.[14,15]

We will encounter many of the basic curves throughout this book. They can all be explained by way of walls, ceilings, and floors affecting the basic linear positive curve with a slope of 1 (+1), which is one of the basic tenets of my model. The walls start or stop the continuation of the linear curve (beginning and end walls), floors set a minimum level that cannot be transgressed and that divert the direction of the basic linear curve, ceilings set a maximum level that cannot be transgressed and that divert the direction of the basic linear curve.

It is the presence and the different positions along the linear ascending curve (the **growth curve** of the project) of the walls, ceilings, and floors that create each individual curve type. This makes comparing processes relatively easy because what needs to be investigated is the presence and placement of walls, ceilings, and floors in order to comprehend any process (e.g., with respect to time, costs, or norms of quality).

To give an example, let's take a hypothetical curve,[16] such as Figure 2.6.

Table 2.8 Walls, ceilings, and floors

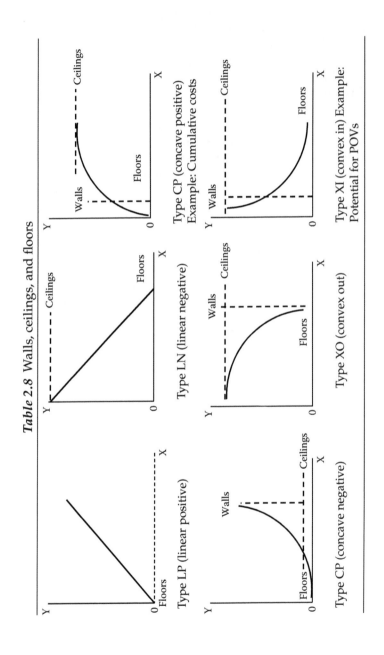

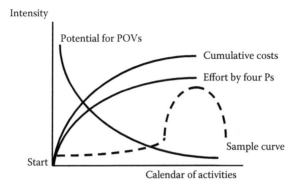

Figure 2.6 A sample hypothetical curve.

This curve is initially quite flat and remains that way for some time; however, much like a growth spurt for a teenager, all elements of the project come to play an active role during the implementation stage as the deadline quickly approaches. Once the project is completed, there is little left to do except close the project. An expert in handling this type of analysis will easily identify curve patterns of processes and determine which of the three constraints to focus on.

What is important here is that parameters set the frame within which the project develops. We will see that walls, ceilings, and floors are important considerations during the life cycle of a project; what matters is to set the parameters' frame of activities that will govern the project. The Building Code is a floor (bottom) parameter: a construction project cannot be approved without abiding by the standards set by it. For the African project, key parameters/requirements would have been the presence of roads as well as water and electrical infrastructures.

In terms of the procedure for producing parameters and setting scales, the following steps have been proposed:[17]

- Step 1: The analyst creates a model after reviewing the literature, conducting interviews, and leading discussion groups.
- Step 2: He/she establishes measuring criteria and places them in a hierarchical order.
- Step 3: He/she conducts a pretest to validate the newly created instrument.
- Step 4: Finally, he/she confirms and administers the questionnaire.

Judging a project without qualitative measurements (added to quantitative measurements) is a sure way to leave POVs uncovered, because it opens the door to faulty interpretations—in addition, at times, people prefer to keep their eyes closed rather than facing the vicissitudes of a project life cycle.[18]

Reverting to the Mervel Farm project, we as analysts would be reviewing what has been written on similar projects; we would conduct interviews with the current promoters and with experienced managers of similar projects completed in the recent past, and we would have a focus group with some of the neighbors. We would then select and classify criteria, thus allowing us to assess whether the project is viable, and successfully so. These criteria would include items such as acceptance by the local community, commitment by the city to upgrade the parking lot and traffic-related infrastructures, management experience, as well as the presence of firm contracts.

A questionnaire would be generated with the aim to measure the overall likely demand for the products and services offered by the promoters. Once validated, it would be distributed to a sample of 1000 local residents and business owners in order to run statistical analyses. In the best-case scenario, everyone would be happy. In the near-worst case, most would reject the idea. In a doomed scenario, the analyst would conclude that the project is feasible and profitable when in reality all kinds of POVs would sprout as the project takes place. Such catastrophic scenarios are possible. Let's imagine bad weather developing, panic kicking in, and so on. It would be normal for the participants[19] who fill out the questionnaire not to have forecasted such problems—this was not their responsibility. Their viewpoint may already have been biased: they responded to the questionnaire assuming the project had been properly planned. The promoters could always tweak the information to make it look like all is fine (overly optimistic), even after having interviewed some experts in the field.

The five analytical frames, including the parametric frame, come in handy because they allow the analyst to examine the project from different angles so that the POVs that people do not think of, do not wish to admit, or else ignore can be unveiled.

Recall that I mentioned parallelograms and boxes when dealing with process modeling. Boxes come in two types: parameters and observables (we will discuss observables in Chapters 5 and 6 on People). Parameters are quantitative criteria that are used to indicate what the value of a parallelogram is. A boxed value can point to a wall, in which case it indicates that the value of this wall is, say, July 16, 1976. The value can also be a range. For example, we will see that the ideal ratio between management control and transparency when it comes to Power is k, that is, 1.3; in general, the ratio remains functional between the values of 1.0 and 1.8. So a box pointing to the construct Control/Transparency could contain the value range of $[1.0 < x < 1.8]$.

Indeed, when preparing a project and when the transformation phase is not developing along a critical path, it may be useful to allocate an **operating margin**.[20] The operating margin is a range that the project manager is willing to allocate to a particular activity; say, for example, finishing the carpentry inside a room with an allocation of 1–2 days. Figure 2.7 illustrates my point.

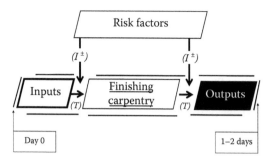

Figure 2.7 Boxes and lines.

Now, suppose we wish to explain the fact that the birth of a particular project is influenced by an initial investment that varies between US$20,000 and US$25,000, no less, but no more (in which case, that money would have a better opportunity of earning interest somewhere else, for example, in another project). We would have: Initial investment $\downarrow$ US$20,000 $\overline{\uparrow}$ US$25,000 influences positively $(I^+) \rightarrow$ project commencement $(|\overline{\uparrow})$. The codes are shown in Table 2.9.

Table 2.9 Codes for the parametric frame

Parametric coding	Meaning
$\underline{\downarrow}$	Minimum
$\overline{\uparrow}$	Maximum
$|\overline{\uparrow}$	Beginning wall
$|\uparrow$	Beginning stage
$|\underline{\downarrow}$	End wall
$|\downarrow$	End stage
$\underline{\downarrow}$	Floor
$\overline{\uparrow}$	Ceiling

Used jointly with the so-called descriptive and consequent arrows, which we shall see in Chapter 4 on Processes, an entire process can be written using this symbolic language. Let us take the example of the Stradivarius concert. We would have $[(\underline{\downarrow}$ US$20,000 $\overline{\uparrow}$ US$25,000 investment $(I^+) \rightarrow |\overline{\uparrow}$ Project) + (Ticket sales)] $(C^+) \rightarrow$ Project success $|\underline{\downarrow}$. This reads: with a minimum investment of US$20,000 and a maximum investment of US$25,000 coupled with ticket sales, conditions are met to ensure the concert's success. This simple way of describing the project has the advantage of showing how critical steps for a causal link (C) are more critical than a link of influence (I). It also shows the relationships between the process elements (or stages) and has the ability to assign a value (a parameter) to each such element, such as a value of US$25,000 for a ceiling maximum cost (⟰).

It is fundamental that every process element be measured and that steps are quantified; thus, as part of the prefeasibility study, every such element that is disclosed must be accompanied by a parameter, which, in a diagram, would be enclosed in a box. Boxes are not related to time. They refer to a number (or, in the case of psychodynamic analysis, to an observable).

Because the parameter coding system reveals the most critical aspects of any process, it cuts straight to the point and produces a pristine working model. It is a great tool to identify POVs because it outlines at a glance where there are potential weaknesses in the flow of reasoning. In the parametric frame of analysis, the emphasis is on tasks and engineering, rather than on the human or organizational aspects of the project. A project promoter, who would not have a clue about how to set the measurement parameters, would disclose that they are not ready to launch said project. Measurements are tailored to each unique project and hence help "contextualize" it, making it more concrete and easier to adhere to by future team members. The following steps are proposed to complement step 2 of the aforementioned method:[21,22]

- Find parameters that already exist for specific tasks and activities. In the home building industry, for example, tasks are divided among various trades, which are governed by standards (e.g., procedures), code (e.g., the Building Code), as well as laws (e.g., the Safety Act). Some authors emphasize the importance of determining what skills are required and what infrastructures are necessary before launching a project, in addition to consulting with stakeholders such as the local community.[23]
- Find appropriate forms, formal documentation, and procedures manuals, such as project contracts, submission forms, technical specification forms, and so forth (e.g., in a scientific laboratory or chemical plant, a material safety data sheet [MSDS]). A promoter that cannot include at least some of these essential forms with the submission of his project is sending a message that he is not ready to initiate his project.
- Find equivalents (for standards, codes, and laws) should exact forms or adequate parameters not be available on site. Building codes, as an example, vary according to the seismic activity of a given country. It might be useful to know what metrics other countries take against earthquakes, if only to prepare for a construction project in an area presenting seismic activity.
- Find similar projects. One can then determine if past learning can be of use. Seeing where similar projects have encountered difficulty is one way of identifying points of vulnerability (POVs).

In the case of the Mervel Farm project, two similar projects are readily available for comparison. They are the cases of two not-for-profits organizations in two different Canadian cities, which decided to organize a hidden chocolate egg contest in their respective municipal parks for the Easter holidays.

The actual number of parents and children that showed up far exceeded expectations, resulting in a near rampage by overexcited children who got there first while latecomers (or runners!) wound up in tears at having been left with nothing. Is this a lesson that can be learned at the Mervel Farm? Probably. Success, strangely enough, can be a point of vulnerability, if it is not adequately managed.

Similar product, contrasting projects.

A promoter who has no or little backup to justify his project may be initiating POVs without even realizing it; hence, the importance of conducting a prefeasibility study that covers as much material as required, and the relevance of meticulously setting parameters (which are entered in a box when illustrating the project in diagram format).

2.6.1   Points of vulnerability and the parametric frame

My modeling proposition has the advantage of finger-pointing at POVs. We will see in Chapter 4 on Processes that it simplifies all processes by using four types of action: Descriptive (S and F), Longitudinal (T), Influence (I), and Causal (C); and three forms of parameters: walls, ceilings, and floors. We suppose that all of the processes that need to be expressed in a given plan, and that eventually must be examined in a prefeasibility and a feasibility study, can be rendered by way of these simple tools.

As an overview of Chapter 4 on Processes, what needs to be done is to draft a model that expresses the project in its totality (not in its minute details yet) and set the scales that will allow the analyst to evaluate whether each step of the project will likely happen according to the preset norms of quality.

Modeling can help shed light on POVs; these points are already anchored in the project or will develop where the logic of the process fails. To understand how this works, a series of questions must be raised, such as: "Is the link between process element A and process element B really an influence (I^+) link? Could it be a causal link (C)? Has the analyst identified all the elements of the main process elements? Why is there a wall at this particular stage in the transformation process?" And so forth.

In other words, the feasibility experts check whether the reasoning behind the project model makes sense. If it doesn't, if a component of the model cannot be explained in a logical manner with a way of measuring it, then there is likely a POV. A link that is wrongly identified between two process elements (it turns out to be a causal link, C, rather than an influential link, I), a main element that is not well defined or incompletely assessed (so that some production steps have been forgotten), or else a wall that cannot be justified are all indications that the project may crash, in part or in whole. The relevance of that crash is measured when looking at the parametric frame of analysis.

Hence, the prefeasibility study is concerned with properly defining the project, identifying the contextual frame of risks, and establishing the scales and parameters. This will become clearer as we progress through this book. For now, it is enough to remember that projects can be put into models and that it is of foremost importance to identify the parameters (such as the Key Performance Indicators, or KPIs) that will permit its measurement. When working with diagrams, parameters are put in boxes, whereas process steps are put in parallelograms.

2.7　Frame of key success factors and key failure factors

The goal of the feasibility expert is to evaluate methodically whether the proposed project has any chance of meeting its objectives; perhaps they would even like to estimate what the probability of success is. In other words, they must set the parameters that help determine whether a project is a success or a failure. Thus, the last but not least analytical frame is that of Key Success Factors (KSF) and Key Failure Factors (KFF); this is tantamount to measuring a project's robustness (see Appendix 2.1).

There are two difficulties inherent in this terminology. First, it may be hard to differentiate between a KSF[24] and a KFF—indeed, some of the KFF are just the opposite of KSF!—and so, for all intents and purposes, my qualify KSF as referring to everything that enhances strengths and opportunities, and KFF as everything that energizes risks and vulnerabilities.

Also, in my modeling system, so-called factors relate mostly to moderating variables[25] ($I^{\pm}$), but in fact the so-called factors of KSF and KFF are not factors *per se*, but rather process elements of influence (I^{+} or positive influence for KSF and I^{-} or negative influence for KFF). In the literature, so-called KSF and KFF most often consist of strategies, tactics, and/or behaviors that affect the transformation phase of the project either positively or negatively. Thus, the terminology of KSF and KFF may be confusing. Because the standard terminology in about every article and book written about the success and failure of projects is tied to the concepts of KSF and KFF, I adopt this terminology, albeit reluctantly. The reader should keep in mind though that the term "factor" in this context refers to a process element of influence (I) when in fact KSF and KFF are usually strategies rather than factors. Let's settle by assigning the expression "Key Success Fundamentals" and "Key Failure Fundamentals" to KSF and KFF, respectively. Also, please note that I treat the difference between KSF and KFF with some skepticism.

Examining historical projects in order to pinpoint what has worked and what has not is certainly a sound idea, and so is reviewing what scientific authors have thought on the subject. Some scholars are of the opinion that a healthy and safe working environment is an essential basis for an effective work team.[26] Accordingly, a KSF would be for a project leader to provide a healthy and safe environment (see, it's not a factor, it's closer to a strategy!); viewed from another angle, a KFF would be to accept work in an unhealthy and unsafe work environment, such as is found in many underdeveloped countries around the world (yet, many projects are brought to term even though working conditions are unhealthy and unsafe). The same authors also elaborate by claiming that a team's performance is tied to the functionality of interactions between stakeholders, the group's stability, the involvement of people in positions of authority, and the rewards that achievers receive. Other scholars have expressed their views that duplication of work, lack of sharing of responsibilities, shaky coordination, and poor task definition are common causes of problems[27] —put differently, these factors may be a cause (C) for or an influence (I) that leads to failure, that is, they are a KFF if they turn the project into a fiasco. A list of necessary yet insufficient conditions that are strongly associated with perceived success has also been given by other scholars.[28] This list can be matched against some of the eight contexts where risks operate their magic (I also include a reference to the four Ps and to external–internal control). I display this in Table 2.10.[29]

For each one of these elements, their presence or absence may potentially influence the success or failure of the project, so that there is no clean separation between what leads to success and what causes failure (C).

Overall, what seems to happen though is that there is a **dreadful combination:**[30] that of poor planning intertwined with Unfits (uncontrolled Forces of Production, FP_{nc}). Inadequate planning, which entails such problems as porous budgeting and scheduling as well as the reliance on excessive red tape may all mix in a somber manner with team members who display a lack of commitment to established schedules, are not cooperative, send excessive negative feedback, show lack of enthusiastic support for the team or else host dark agendas. Behaviors that would accompany such an organizational state would likely include "buck-passing", a disregard for others, doodling, filing excessive grievances, procrastination, and sabotage. The following could also happen: constant interruptions, enforcing unannounced changes, holding

Table 2.10 Risk contexts, success factors/POVs, and the four Ps

Risks context	Reason for project's perceived success	Four Ps
Organizational	(Internal; ≈ POV) Organization structure suited to the project team	Plan
	(Internal; ≈ POV) Project team participation in scheduling and budgeting	
	(Internal; ≈ POV) Availability of backup strategies*	
Technological	(Internal; ≈ POV) Judicious use of networking techniques*	Processes
Marketing	(External) Frequent feedback from clients	People
	(External) Client's commitment to established schedules, budgets, and performance goals	
	(External) Enthusiastic public support*	
	(External) Minimized number of public/ government agencies involved*	
Legal	(External) Absence of legal encumbrances*	Power
Political	(Internal; ≈ POV) Promoters' desire to build up internal capabilities*	Power
	(Internal; ≈ POV) Frequent feedback from parent organization	
	(Internal; ≈ POV) Flexible parent organization*	
	(Internal; ≈ POV) Adequate control procedures*	
	(Internal; ≈ POV) Promoters' commitment to an established budget and performance goal*	
	(Internal; ≈ POV) Promoters'enthusiasm*	
	(External) Absence of excessive government red tape*	

Note: (External) refers to external influence and hence pure risk and (Internal) refers to potential points of vulnerability POVs.

*What the feasibility analyst can check ahead of time.

too many meetings, misplacing documents, and resorting to excessive paperwork. Does it sound familiar?

In other words, POVs result not only from poor planning, but also from a combination of poor planning and an inadequate workforce (put differently, Unfits).[31] Planning is perfomed by humans; humans can correct most mistakes they make.

The role of People and Power (the human component of my model) cannot be underestimated, as can be seen in Figure 2.8.

Looking at the MOS case, it is the addition of the overly ambitious technological challenges (a retractable roof from a tilted mast that is supposed to work well in the most grueling winter conditions) to the

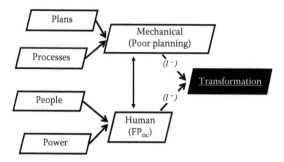

Figure 2.8 Positioning of POVs in the basic input model.

strike-loaded sociocultural environment that is to blame for the delays in finishing the roof.

Some authors offer a handful of what they call the "risk areas" and "Key Success Factors" (KSF) of a project.[32] I extracted those from the source article and reworked the information as follows (please be prepared to read fast!). The star "*" means that the factor could be given consideration in a feasibility study. Cultural or sociocultural [33] "risk factors" are (Plans) a poor acknowledgment of risks* and a lack of analyses*; (Processes) overly rigid systems*; (People) a culture of fear, a lack of training*, and a limited sense of belonging. Cultural or sociocultural KSF are: (Plans) ongoing analyses; (People) an open culture, a culture of knowledge, learning from experience, and trust; (Power) responsible risk management*.

For the stakeholders (People/Power) or psychodynamics (if internal), the risk factors are (Plans) ambiguous contracts*, changes to the norms, a lack of long-term planning*; (Processes) supply delays; (People/Power) a lack of commitment and strained relationships; (Power) a lack of support. For the stakeholders (People/Power) or psychodynamics, the KSF are (People) bonding with a community*, supplier relationships, tight human bonds, efficient communication; (People/Power) a partnership-like strategy*; (Power) collaboration with the regulators*, facilitating commitment, as well as managing risks and competencies.

For leadership, the risk factors are (Power) an authoritarian system*, lack of communication with external stakeholders, lack of support from the management, and poor management–employee relationships. The KSF are (Plans) focus on a project*; (People) team collaboration; (Power) external networking* and joint decision-making.

For risk management, risk factors are (Plans) an inefficient matrix*; (People) a lack of multilevel collaboration, a poor team spirit, and poor teamwork. KSF are (Plans) creating multifunctional teams* and a reinforced organizational matrix*; (People) easing communication and fostering a sense of belonging.

For organizational risk factors, the following are proposed: (Plans) a lack of risk countermeasures*, poor levels of analysis*, and poor risk identification*; (Processes) a lack of procedures*. KSF are (Plans) a crisis response plan*, a formal risk management plan*, risk assessment*; (Processes) continuous improvement.

For Processes, risk factors are (Plans) insufficient preparation*, lack of financial resources*, unrealistic calendar*, and poor planning*; (People) poor teams; (People/Power) a lack of commitment; (Power) contractual disputes and a lack of control. KSF are (Plans) a budget*, a contingency plan*, a formal plan*, an in-depth evaluation of the vision stage*, a realistic calendar*; (People/Power) sharing and commitment; (Power) control measure improvements*.

For Technology, risk factors are (Plans) poorly defined canvasses*; (Processes) excessive reliance on technology in order to solve problems and moving standards; (People/Power) lack of expertise*. For KSF, the following are listed: (Plans) a better definition of Work Breakdown Structures (WBS)*; (Processes) proven standards* and proven technology*; (People/Power) acquiring competencies.

The reader is entitled to go drink a glass of water!

In these lists, I observe that processes account for little in the attempt by the authors to find the KSF. Yet, processes play a vital role in the economy; a prime example is that of the transformation of iron ore by way of the Bessemer and open-hearth processing technologies that saw the rise of the iron and steel industry, which became a major force in industrialization (railroads, buildings) efforts around the world (especially in the United States and Great Britain). I also note once more that the merging of poor planning and Unfits is a recipe for disaster. On this topic, another scholar[34] finds a positive relationship between project success and the project management method, which includes, in his terms, knowledge area tools, process capability, and profile techniques. It is also worth mentioning that the management of costs is barely mentioned (under "budget") in the lists; references to calendars of activities (e.g., WBS) and to the four Ps are, however, made plentifully, if not directly then indirectly.

Other authors also have their opinions on KSF and KFF and how to label them. Some scholars[35] identify the following "promoting" and "inhibiting" factors, which I extracted from their writings and reclassified according to my four Ps model. For the "Promoting" factors (I^+), we have: (Plans) alignment of goals and objectives* and a well-defined charter*; (Processes) agility; (People) diverse and balanced teams, an effective reward system*, and good communication; (Power) empowerment. For the "Inhibiting factors" (I^-), the following are listed: (Plans) discrepant priorities*, poorly defined roles and responsibilities*, and confusing resource requirements*; (Processes) contextual barriers; (People) limited training; (Power) bureaucracy and poor leadership.

Our observation of this list is that, just as was the case with the previous list, processes are scarcely listed as a source of either KSF or KFF. A report that is often cited[36] adds a level of importance (by %) to success factors; these include executive management support (15%), experienced project managers (14%), the clarity of business objectives (14%), standards infrastructures (5%), and formal methodology (5%).

Interestingly enough, "formal methodology" is mentioned, which tends to show that my argument in favor of proper modeling techniques makes sense (as touched on in the previous section on parameters). Without going into too much detail, some scholars list the following: (1) (Plans) meticulous WBS and a clearly defined mission*; (2) (Processes) (nothing listed); (3) (People) client acceptance, client consultation (market analysis), communication, strict personnel selection, and technical competencies; (4) (Power) monitoring and feedback, top management support, and troubleshooting. Some other authors allude in particular to factors that are people related or processes-and-tools related.[37] A recent study in the IT sector found that the (internal) KFF that cramp the relationship between the client and the vendor relate to psychological constructs (biases, commitment, morale, and support), structural components (lack or poor definition of contracts, costs, criteria, documentation, expertise, plans, requirements, roles, and responsibilities), and functional components (auditing, change management, communication, customization, development methodology, knowledge transfer, and staffing). These internal vulnerabilities are likely exacerbated by financial instability.[38]

As the reader can see, the subject of KSF and KFF is quite overwhelming. Let's try to ground some of the ideas into actual projects.

2.7.1 NSTP (Sydney) versus Environ (Europe) megaproject

A view of public–private megaprojects is provided by Sydney's Northside Storage Tunnel Project (NSTP) and the Environ megaproject,[39] out of which I have extracted what could be called KSF and KFF, as shown in Table 2.11 (again, factors that are marked with an * can be assessed during a prefeasibility study, and again, I classify the themes found in the source article according to the four Ps).

Again, I note that processes receive little blame for failure and little applause for project success. Yet again, it is most likely that it is the interaction between poor planning and Unfits that is to be regarded as the main concern.

Put differently, an articulate plan (Plans) and a positive human input (People and Power) play a fundamental role, even in a preliminary stage

Table 2.11 The NSTP (Sydney) versus Environ (Europe) megaprojects

KSF	Four Ps	KFF	Four Ps
		NSTP (Sydney)	
Clearly defined mission*	Plan	Tight schedule	Plan
Clearly defined goals*		'Us against them'	People
Match: Employee/tasks	Processes	Slow decision-making	Power
Sharing of risks		Divergent political interests	
Solid PPP*		—	—
Collaboration	People	—	—
Community involvement		—	—
Excellent communication		—	—
Shared commitment to success	People/ Power	—	—
Feedback	Power	—	—
		Environ megaproject	
Clearly defined mandate*	Plan	Impacts undefined*	Plan
Flexibility	Processes	Ambiguous calendar of tasks	
Based on objectives*		Different interpretations of the project's nature	
Project accepted by clients	People	Poor match employee/tasks	
—		Lack of clarity	Plan/Processes
—		Conflict laden	People/Power
Traditional culture		Shortage of competencies	
Value of entrepreneurship		Poor cooperation	
Centralized management*	Power	Shortage of expertise	
Collaboration between PPP		Poor communication	
—		Lack of involvement	
—		Poor conflict management	Power

*Factors that can be assessed during a prefeasibility study.

of a project. With regard to Power, some authors note that the chances of delivering the project as per the initial specifications are fourfold when directed with advanced managerial techniques, while it increases only by nearly twofold when it only receives top management support.[40] This is in line with our previous observation that the type of management resorted to in a project is instrumental: obviously, advanced management techniques (those that tackle POVs effectively when confronted by Unfits) are crucial to the performance of the projects. According to some experts, it is important that employees enjoy the freedom to pose questions while management should have sufficient experience to provide answers.

Dealing with risks, infrastructures, opportunities, and management techniques with respect to the Unfits is a fine balancing act. To any project manager, this will resonate: day in and day out, he has to address risks, check the infrastructures, make sure that his staff performs to the expected level, and reevaluate and capitalize on opportunities.

2.7.2   Summary of KSF and KFF

Table 2.12 summarizes Dominant and Contingency strategies (DS and CS) based on my literature review.[41] These strategies (which many authors would call KSF) subdue the influence of POVs.

I make a number of observations:

1. We have now specified the meaning of Dominant and Contingency strategies (DS + CS) and assume that the items in the lists are functional variables (*F*), that is, these characteristics are an expression of Dominant and Contingency strategies. They are not all *sine qua non* conditions, and some of these characteristics work when paired with others (colinearity); for example, Trust and Collaboration probably go hand in hand.

2. A number of the characteristics can be checked during the prefeasibility study phase (they are marked with an *) but many cannot.

3. A number of concepts reoccur: Agility, Clarity, Collaboration, Commitment, Communication, Competencies, Control, Culture, Efficacy, Efficiency, Matching, Sharing, savvy Conflict management, and Supportive management. In fact, when we reclassify these terms more tightly we have

   a. Plans: Having clarity (transparency) and control.
   b. Processes: Showing efficacy and efficiency.
   c. People: Being collaborative (including communicating and sharing), committed, competent, and proactive.
   d. Power: Being agile, competent (including the use of advanced conflict management techniques), and supportive (including communicating).

4. We will see how these terms become handy in the sections on Processes, People, and Power (Chapters 4 through 7). I estimate that I have given, in a nutshell, what seems to be KSF in most projects. Successful projects occur when plans are comprehensible and include control mechanisms; when processes display efficacy and efficiency; when people collaborate, have competencies, and adopt a proactive attitude; and when leaders resort to agile management, are competent, and supportive. That sounds about right.

5. Indeed, Contingency strategies are listed, though without being recognized as a contingency strategy by many authors. Contingency strategies are linked to the establishment of backup plans, contingency

Table 2.12 Dominant and Contingency strategies

Plan	
Well-defined charter, objectives, impacts, and roles*	Clarity
Access to financial resources*	Efficacy
Clear resource requirements*	
Establishment of norms*	
Realistic calendar*	
Sound budget*	
Processes	
Agility	Agility
Feedback	Communication
Formal methodology*	Efficiency
No time delays	
Proven standards and procedures*	
Proven technology*	
Solid infrastructures*	
Match employee/tasks	Matching
Sharing of risks	Sharing
People	
Team work	Collaboration
Commitment	Commitment
Good communication	Communication
Skilled staff	Competencies
Culture of knowledge	Culture
Learning from past experience	
No culture of fear	
Open culture	
Sense of belonging	
Sharing	
Trust	Trust
Diverse and balanced teams	Efficiency
Power	
Nonbureaucratic	Agility
Frequent feedback from parent organization	Communication
Experienced project management*	Competencies
Leadership	Control
Control measures improvements	
Adequate control procedures*	
Solid decision-making	Efficiency decision-making

(Continued)

Table 2.12 (Continued) Dominant and Contingency strategies

Empowerment	Sharing
Excellent conflict management	Savvy conflict management
Good management–employee relationships	
Management support	Supportive

Contingency strategy

Backup strategies*
Contingency plan*
Crisis response plan*
Formal risk management plan*

plans, crisis response plans, and formal risk management plans. In summary, Contingency strategies are related to risk management.
6. There is no reference to robust management, but we can assert that robust management develops when Dominant and Contingency strategies are in action. We have

$$\text{Robust management} = \text{Dominant strategy} + \text{Contingency strategy} \quad (2.2)$$

I illustrate my findings in Figure 2.9.

Figure 2.9 Summary of Dominant and Contingency strategies.

Table 2.13 can be used to assess the frame of so-called KSF in a prefeasibility study.

I pretend that all of these are elements that enhance a project's strengths and opportunities; hence, I label them as KSF. A KFF would be the dreadful combination of a poor plan and Unfits. However, the reverse of this combination (an excellent plan and controlled Forces of Production, FP_c) could not be assumed to lead to success because if these controlled Forces of Production lack the experience and the talent to complete the project, it will probably go nowhere. Similarly, the fact that all stakeholders are in agreement on a project does not guarantee success (many things can go wrong, including some secluded POVs surfacing during implementation); however, divergent interests (surreptitious or not) among the

Table 2.13 Assessment of the frame of KSF

Dominant strategy (DS)	
Plan	
Well-defined charter, objectives, impacts, and roles*	Clarity
Access to financial resources*	Efficacy
Establishment of norms*	
Realistic calendar*	
Sound budget*	
Processes	
Formal methodology*	Efficiency
Solid infrastructures*	
Power	
Experienced project management*	Competencies
Contingency strategy (CS)	
Preliminary risk management plan*	

key stakeholders and hidden agendas certainly carve a road to failure (hence, the importance of transparency as we shall see in Chapter 7).

I believe that three of the KFF are

- The dreadful combination of poor planning and Unfits (and bad apples)
- Divergent interests among stakeholders
- Management blindness and overoptimism (see Chapter 7)

As can be judged, these are not at all the opposite of KSF; they stand on their own. If you want to make sure that your project goes amiss, regardless of the KSF listed in Table 2.13, simply stick to the KFF!

2.7.3 *Prefeasibility and the Mervel Farm project*

In the case of the Mervel Farm project, the local business development center that did the prefeasibility analysis identified the criteria picked to ponder whether public funds would be distributed to the promoters or not, as follows: (1) Plan: a realistic implementation plan; (2) People: existence of contracts with clients or else formal commitments from clients; (3) Processes: pragmatic work breakdown structures; (4) Power: respect of rules and regulations, such as zoning and management's operation experience.

The local development center deemed that any pitfalls among these criteria would disqualify any proposed project. Other criteria (given the mandate of the government-funded business development center) were also taken into consideration, but to a lesser degree, such as economic diversification, employment creation, integration within the community, matching the funding government's goals, positive impacts on the City of Pierreville, quality of innovation, sustainability, and value creation.

I devised a model based on my findings as per Figure 2.10 (recall that walls are timely conditions for operations; in this case, criteria for engaging further in the submitted project).

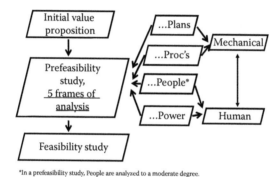

*In a prefeasibility study, People are analyzed to a moderate degree.

Figure 2.10 A model for a prefeasibility study.

Charts that illustrate the link between prefeasibility and feasibility studies have been offered. One such chart begins with a project idea, followed by an informal plan, a prefeasibility study, which if favorable, then leads to a feasibility study.[42] I believe that the initial value proposition summarizes the preliminary steps quite well, before the prefeasibility study gets under way.

Let us summarize the lessons learned so far by way of Table 2.14 that highlights the most important elements of the five analytical frames.

There may be some redundancy between some of the components of Table 2.14 (e.g., a well-defined plan in both potentiality and DS) but given that each frame attempts to measure the same single project (from different correlated angles), this is deemed acceptable. In this case, it reinforces my argument that proper modeling of the project is critical if the analyst wants to uncover POVs.

2.8   Conclusion to Chapter 2

An analyst doing a prefeasibility analysis is content with taking a broad view of the proposed project.

Table 2.14 Summary of the five frames of a prefeasibility analysis

		Frame of...		
Definition	Contextual risks	Potentiality	Parameters	DS and CS[49]
Plan	Financial	Scope	Walls	Financial resources
—	Organizational	Well-articulated plan	Ceilings	Articulate budget
Processes	Environmental	Change process	Floors	Well-defined plan
—	Technological	Measures of success	—	Realistic calendar
People	Marketing	Group of managers	—	Norms of quality
—	Sociocultural	Monitoring system	—	Methodology
Power	Legal	—	—	Infrastructures
—	Political	—	—	Experience
Characteristics	—	—	—	Risk and POVs management

The analyst bases their judgment on the afore-mentioned five frames of analysis and can use the prefeasibility template provided in Appendix 2.2. As we shall see, these frames will also be of use in the feasibility study.

The prefeasibility analysis is highly subjective. The expert offers an opinion based on a general impression by seeking relevant information, but without calling into question the minutiae of the proposed project. They are not afraid to look for what can go wrong.

We can summarize what has been learned so far by way of a visual model that does not take into account all of the links that exist between the various concepts, but that simply conveys a general sense of

Playing devil's advocate.

what is happening with respect to the project. We will define, further along in this book, the meaning of the triangle at the center (refer to Figure 2.11).

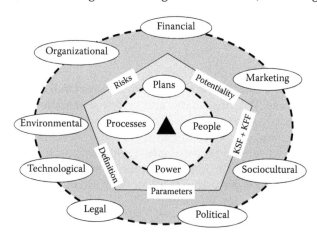

Figure 2.11 A simplified model in development.

If we were to analyze the Mervel Farm project in the usual way (without due consideration for POVs), we would probably approve it because it resonates well in terms of revitalizing a core downtown area. However, if we were to look for POVs using our method, we would identify the following potential problems: the absence of an accountant and of a lawyer/notary; a challenging ideology; difficulty in growing farm products; heavy reliance on volunteers; a lack of any defined key performance indicators, or KPIs; a lack of human resources; a lack of tourist attractions; limited use, with summer being the most intensive period; the need for an architect (costly); the need for specific agricultural competencies (costly/hard to find); not enough sales versus costs (high dependence on financing); poor financing; a poor promotional strategy; possibly poor crop profitability; and sales that are done exclusively on site. As for risks, these would include a dependence on the weather, incomplete infrastructures (parking), the need for specific approvals from the National Capital Commission and the City of Pierreville, and possibly adverse reactions from neighbors.

You may think of a prefeasibility study as the act of Caesar and Cleopatra being engaged. They would commit with the best of intentions but they could always withdraw from their promise; although this would be heartbreaking, it would nevertheless be less costly than a quasi-inevitable upcoming nasty divorce. Caesar would lose his investment in the gold ring inlaid with diamonds and rubies, but he would get to keep his furniture, his vessels, and his armies!

Chapter 3 examines the notion of feasibility. I discuss the criteria associated with feasibility study analyses and present the concept of the "triple constraints".

2.9   *What we have learned about POVs: Chapter 2*

POVs…

1. Can appear in any of the eight different contextual risk areas.
2. May include: An unrealistic and poorly articulated plan, diffuse measures of success, an inadequate change process, incompetent managers, and a lack of safeguards to prevent derailment.
3. May be uncovered by resorting to qualitative measurements.
4. May be uncovered through proper modeling.
5. Exist because people do not think of them, do not wish to admit that they exist, or else ignore them.
6. Often result from poor planning and an inadequate (if not hostile) workforce.
7. Can be subdued in a prefeasibility study by checking for: (Plans) access to financial resources, a realistic calendar, a sound budget, detailed resource requirements, the establishment of norms, a well-defined charter, objectives, impacts, and roles; (Processes) a formal methodology, a reliance on solid infrastructures, the use of proven standards and procedures, and the use of proven technology; (Power) adequate control procedures and resorting to experienced project managers.

2.10   *Key managerial considerations: Chapter 2*

1. All four Ps must be evaluated in a prefeasibility study.
2. Avoid poor planning along with the parallel presence of Unfits.

3. Use all five frames of analysis as a way of shedding light to uncover dark spots.
4. Aim for a robust strategy.

2.11   Case study Chapter 2: Recycl'Art

Recycl'Art is a venue for exhibiting art made out of recyclable material. The concept originated in a remote village in the province of Québec, Canada. At the end of December 2014, the board of directors met with a new project in mind: to relaunch Recycl'Art in Gatineau, which is a city adjacent to the national capital—Ottawa. Ottawa is a city known all over the world that attracts tourists by the thousands, including on Canada Day, July 1. The cultural core of the City of Ottawa sits within walking distance of its neighbor. Most people wander in the ByWard market to buy local farm produce or to wine and dine, but then cross the bridge to enjoy the numerous parks and outdoor attractions (as well as a casino) that Gatineau has to offer.

A number of challenges awaited the enthusiastic promoters, including (1) ensuring that the event would receive support from the local community, (2) obtaining financing from the City of Gatineau, (3) preparing a five-year plan, and (4) finding a catchy slogan.

The origin of the Recycl'Art Festival dates back to 2004, in Montpellier, Québec. Being in a remote village, it did not attract enough tourists to make the event viable, hence the idea of trying it out with the 500,000 inhabitants that comprise the City of Gatineau. A myriad of visual artists exhibited their art made out of recycled material, which included wood, metal, plastic, rubber, and paper; in short, anything that is found on dump sites or in garbage bins and that can be remodeled to create some figurative or abstract art. One key structure that had people and newspapers talking was the assembly of a totem-like structure composed of scrap laundry and drying machines, which, by its mere height and imposing presence, managed to catch people's attention while conveying a message against overconsumption. Anything that can serve as a source of inspiration has an appeal to artists or would-be artists, be it nature, urban life, industry, or poetry. The ministry of culture strongly supported the project, for it was in line with its goal of "democratizing" visual arts.

Yet, there was a need to find financial sponsors. These eventually included a local gallery, local universities and high schools, the City of Gatineau of course, the National Capital Commission (Ottawa being just across the river from Gatineau), as well as various businesses.

To top the offer of an outdoor venue where recycled art would be exposed along pathways and park grounds, the promoters examined the idea of hosting specialized exhibitions with international guest artists, creative workshops, guided tours, as well as a sculpture fair.

The initial problem (lack of tourists in the remote village of Montpellier) turned into an opportunity for the City of Gatineau. The project was launched in the summer of 2015, and attracted a vast array of local and foreign tourists.

2.11.1   *Appendix A of Case 2: Recycl'Art*[43]

2.11.2   Questions related to Case 2: Recycl'Art

1. Discuss what need triggered the move to a different region.
2. Give the list of the key stakeholders and classify them according to the terminology favored in this book (e.g., suppliers).
3. Discuss how a problem became an opportunity.
4. Discuss the nature of the opportunity.
5. Discuss how the board of directors resorted to robust management.

2.12   Appendix 2.1: Assumed KSF or KFF based on a modest literature review[44]

Assumed KSF or KFF	Harmonized categorization
Acknowledgment of risks*	Contingency strategy
Availability of backup strategies*	
Contingency plan*	
Crisis response plan*	
Formal risk management plan	
Managing risks and competencies*	
Presence of risk countermeasures*	
Responsible risk management*	
Risk assessment*	
Risk identification*	
Client's lack of commitment to established schedules, budget, and performance goal	External risks
Excessive government red tape*	
Lack of feedback from client	
Large number of public/government agencies involved*	
Legal encumbrances*	
Project accepted by clients	
Poor enthusiastic public support	
Ambiguous calendar of tasks	Plan: Clarity
Ambiguous contracts*	
Better definition of WBS	
Well-defined charter*	
Clear business objectives*	
Clearly defined mandate	
Clearly defined goals	
Clearly defined mission	
Different interpretations of project's nature	

(Continued)

Assumed KSF or KFF	Harmonized categorization
Impacts undefined	
Lack of clarity	
Poorly defined canvasses*	
Poorly defined roles and responsibilities*	
Poor planning*	
Partnership-like strategy*	Plan: Collaboration
Lack of financial resources*	Plan: Efficacy
Alignment of goals and objectives*	Plan: Efficiency
Budget*	
Changes to norms	
Contextual barriers	
Formal plan*	
In-depth evaluation at the vision stage	
Insufficient preparation*	
Lack of analyses*	
Unrealistic calendar*	
Ongoing analyses	
Organization structure suited to the project team	
Poor levels of analysis*	
Realistic calendar*	
Reinforced organizational matrix*	
Unclear resource requirements*	
Tight schedule	Plan: Efficient planning
Focus on project*	Plan: Focus
Minimized scope	
Inefficient matrix*	Plan: Inefficiency
Based on objectives	Plan: Objectives
Lack of long-term planning*	Plan: Planning
Agile requirements process	Process: Agility
Agility	
Flexibility	
Overly rigid systems*	
Feedback	Process: Communication
Moving standards	Processes: Efficiency
Excessive reliance on technology to solve problems	
Proven standards*	
Proven technology*	
Lack of procedures*	

(Continued)

Assumed KSF or KFF	Harmonized categorization
Continuous improvement	Processes: Efficiency and efficacy
Standard infrastructure	Processes: Efficient infrastructures
Formal methodology	Processes: Efficient methodology
Reliable estimates	Processes: Reliability
Effective reward system	Processes: Rewarding
Sharing of risks	Process: Sharing
Solid alliance Private–Public Partnerships (PPP)	
Supply delays	Processes: Timeliness
Match employee/tasks	Process work allowance
Poor match employee/tasks	
Adversarial relationships	People: Collaboration
Bonding with community*	
Based on collaboration	
Collaboration between PPPs	
Collaboration with regulators*	
Conflict laden	
Joint decision-making	
Lack of multilevel collaboration	
Poor collaboration	
Project team participation in scheduling and budgeting	
Poor team work	
Strained relationships	
Team collaboration	
Suppliers' relationships	
Community involvement	People: Commitment
Creating multifunctional teams*	
Facilitating commitment	
Lack of commitment	
Shared commitment to success	
Lack of involvement	
User's involvement	
Easing communication	People: Communication
Efficient communication	
Excellent communication	
Good communication	
Lack of communication with external stakeholders	
Poor communication	

(*Continued*)

Assumed KSF or KFF	Harmonized categorization
Acquiring competencies	People: Competencies
Lack of expertise*	
Lack of training*	
Limited training	
Poor teams	
Poor team spirit	
Skilled staff	
Shortage of competencies[45]	
Shortage of expertise	
Diverse and balanced team	People: Efficiency
Sharing	People: Sharing
Sharing and commitment	
Culture of fear	People: Culture
Culture of knowledge	
Learning from past experience	
Limited sense of belonging	
Open culture	
Sense of belonging	
Sharing and commitment	
Tight human bonds	
Traditional culture	
Trust	People: Positive culture
"Us" against "them"	
Value of entrepreneurship	
Authoritarian system*	Power: Agility
Flexible parent organization*	
Promoters' commitment to established budget and performance goal*	Power: Budgeting
External networking*	Power: Communication
Frequent feedback from parent organization	
Judicious use of networking techniques*	
Experienced project management	Power: Competencies
Adequate control procedures*	Power: Efficiency
Bureaucracy	
Centralized management	
Contractual disputes	
Control measures improvements*	
Discrepant priorities	
Lack of control	

(Continued)

Assumed KSF or KFF	Harmonized categorization
Poor leadership	
Poor management–employee relationships	
Slow decision-making	Power: Efficient decision-making
Divergent political interests	Power: Efficient communication
Empowerment	Power: Sharing
Poor conflict management	Power: Sound conflict management
Executive management support	Power: Support
Lack of support	
Lack of support from management	
Promoters' desire to build up internal capabilities*	
Promoters'enthusiasm*	Power: Supportive

2.13   Appendix 2.2: Prefeasibility study template

Author: _____

Position: _____

Date: _____

Care of: _____

PROJECT

Name of project:_____

Goal: _____

Initial value proposition of a project:_____

Calendar: Start: _____ End: _____

Estimated cost (US$): _____

Main challenges (norms of quality to be met):_____

General description: _____

Frames of analysis: (1) frame of definition, (2) contextual frame of risks, (3) frame of potentiality, (4) parametric frame, and (5) KSF and KFF

FRAME OF DEFINITION

Characteristics of the project (structural elements)	Comment
Responds to an opportunity	
Guarantees stakeholder involvement	
Is unique	
Innovative	
Is subject to time constraint	
Requires some investments	
Contains challenges	
Each characteristic present?	

Functional Elements

Generates deliverables (product, service, process, research)

Creates formalized knowledge

Has some impacts (+ and −)

At least two elements present?

Conclusion (is the project well defined?): _____

CONTEXTUAL FRAME OF RISKS

Type of contextual risk	Four Ps	Evaluation on a scale of 0 (no risks) to 10 (highly risky)
Financial	Plan	
Organizational		
Environmental	Process	
Technological		
Marketing	People	
Sociocultural		
Legal	Power	
Political		
Average		
Prefeasibility index		10−(…) => %

Conclusion: _____

FRAME OF POTENTIALITY

Yes/No	List	Four Ps
	Scope: Is the plan realistic?	Plan
	Well-articulated plan: Is it so?	
	Change processes: Are they adequate?	Processes
	Measures of success[46]: Are they well established?	
	Group of managers: Have the best managers been chosen in the circumstances?	Power
	Monitoring system: Have the necessary safeguards been put in place to prevent derailment?	

Conclusion: _____

PARAMETRIC FRAME

Parameter	Comment
KPIs	
Norms of quality	
Needs analysis (assessment)	
Existing parameters, codes, laws, and so on	
Similar projects	

Conclusion: _____

KSF and KFF (robustness)

KSF		
Dominant strategy (DS)		
Yes/No	**Plan**	
	Well-defined charter, objectives, impacts, and roles	Clarity
	Access to financial resources	Efficacy
	Articulate budget	
	Establishment of norms	
	Realistic calendar	
Yes/No	**Processes**	
	Formal methodology	Efficiency
	Solid infrastructures	
Yes/No	**Power**	
	Experienced project management	Competencies

Endnotes

1. As set by the S and F process elements that we will see in Chapter 4 on Processes.
2. Neuvel et al. (2015).
3. Deng et al. (2014).
4. Modeling will be seen in Chapter 4 on Processes.
5. Mesly (2010, 2011, 2013, 2014, 2015a–c).
6. We can be more specific and pretend that risks affect mostly time, and that POVs are more intimately related to costs, so that the one constraint that is left as a condition (the "|" sign or "given") is actually norms of quality. Hence, I would say that risks (as they affect time) are in an inverse relationship with POVs (which imply costs) given norms of quality, as affected by k (which, as a psychological measure, is an expression of perceived threat).
7. The [risks $(I^-) \rightarrow$ project] renders in words and symbols what would otherwise be an illustration of the process, without any chances of confusion.

When risks increase, it follows that the project's chances of being successful diminish because there is a negative bond between the two concepts.

8. Not in the control of the three promoters and of their company.
9. Recall that we estimate the presence of POVs to be in the neighborhood of 4%, so with a 96% feasibility index, the probability is that only the POVs that are due to the "acts of God" persist.
10. For the value of 80%, I use the common Pareto number: 20/80.
11. del Caño (1992).
12. Tardif, P.M. Interviewed April 2016.
13. In a nearby similar event, this actually took place a couple of years before.
14. Cooke and Tate (2010).
15. See also Sterman (2000).
16. Curves presented in this book are stylized to simplify drawing.
17. Xia and Lee (2005).
18. We will discuss the project's life cycle at length in Chapter 4 on Processes.
19. The term "participants" is common when referring to marketing surveys.
20. This cannot take place, of course, along the critical path.
21. Xia and Lee (2005).
22. See also the concept of "organizational process assets" in the PMBOK.
23. Schultz et al. (2011, p. 17).
24. Most of the time, I use KSF and KFF in the plural form.
25. Moderating variables exercise a moderating effect, translated by the symbol ($I^{\pm}$). A moderating variable has two types of influences, a positive or a negative one depending on circumstances.
26. Bstieler and Hemmert (2010, pp. 485–487).
27. Gould and Joyce (2000, p. 150).
28. Cleland and King Editors (1988, p. 905). See also Pinto and Mantel (1990, p. 270).
29. Items listed marked with an * can be assessed when doing a prefeasibility study.
30. This dreadful combination is certainly a negative force being exercised upon the project.
31. This is definitely a Key Failure Factor/Fundamental (KFF).
32. Yeo and Ren (2009, pp. 282–283).
33. In fact, in this list on culture, all elements are internal, so really, they should be considered potential POVs and not risks.
34. Chen (2015).
35. Martin (2009).
36. The 2003 Standish report.
37. Joslin and Müller (2015).
38. See Liu and Yuliani (2016).
39. Van Marrewijk et al. (2008). The Sydney (Australia) North Side Tunnel Project (NSTP) was a 12.4-mile (20 km) tunnel that was part of the 2000 Olympic infrastructure. The Environ project was a large European infrastructure project, a PPP that took place from 1990 to 2006.
40. Berssaneti and Carvalho (2015).
41. I do not pretend to have the final word on the process elements that ensure success; these may change, I suppose, with the size of the project, its nature as well as with other considerations. This is a first step in identifying Dominant and Contingency positions.

42. Farrell (1995).
43. Visual material generously provided by Recycl'Art 2015. A special thanks to Gaston Therrien.
44. Some of the items appear redundant; I have kept the original terminology used by various authors.
45. See Loufrani-Fadida (2008).
46. This will be detailed in the KSF/KFF frame.
47. At this point, the processes are looked at in general terms. Details will be examined in the feasibility study (Chapter 3).
48. This will be detailed in the KSF/KFF frame.
49. I now refer to DS and CS instead of KSF and KFF.

chapter three

Plan—Project feasibility study

So it is, said the Wise Eagle, that it takes quality time
to generate unnecessary costs.

Darloz

3.1   Introduction to Chapter 3

A prefeasibility study seeks to validate a project and to expose organizational spots or problems that may condemn it before it sees the light of day, using a limited number of analytical tools. A prefeasibility study does not concentrate on the day-to-day operations; in other words, it doesn't look into the specifics of tasks or the team member composition. This is something that the feasibility study does. From this perspective, a feasibility study is very pragmatic; once we have defined the project in approximate terms using the five frames of analysis, we can then examine how this is going to translate into daily operations, most evident during the transformation phase.

Should the prefeasibility study be conclusive, the project promoter is invited to feed in more information to the analyst. The feasibility expert scrutinizes all the possible contextual risks (environmental, legal, political, etc.) that trigger or amplify Existing vulnerabilities. Some authors have echoed the idea that a feasibility study consists of a thorough analysis of the different project components,[1] including assessing its risks, identifying its key components, and measuring its impacts in order to reach a "go," "no go," or "put on hold" decision.

As an example, let's take the case of the replacement of the Champlain Bridge that connects the island of Montréal to its south shore and the United States and which is nearing the end of its useful life. The 4.2-billion-dollar (C$) project will inevitably generate a perimeter of intense noise (machinery, etc.) and dust in an urban area that is quite populated. Natural ecosystems are threatened. A thorough environmental feasibility study is required. A feasibility analyst could linger on whether or not the construction is feasible, taking into account dust and displacements of rare animal and plant species as well as maximum noise levels. Of course, that would not be enough in this case: other aspects such as financial and technical concerns would also have to be addressed.

This chapter addresses the notion of project feasibility. It discusses in more detail the concept of PRO (Pessimistic, Realistic, and Optimistic) scenarios, the triple constraints. It also introduces the concept of POW (Product, Organization, and Work) breakdown structures, and makes additional comments on the notion of POVs (points of vulnerability).

3.2   Some characteristics of feasibility studies

Sound feasibility studies include the five analytical frames of the prefeasibility study in their introductory sections, with, of course, more attention to detail and to People, especially management and team members (the Forces of Production, FPs). The costs of such studies vary; some authors write that feasibility analyses for small projects may cost US$100,000 while the bill may inflate to more than US$1MM for large projects.[2]

There is no standard format for feasibility studies, as each expert firm or analyst tends to adapt the format to the particular project to his own style. We are not competing with existing prefeasibility or feasibility forms or formats; we are merely seeking to develop tools for uncovering points of vulnerability. The theory and tools that I have provided so far and that we will continue to discuss throughout this book are designed to help the analyst to make up his mind regarding the future of a project. It is much better to spend time dealing with current situation than having to react after the fact. Often, a POV exists when one thinks that there's none. The Chinese constructed the Great Wall thinking this would stop any possible attack from the northern tribes, but all it took was one person to open one of the fortified gates to the enemy to render futile the entire 13,000 miles of rock and sweat. Similarly, the French conceived the Maginot Line after WWI believing it was the best defense against their Eastern neighbors, but it took a curt ride through the Ardennes Woods during WWII for the Nazis to take over the country. In both cases, marvelous (and costly) undertakings contained an inherent point of vulnerability while an entire population felt confident it was shielded as never before.

Some authors have put forth some specific layouts for feasibility studies. The following sequence has been proposed[3] : (1) expressing the vision of the project; (2) conducting an informal analysis (which I called the "initial value proposition" of a given project); (3) conducting a prefeasibility analysis (which we reviewed in Chapter 2); (4) conducting a market analysis (this is part of a marketing assessment of a feasibility study, which aims to prove that there is indeed an opportunity); (5) doing a technical analysis (again, this is one of the eight risks areas that can be explored); and (6) performing a financial analysis (mandatory in any project).

As mentioned in the previous chapter, all feasibility studies do not necessarily have to examine every angle of the contextual frame of risks. For example, a project may not require extensive technical analysis; however, every project involves humans and as such, a close look at the makeup of the management force and team is imperative. Who wants Unfits (uncontrolled Forces of Production, FP_{nc}) on his team, whether among the workforce or in management? In short, a feasibility study addresses each of the four Ps (Plans, Processes, People, and Power), including People and Power since they are, in the end, the ones accountable for ensuring that norms of quality are met.

Contextual risks may overlap. A marketing campaign must respect legally binding rules: the feasibility analyst will scrutinize the Consumer Protection Act, laws relating to trademarks, signage and labeling, and so forth. If the study is based on technical specifications, the expert may emphasize contracts, health and safety standards, industry standards, laws relating to product safety, weights and measures, and so on. In the case of an environmental study, bylaws and zoning regulations will be inescapable. Laws on taxation (financial risk) are, of course, inescapable.[4] The contextual frame of risks can be complex, which makes it that much more necessary to ensure a project is least vulnerable.

While the prefeasibility study sets some parameters (when assessing the parametric frame), the feasibility study digs further into the subject and verifies that all key stakeholders (People and Power) are taken into consideration. For example, in a Private–Public Partnership structure, stakes will change depending on the point of view (Table 3.1).

While there are similar ambitions (e.g., profitability for both the private and government sectors), there are important divergences of interests (e.g., political interest). Hence, the value of the project will be assessed differently depending on the project and the stakeholders. This provides a fertile ground for external "bad apples" and internal Unfits (uncontrolled

Table 3.1 Criteria for participation in PPP depending on the stakeholder

Private sector	Government	Community
Financial appeal	Acceptable mandate	Acceptable standards
Free from political influences	Contractual flexibility	General approbation
Strong business ties	Political interest	Respect of infrastructures
Profitability	Profitability	Reasonable charges
Stability	Reasonable charges	Stability

Forces of Production, FP_{nc}) to surface: this is regularly seen when a project that does not please a community is pushed forward or when employees start expressing discontent. The sociocultural clash between the community and management may invite some internal team members (Forces of Production) to behave contrary to the well-being of the organization. This is in line with the observation made in Chapter 2; having a poor plan while operating with Unfits (uncontrolled Forces of Production, FP_{nc}) is a recipe for disaster. Building on these observations, we can conclude that the four Ps, as they interact with each other in the transformation phase, are critical. Anything that jeopardizes the harmony between the four Ps begets a POV.

In a feasibility study, as mentioned, People are a core element; the same comment applies to Plans, Processes, and Power. The *Project Management Body of Knowledge* (PMBOK) remains quite general in terms of People, as it focuses mostly on processes, being engineering oriented. However, projects are conceived, initiated, managed, completed, enjoyed by individuals.

A general definition of a project feasibility study is as follows:

3.2.1   *Definition of project feasibility study*

A project feasibility report is a comprehensive study which examines in detail the five frames of analysis of a given project in consideration of the four Ps, its risks, POVs, and its constraints (calendar, costs, and norms of quality) in order to determine whether it should go ahead, be redesigned, or else totally abandoned.

This is illustrated in Figure 3.1.

In the Québec Multifunctional Amphitheatre (QMA) case, a color-coding system was chosen to express changes to the project chart in the *Programme fonctionnel et technologique* (PFT) as follows:

- Green: Proposed changes are accepted conditional to changes.
- Yellow: Proposed changes must be further specified/explained and then validated by the City of Québec.
- Red: Proposed changes refused by the City of Québec.

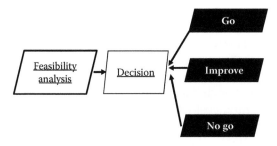

Figure 3.1 Feasibility choices.

3.3 Triple constraints

As seen in Chapter 2, there are walls, ceilings, and floors to take into account in any project worth this name. The **triple constraints**,[5] which I sometimes call the "Bermuda Triangle" refers to the quasi-mutually exclusive requirements to meet set deadlines, to abide by forecasted costs, and to respect quality criteria (with quality being the result of the four Ps working tightly together in order to meet these criteria).[6] It is by measuring the time line (risks affect the time line more directly), costs (POVs affect the costs more directly), and norms of quality (errors or deviations from standards affect quality more directly) that project performance can be evaluated. There is no need for other measurements, even though some authors add elements such as customer satisfaction or scope. Time line, costs, and norms of quality can be assessed objectively, consistently, and precisely; this is not the case for customer satisfaction, for example, which is eminently subjective. Costs in particular can be measured along three axes: planned, actual, and forecasted.[7] As put forth by some authors: "The traditional definition of project management is the accomplishment of a stated objective on time, within budget, to technical specification."[8] I fully agree.

The PMBOK ascribes the term "scope" while others privilege the word "performance"[9] to what I refer to as "norms of quality". Other authors use the term "quality". From my perspective, in order to remain consistent with my logic, the three constraints have to be exactly that: **constraints**, limits that cannot be transgressed or stretched without dire consequences. A scope is not a constraint; however, norms of quality certainly are, and so are, albeit indirectly and in a large sense, the four Ps, since they interact in the transformation phase, and since they are accountable for meeting, failing to meet, or exceeding the norms of quality.

An example of time, costs, and norms of quality interacting with each other is the preparation of a resin gel with a hardener: too much hardener compromises the hardening process, as the act of adding sufficient hardener is bound by time. The laminating job of, say, a boat made of fiberglass has to be completed before the gel hardens, and yet, the laminator has to ensure that all of the air bubbles are removed.

Typically, walls refer to calendars of activities, which have a beginning and an end. Floors are the absolute minimum quality standards that must be met while ceilings correspond more naturally to maximum allowable costs. As seen before, ranges can be put in each of these coding elements: a maximum cost (ceiling) does not preclude the existence of a range of acceptable values for costs, say from US$35,000 to US$50,000, but it states that US$50,000 is the maximum amount that can be reached without sounding the alarm bell.[10]

I like to refer to this view of things as the "Bermuda Triangle" because the Bermuda region is well known for being a dangerous zone with inexplicable aircraft and vessel disappearances. In project management, there is also an area of disasters (chaos, which we will study in Chapter 7) so to speak, defined by the outer limits of the constraints of a schedule, costs, and norms of quality.

One way to go about an investigation into the feasibility of a project is to pose questions that dig further and further into potential problems while including each of the three constraints—calendars, costs, and norms of quality. For example, for each major task listed during a transformation phase[11] (vision, planning, mobilization,[12] deployment, and completion/evaluation), the questions could be: "Are the right people available to accomplish them?" If yes, "Can they complete the task in time?" And if so, "Can they be afforded?" The same procedure can apply for each of the four Ps. Given a specific task, the analyst may ask: "Is the person in authority able to provide the necessary guidance?" If yes, "What salary does he command?"

The "triple constraints" concept is composed of three axes of potential vulnerability: time, costs, and norms of quality. The problem is that it is nearly impossible to abide by the limits of these three constraints at once. Inevitably, if a project is deferred, costs will be incurred. If a deadline is to be achieved in a hurry and contrary to plan, norms of quality may have to be compromised (sometimes in a dishonest fashion).

Recall that each project is a closed dynamic system that is assumed to evolve in a linear, ascending fashion (this is one of my many hypotheses). If the system is bounded, it also necessarily has a ceiling and a floor. This is represented by upper and lower barriers of sorts. I postulate that this results in a convex and a concave curve that delimits the chaos area. The ensuing image is shown in Figure 3.2.

We will see in Chapter 7 how to measure these limits. I will also explain then why there is a concave and a convex limit. This overall "football" shape epitomizes the fact that, if we were to put time, costs, and norms of quality on three orthogonal axes, the growth curve of the project (linear and ascending) would have to bend in order to accommodate each

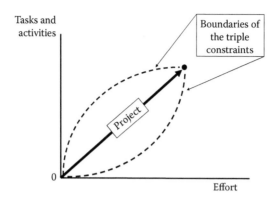

Figure 3.2 The linear growth curve and its boundaries.

element. Outside the limits of such a "football", things tend to get out of control. Moreover, you don't want to see an employee the size and the strength of a 6-foot-5, 325-pound professional football middle linebacker out of control, do you? Maybe on your enemy's field, but not on yours anyway!

Within the "football", less than perfect performance is achieved at any point except right in the middle, where compliance with the demands of time, costs, and norms of quality is granted. Because the system is closed (limited time line, maximum costs allowance, preset norms of quality), what goes wrong within it cannot readily escape outside of its frame and is obligatorily self-contained, which allows the problems to compound quickly and to cause havoc.

Victor Hugo (2001), who is well known as the author of *Les Misérables* and *The Hunchback of Notre Dame*, provides an eloquent description of chaos in his book *Quatrevingt treize* ("Ninety-three"). The scene takes place aboard a warship in the nineteenth century. Inside the ship, cannons are aligned to face the enemy. Cannons are normally fixed solidly to the floor with a series of lock mechanisms designed to resist the impact of firing the bullets. However, the crew responsible for securing the cannons has failed to do so with one particular cannon. As a result, in a sea that is agitated by powerful waves, the cannon rolls loose on the ship, punching holes in its various walls, crushing crew members who happen to be in its way as it waltzes without apparent direction. This is truly a "loose cannon". Hugo writes[13]:

> "You cannot kill it, it is dead by nature. Yet, it has
> a life of its own. It has a sinister life that is injected
> into it from infinity. The floor, moved by dreadful
> waves and hellacious winds, is causing it to bounce
> around and lurch without coherence."

Questions are posed: "Is this the toy of a terrible tragedy? How to stop this monster from causing a shipwreck? How to predict its next move?" Observations are made: "It almost seems like it has ideas and plans of its own, which keep changing. It fights its own movement, moves forward, backward, hits to the right, hits to the left; it escapes, beats the crew's anticipations, waits, and destroys." Impacts are laid out: "The vessel's guts are held captive by an ongoing thunder.... Apocalypse is all around and all within, with no means of putting an end to it."

This powerful image illustrates how a team can become a miasma of people not doing their job right. Things start to fall outside the plan set for work tasks, costs, and norms of quality. It really portrays what an Unfit pushed to the extreme can do to a project, especially when the plan has gone awry. Of note, Unfits can exist among management as well: project managers too busy navel-gazing don't get the job done or else antagonize their workforce.

Projects that deliver on time, within budget, and that have risen to the quality challenges are not the majority, as we know. My vision of a project links the four Ps to the triple constraints[14]; if we were not to do this, we would assume that costs, for example, are independent of what People (e.g., the team members) do, which would be turning a blind eye on the reality. In a closed dynamic system, POVs are prompt to inflate—the Bermuda Triangle lives up to its reputation! Hugo's description is a vivid expression of what happens when there are no exits.

The project is indeed walled off by the triple constraints. There cannot be a project without each one of its elements being present: calendar, costs, and norms of quality. This kind of dilemma exists in chess. According to Kasparov (2006), former world champion, there are three things to evaluate in order to win: (1) time (games are timed by a clock), (2) position (the strategic position of pieces on the check board), and (3) material (the number and quality of the pieces, with pieces being valued differently and acquiring different values depending on their position). A player may have a winning position, but if he's got only one second left on his time allowance, he'll lose on time. If he takes too much time to cogitate about his next move, he'll end up panicking and sacrificing his strategic advantage, as he'll run out of time later in the game. If he plays a foolish move, he'll have to spend time considering how to redeem his player's sin. He may sacrifice a piece in order to gain a positional advantage. The entire game is built around one key behavioral concept: targeting the opponent's vulnerability, finding a POV in the opponent's game, to act swiftly in due time in order to defeat him.

The same dynamic takes place in project management: time, costs, and norms of quality act as time, position, and material in chess, and these three axes of interest can initiate or activate POVs. Chess world champions master the art and science of conciliating time, material, and position while capitalizing on their opponent's vulnerability. Great managers handle the triple constraints by minimizing the emergence of POVs. While a prefeasibility study is not overly concerned with the triple constraints, it is an essential component of any feasibility study.

Note that even if a project has escaped the dangers of the Bermuda Triangle, this does not guarantee a commercial feat: the high-speed train linking Shanghai to the city's airport is a painful example.[15] It has so far proved to be a great technical achievement, but a disastrous commercial venture.

The Montréal Olympic Stadium (MOS) had its own challenges with respect to the triple constraints. Table 3.2 points to the extent of the problem that the City of Montréal faced in 1976.

Table 3.2 The triple constraints and the MOS

Triple constraints	Comment	Criteria met or not?
Calendar	Started: April 28, 1973 Completed: 1987	?
Costs	Initially: C$134 MM At the end: C$1.61 Bil.	?
Norms of quality	Uncompleted roof	?
Success or not?	Hmmm! Debatable.	

Recall that we assumed that a project is represented by an ascending linear curve with a slope of 1 (+1) and that anything that bends this curve is related to walls, ceilings, and floors. By looking at the shape of a curve, the feasibility expert is able to determine which of the axes of the triple constraints is implicated in the development of the project. As an example, we have seen a sample curve in Figure 2.6. Evidently, there is a ceiling that prevents that curve from adopting a linear ascending shape; ceilings are associated with costs. Hence, the main concern may have initially been to maintain costs under control. However, as the project got closer to its end wall, this wall gained in importance: the situation became more critical. This is expressed by the sudden rise of the curve: the project had no concern with costs at that point, but merely only with time—it was a wall that moved the curve upward. Table 3.3 shows the analytical process we just went through.

Table 3.3 Analytical process using the triple constraints

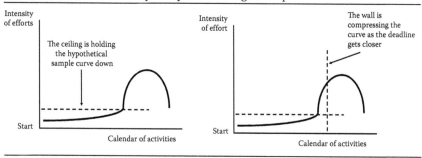

I believe that this way of analyzing a project adds value to the notion of the triple constraints that is a core element of any book on project management. It allows the feasibility analyst to focus on which constraint is most active at any particular stage of the project, providing he knows, at least roughly, what the growth curve looks like for that project.[16] Pervasive POVs have a tendency to throw the project outside the "football" formed by the three constraints, into an area that I call "chaos" and that we will, as mentioned, analyze further in Chapter 7.

While the prefeasibility study examines mostly Plans, and partly Processes and Power, the feasibility study takes a hard look at all four Ps; however, the attention centers on their interaction in the transformation phase. Additionally, the four Ps as they interact during this transformation phase are bounded by the triple constraints.

Let's continue to detail our working model as shown in Figure 3.3.[17]

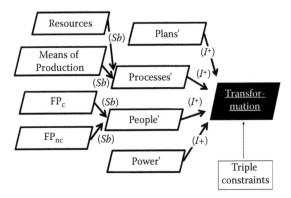

Figure 3.3 The four Ps and transformation; an evolving model.

To indicate that "transformation" is in this case an action and not an object or a concept, we underline the word. *I* stands for "influence": for example, Power has an influence on the transformation phase, which is obvious (here, the influence is assumed to be positive (I^+)). This way of modeling has the advantage of simplifying the entire process to make it easier to understand (in particular as more and more process elements are brought into the model).

Normally, the influence should be positive so that (*I*) should really read (I^+), but in fact, the influence may be negative (I^-), especially when the process is beleaguered by POVs activated by the presence of risks. Note also that (People) is composed of two process elements: "Fits" (Forces of Production under control, FP_c) and "Unfits" (uncontrolled Forces of Production, FP_{nc}). In the previous example, we assume that these two sub-elements are measured using a binary scale (present/not present). Finally, note that Processes are structurally composed of Resources and Means of Production. Resources include energy,[18] infrastructures, materials, and money. Means of Production include building, equipment, machinery, and measuring instruments. This is illustrated in Figure 3.4.

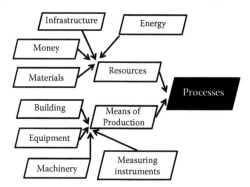

Figure 3.4 Transformation inputs.

It can be observed that the transformation phase promotes the recourse to resources by Forces of Production (FP) using Means of Production (MP) anchored in a predetermined plan and as directed by some line of authority. Of course, this entire process is bound by a beginning wall (start date) and an end wall (completion date), a ceiling (costs), and a floor (norms of quality). We will see when we discuss Processes how, in particular, resources, management and team members (Forces of Production), and Means of Production interact.[19]

We need to review each of the three constraints individually. Before we do this, however, we shall examine the PRO (Pessimistic–Realistic–Optimistic) tool and introduce a new concept: the POW, (Product–Organization–Work breakdown structure), because both tools come in handy when analyzing POVs.

3.4   The PRO system

In the general introduction, we have seen how important it is to examine everything that can go wrong with a project if we are to uncover POVs. The example of the oil filter was given.

In the PMBOK[20] and other similar texts, three scenarios are proposed:

- Pessimistic (P). Managers are preparing for the worst while having the heart to finish the project.
- Realistic (R). Managers regard the project in terms of available resources and the realistic possibilities of problematic occurrences, such as interruptions.

- Optimistic (O). Managers decide on the project duration in the best of scenarios, virtually ignoring any possible hurdles (a little bit like saying "Yes, I do" when exchanging vows at a wedding!).

The PERT (program evaluation and review technique) method as well as PMBOK proposes an expected cost (EC) measurement using a weighted average of these three estimates,[21] as follows:

$$\text{EFC} = \frac{P + 4R + O}{6} \tag{3.1}$$

Where EFC stands for "expected final costs."

This measure is very subjective, as is a large part of the prefeasibility analysis. There is nothing wrong with that; what I wish to do here is to offer some guidelines that may help identify POVs, even if some judgmental effort is encouraged. Because we are seeking to identify POVs, we are not merely concerned with optimistic scenarios; we are more attracted by everything that can go wrong. However, given what we have said in the previous chapters, that is, that POVs account for approximately 4% of any project (delays, cost overruns, and quality issues), it may be worth reconsidering Equation 3.1. In this equation, the pessimistic scenario accounts for one-sixth or 17% of the project [(P + 4R + O)/6]. As chief project directors, we would be quite upset if a project manager were to come to us and tell us "I can work on this project, but I need a 17% leeway on calendars, costs, or norms of quality." In other words, let's suppose the cost of building your house is US$60,000; the constructor would try to sell it to you by stating that some US$10,000 may be lost in the process. You would certainly think that the builder has not done or will not do his job right or that he is not capable of doing his job, and you would be convinced that the project would suffer sooner rather than later. So, from the point of view of a project feasibility analysis, I somewhat object to Equation 3.1 for which I have found no convincing proofs. In the introductory chapter, we saw that approximately 30% of projects fully succeed; yet, that doesn't mean that 70% of projects are failures—some turn into nightmares or white elephants, true, but not all. It simply states that approximately 30% respect their initial plan in full. The QMA is one example of such an achievement.

In essence, I propose to replace the weights suggested in Equation 3.1 as follows—estimating that about 70% of projects are, if not completely successful, at least successful in some measure despite some shortcomings:*

$$\text{EFC} = 0.04\,P + 0.70\,R + 0.26\,O \tag{3.2}$$

---

* This can be simplified to EFC = 5P + 70R + 25O.

This seems to make intuitive sense from a managerial point of view. If we were investors in a given project, we'd like to see the project's promoter admit that there is a slim chance that things could go wrong (4%–5%), that he has done all he could to guarantee that the project will develop as planned, but that there is even a chance that it may go better than planned, although this chance is not as strong as the one expressing how things are realistically expected to happen. Accordingly, we'd say that the manager is seeing reality with common sense and we'd be interested in investing. An overly optimistic forecast would raise some flags: perhaps, the promoters are ignoring some POVs? My proposed formula is a far cry from the one proposed in PMBOK and various such books; however, I do not believe any manager would like to work with a set of circumstances where things can go wrong 17% of the time. This can only be true if POVs have not been addressed from the start. Since the current success rate of projects is estimated at 30%, this means that there is theoretically room for improvement by way of tackling POVs, by more than twofold. The reader will have guessed it: I are trying to make the assessment of scenarios more in line with common sense and to highlight the importance of dealing with POVs. This is shown in Figure 3.5.

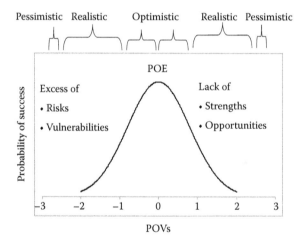

Figure 3.5 A stylized view at PRO.

Viewed from this angle, Equation 3.2 seems to correspond to a normal situation, even if some readers may question the logic that I adopted to arrive at it. I would venture to say that Equation 3.1 should be parsimoniously used because it underscores some poor management practices and because it does not seem to reflect a normal situation, which would be expected in any project undertaking. To the right of the mean, we could postulate that projects were lacking Key Success Factors (KSF) (strengths and opportunities), and more and more so as the variance escalates (e.g., there were no or few negative forces but the management team just wasn't performing enough); to the left, we could assume that projects failed more and more as the negative variances inflated because they were under the influence of Key Failure Factors (KFF) (e.g., there were dreadful combinations of poor planning and Unfits—uncontrolled Forces of Production). KFF include risks and vulnerabilities. That's one way of looking at the expected final costs; from a managerial point of view, it seems to make sense. It says that a manager should face risks and reduce vulnerabilities appropriately, and build on strengths while answering adequately to opportunities; by doing so, a manager builds a strong realistic scenario for the project. He is robust.

3.5   The POW system

The fact that a promoter cannot measure his endeavor or else that he doesn't care enough about finding ways to assess them other than financially or operationally is a strong warning sign that POVs will bloom. A project is a human effort. Discounting this is falling for a dream scenario when in fact the opposite is likely to happen. There is no better control on quality than by preparing for the worst, and quality is something we aim for in a project. Recall that the prefeasibility study has a look at the parameters and at KSF; in a feasibility study, we review them within the context of the process "transformation-deliverable" to determine if the project promises still hold true. To do this, a number of documents are necessary. They help in sharing the vision of the project, they are[22] the following:

1. The project charter, which authorizes the project formally.
2. The statement of the project's content, which describes the tasks (part of the transformation stage in our model) and the project deliverables (part of the output stage in our model).
3. The project management plan, which explains how the work will be performed, and more particularly, how the preparation, integration, coordination, and control of each of the project's steps will be conducted.

A promoter should be able to provide the necessary information, including a budget, a list of work tasks, a plan for the management of anticipated resources, estimates of total work, the anticipated schedule, the expected deliverables, as well as his take on KPIs (Key Performance Indicators) and KSF[23] (Key Success Fundamentals). From this, the analyst can outline the Dominant and Contingency Strategies (DS, CS).

Using terminology commonly found in project management, we could roughly group these elements into three sections,[24] which I refer to

Table 3.4 The POW

Item	POW
Key performance indicators (KPIs) are identified	PBS
Deliverables are well defined	
Management of resources is well planned	OBS
DS and CS are established	
Budget is completed and realistic	
Current tasks are explicitly laid out	WBS
Calendar of tasks/activities is established	
Forecasted work is done	

as POW (Table 3.4). The POW concept is simple: an organization (O) works (W) to produce a deliverable (or else, a product is made by an organization by means of work—*P* for "product"):

1. Product Breakdown Structure (PBS), which includes resources as inputs (Processes), hence the importance of clearly defining the deliverables from the start.[25]
2. Organization Breakdown Structure (OBS), which relates in particular to talent within the organizational structure as well as the delegation of Power (authority/line of command).
3. Work Breakdown Structure (WBS), which examines tasks jointly with responsibilities.

I propose the core minimal efforts related to the POW concept as indicated in Table 3.4.[26]

A project promoter that is unable or unwilling to provide a POW structure is probably not serious about the project; it would be unwise to pursue with the feasibility study.

As can be seen, the POW concept gives the analyst a closer look at the project, with an emphasis on its processes, and more particularly, on the transformation–deliverable axis. With the POW concept, one answers the questions: "What is going to be achieved? How are the achievements going to be measured? When? At what costs?" Often, POVs develop because there are no measurements; this prevents the manager from doing the follow-up on, say, tasks, so that he cannot judge the progress being made. Additionally, if an employee is not evaluated on the quality of his work, what will he care? Soon enough, he'll let his eagerness to achieve high standards fade away and will become sloppier. This is human nature, at least for many people. The POW concept provides a very succinct means of saying: "What is to be done can be done (what is promised is what you'll get)."

Let us take the oil container as an example. The promoters can probably very easily come up with a list and flow of tasks, with promises of deliverables and time for completion. The real question is: "Given the optimistic scenarios they provide, what measurements of performance and success can be used?" A feasibility analyst would challenge the promoters and find out that there are few realistic answers to these two questions. The same logic applies to the Mervel Farm project: as soon as performance indicators and KSF are demanded from the promoters with respect to the transformation phase (turning the abandoned farm into a tourist venue), the project seems less realizable, sad as it is for them.

3.6   PRO-POW

The PRO system can be merged with the POW analysis, as shown in Figure 3.6.

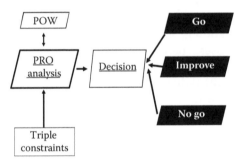

Figure 3.6 PRO-POW.

I refer to this as the PRO-POW system, which is the heart of a feasibility analysis,[27] and which I will expand on even further as we go. Everything that cannot be managed within this framework falls into the area of disasters (chaos), because constraints (time, costs, and norms of quality) will affect the scenarios (PRO) and at least one element of products, organization, and work (POW). Put simply, the analyst looks at the organization that works to produce the deliverable through the lenses of the three scenarios: realistic, optimistic, and especially pessimistic. The analyst would rightfully pose the question: "Given the most pessimistic scenario, for example, what would happen to the Mervel Farm project in terms of deliverables, organization, and tasks?" In fact, he could refine his investigation even further in very practical terms: "Given the most pessimistic scenario, that is, given that time, costs, and norms of quality go awry, what will be the exact consequences on Mervel Farm's deliverables, organization, and tasks?" An answer could be: "The Mervel Farm project will have to be put off for one year, its promoters will have to request an emergency fund of C$35,000, then the project would see two of its leaders quit, the one manager left would have to revise tasks and activity flows, and to reduce the offer to a more manageable level." Any change in the POW structure is likely to influence the calendar (time), costs, and norms of quality (with the four Ps being mandatorily involved); these changes, in turn, will affect the POW. Daily project operations that take place within the confines of the constraints are likely to produce a successful project, yet any changes to the constraints are likely to affect the daily operations. This is truly a complex dynamic system.

Let's now turn our attention to each one of the three constraints.

3.7   Calendar

Terminology used in project management refers to a "calendar" but in fact, it really is a **calendar of tasks and activities** that matters during the transformation phase.

In general, the calendar is initially the simplest of the three constraints to assess. Cost analyses, marketing research, and technical evaluation that lead to the establishment of norms of quality require time, money, and effort. Examining similar past projects, even roughly, can assist in determining a realistic schedule.

The PMBOK guide divides the calendar into five stages, one adding up on top of the previous ones over time, which I label as (1) "vision", (2) "planning", (3) "mobilization", (4) "deployment", and (5) "completion and evaluation".

The vision stage includes the creation of the project charter, the project scope statement, and the project management plan.

Some authors resort to the notion of "prephase" also sometimes called "incubation"; this is when the prefeasibility study takes place.

I encompass incubation within the vision stage. At this point, there is no substantial commitment toward the project, as opposed to the planning stage, when inputs are beginning to enter into the **transformation chain**, which will eventually generate outputs.

Let's replace the term "execution" commonly used in many project management models, with the term "mobilization" because it describes best the

Keeping an eye on the source of light.

fact that passive elements (e.g., material resources, MP) and active elements (e.g., Fits, FP_c) are fed as inputs into the transformation chain. I also add the term "evaluation" in the final stage (compared with most project management models, which omit this important step) because a project must by all means be evaluated at the end, if it were only to produce the Book of Knowledge associated with it, or else, some form of it.

The five **stages**[28] that form a project's life cycle are captured in Figure 3.7.

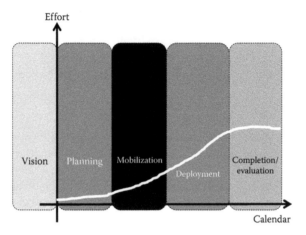

Figure 3.7 A project life cycle.

We can also illustrate this according to our modeling method, shown in Figure 3.8.

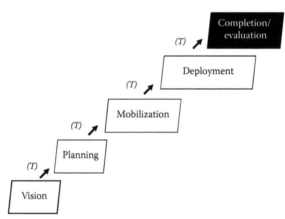

Figure 3.8 The transformation stages.

In Figure 3.8, the shape of the curve [effort=f (time)] (found in the PMBOK among other sources) contains a floor during the vision stage (when norms of quality are set) and two ceilings (when costs are imposing their weight onto the project); one during the planning stage and one at the beginning of the completion stage. The first ceiling (there is a ceiling when the curve is not a straight ascending curve) indicates that there is something preventing the growth of the project along a linear ascending curve with a slope of 1: indeed, this means that all the resources (more generally, the four Ps) are not yet in place and not ready to move into the transformation phase. Because it is a ceiling, this is intimately linked to costs. Assessed differently, the first ceiling means that the money needed to engage the inputs (e.g., resources, FPc) has not been released yet or else is not available yet.

At the planning stage, POVs are theoretical in nature; they are not active yet. They come to life once the curve adopts the linear ascending shape all the way up, theoretically, to the next ceiling. The second ceiling is an indication that the project has reached its maturity or, technically speaking, the maximum involvement of the four Ps. Because this is a ceiling, it is, again, intimately related to costs. This means that the project has neared its total budget; if it were not the case, some sacrifices would have to be made, most likely in terms of quality.

Note that this curve cannot possibly represent the Total Costs (as some project management models erroneously claim), because it has a descending shape at the end; rather, it represents the efforts needed to run each stage of the project.

Hence, from a calendar's point of view, POVs represent a source of clear and present danger during the stages of mobilization and deployment. This is when they are most invigorated. Of course, projects often have substages so that there are various project cycles within the entire project. In addition, some POVs are sometimes discovered after the project is completed—when part of a building collapses, for example. Identifying POVs requires the unfolding of each stage (vision, planning, mobilization, deployment, and completion/evaluation) under a pessimistic scenario. A good way to start is to review the list of tasks (the W in WBS). The vision stage, for example, can be subdivided as follows: the evaluation of the concept, identification of the prerequisites, design, and validation from potential clients and project investors. Market validation is a crucial marketing effort because projects must respond to an opportunity. Let us take the example of the Stradivarius musical concert as per Table 3.5.

Table 3.5 Calendar of a concert

Stage	Start date event	End date event	Number of days
Vision/Planning			
Mobilization	First meeting with musicians		
Deployment	First concert		
Completion and evaluation			

The feasibility analyst and the manager on site (during the actual implementation of the project) will speculate about what could go wrong as the concert event approaches. The analyst might ask: "Given what could go wrong, how realistic is the time line for the concert?"

Let us move tangentially to consider the story of Moby-Dick.[29] The captain has set up a shrouded agenda to go chase a white whale that has amputated his leg in a previous encounter. The ship must be prepared for the journey: a Plan is set as to where the ship will sail and what material and human resources must be on board. Processes are established: undocking the vessel, sailing, and surveying the sea. Team members (Forces of Production), such as the cooks and the harpooners, are put to work. The agenda has changed by an effect of surprise, from chasing whales for their oil to exacting revenge on a particular whale: Moby-Dick. All four Ps are in place to do what they are supposed to do, but in the end everything goes awry because there was a secret plan, which by itself was a POV driving the mission to catastrophe (the captain was the Unfit in this case). The initial question related to the project would have been: "Given that the captain is nowhere to be seen during preparation and departure, what could go wrong?" Most certainly, something related to the captain.

When it comes to time, there are two kinds of flows to be cognizant of: **relative and absolute**, or, put differently, conditional and unconditional. Technically speaking, the conditional one is a cause-and-effect relationship, whereas the unconditional one is purely based on the passage of time. The difference is fundamental, but seldom understood by some practitioners. The relative time is the path that is designed for the project (the full transformation chain), and that can become critical as the concert date gets closer—it is created by humans by way of calendars of tasks and activities; the **absolute time** is simply the normal passage of time as

Sun shining on the QMA.

set by nature; it has a steady pace as opposed to relative time, the pace of which can change depending on circumstances.[30] In the case of the QMA, for example, the pouring of the concrete for the foundation was postponed by roughly one month because of poor weather: absolute time kept ticking while relative time had to adapt. Einstein would inarguably agree!

The feasibility analyst ensures that tasks and activities deemed essential to the project at each of the five stages are recognized. He must verify that the units of measurement (e.g., currency, day, hour, and quality metrics) are identified and precise. He explores the types of links that bind tasks together. This will give him a good view of the project.

By making a succinct list of tasks/activities included in each of the five stages of the project and deciding on their types of links, the analyst will be able to identify the most vulnerable points. Causal links can lead to the most dramatic negative impact on a project; however, as the project travels along its critical path and nears its end, the time factor becomes more and more acute. Finally, under certain circumstances—such as the dramatic gush of the price of steel during construction of the Montréal Olympic Stadium, MOS—some process elements may have a tremendous influence on the project. Generally speaking, influence process elements (those having an influence on the project) are more critical during the mobilization stage, causal process elements (those exerting a cause-and-effect pressure on the project) during the implementation stage of the project, and the longitudinal process elements (those strictly time related) toward the end of the project as it approaches its deadline. The definition of the project contains POVs at the planning stage.[31] We can illustrate this as shown in Figure 3.9.

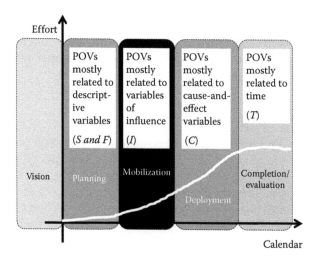

Figure 3.9 General importance of types of links during the project's life cycle.

POVs are thus temporally spaced along the calendar of tasks and activities. This observation may provide an additional clue that the analyst can use in his effort to uncover them. POVs' prevalence is related to proper Description during the planning stage[32], their Influence during mobilization, their level of Causality during implementation, and their Timeliness during the completion of the project. In conclusion, the mobilization and implementation stages[33] are when the POVs are theoretically most potent.

From these observations, I draw the second law of project feasibility:

3.7.1 Second law of project feasibility: Dependencies

The stronger the dependencies between the project's tasks, the more vulnerable the project is. High task interdependence generates high potential vulnerability.

Here, dependencies refer to the critical nature of the bond between process elements. From this perspective, a project is theoretically most vulnerable during the implementation stage. At this stage, there is a potential for all POVs past and present (whether they relate to Description, Influence, Causality, or Time) to become active. The Challenger shuttle explosion on January 28, 1986, which happened 73 s after takeoff, while it was traveling at a very high speed, provides a painful example. Seven crew members perished. The catastrophe was eventually traced back to a ruptured O-ring on one of the rocket boosters attached to the main tank of hydrogen, which is, of course, a highly explosive source of energy. The lives of seven astronauts were partly dependent on a simple O-ring, the cost of which was altogether minimal compared with the total cost of the shuttle and mission: there was a very high POV at that particular joint between the tank and the shuttle. Most POVs regretfully act up during the implementation stage.

The second law of feasibility must be acknowledged when making the final assessment of a project's feasibility.

3.8   Costs

When doing a feasibility study, all costs are *de facto* estimates. There is little room here to be optimistic and in fact, the best strategy is to rely on a pessimistic scenario. Costs are to be spread along the five stages of the project's life cycle. Seventy-five percent of estimated costs are usually set in the vision stage, but only 5% are actually incurred at that time.

Table 3.6 offers an example of real costs, according to each stage.

Table 3.6 An example of a project's cost structure

Stage	Real costs (in *US$*)[41]	Official documentation
Vision	Getting married	–
Planning	Sending cards	Cards and contracts
Mobilization	Setting the stage	–
Deployment	Party time	–
Completion/evaluation	Well, you decide!	"Thank you" cards

Most project management books[34] illustrate the major project stages' costs as shown in Figure 3.10.

A number of observations can be made with respect to the cost structures shown in Figure 3.10. First, the more time elapses, the more

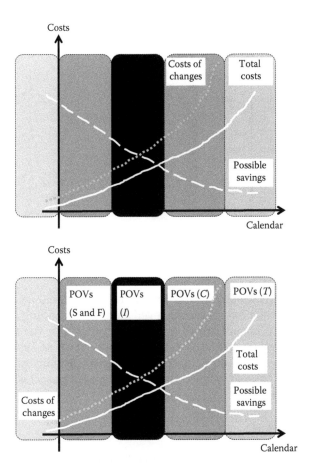

Figure 3.10 Costs by transformation stages.

expensive the changes are; secondly, the savings seem to reach a floor toward the end of the life cycle; thirdly, the situation becomes increasingly critical,[35] starting from the point where the savings curve meets the cost of the change curve, midway in the mobilization stage; and fourthly, the cost of changes forms a near straight linear ascending curve during the mobilization–deployment stages. For this reason, the assumption I initially set that all projects follow a straight linear ascending evolutionary curve is associated with the *relative* cost of changes. The **relative cost of changes** is the cost of changes relative to the budget that has been expended so far during the project. At the beginning, a change, say in the planning, may not represent a high cost *per se*, but in comparison to all the costs incurred at that time, it may actually be small. The relative and punctual curves of the cost of changes are indicated in Figure 3.11.

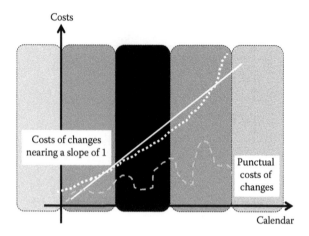

Figure 3.11 Relative and punctual costs of changes.

Thus, the fundamental basis for the feasibility analyst, that is, the assumption that all projects are deployed along an ascending linear curve with a slope of 1, actually rests on the observation that the relative cost of changes evolves in a quasi-ascending linear curve. At least, that's my working hypothesis. Looking at a project, from a feasibility point of view, is looking at the cost of changes. This makes sense: a project is fundamentally a change process.

3.9   Norms of quality

We have already defined perceived quality as being [Perceived quality = (Functionality + Design)/Costs]. We have seen that Functionality and Design can be linked to Short strategies, which are strategies that need to be implemented when everything else (Dominant and Contingency strategies) fails; this requires imaginative solutions, that is, some level of innovation.

Judging from this perspective, norms of quality and costs are interconnected. The establishment of norms of quality as constraints is related to technical expertise. At the core of projects, indeed, lays a technical challenge, as seen when we defined the characteristics of projects. The three elements of the triple constraints are not independent from each other, but are, rather, inextricably linked, with each one taking on a unique role. The four Ps play a critical role in the triple constraints. Let us take the example of People: the quality of people employed in a project is equal to their capacity to be functional, to abide by the design of their tasks, as reduced by the costs to employ them, given the time line set for them to complete their tasks. The final value of a project can be seen as the quality achieved over costs, given the time allotted to complete it.

This makes intuitive sense to a project manager. A proper organization, the delivery of quality deliverables and well-structured tasks seem to be ideal conditions for the achievement of the project, which is itself a function of the four Ps. The reverse, however, may not be true. Aiming for quality of the four Ps may not translate into quality in terms of the POW concept for a single reason: the presence of POVs. It may happen, for example, that a project respects the high standards of the four Ps' qualities, but this was achieved only because POVs were not triggered by external risks, which would have together adulterated the quality of the POW structure. Hence, the active presence of POVs is what determines if a project will succeed. From this perspective, a feasibility analysis should look at the POW elements and the POVs altogether, because POVs can compromise the products, the organization, or the work (POW). No matter how hard people work, for example, it will not affect the POW elements until that work focuses on solving issues caused by POVs.

3.10   Summative triangle

Glancing at the big picture is just as important as screening the details. The devil hides in the details, for sure, but missing the forest because the analyst is too close to the trees may lead him down the wrong path. One way of expressing the big picture is to use the **summative triangle**, which merges the three axes of the triple constraints with the five stages of the project's life cycle, as shown in Figure 3.12.

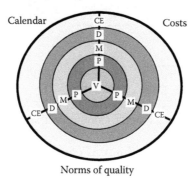

Figure 3.12 The summative triangle.

This summative "triangle" makes sense when units of measurement are specified, as in Figure 3.13.

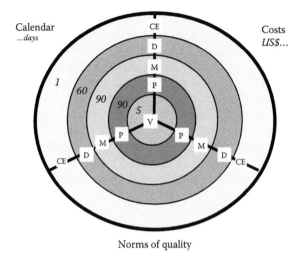

Norms of quality

Figure 3.13 The summative triangle with units of measurements.

It is most advisable for the feasibility analyst to compare the project with similar projects that have already been completed, using the summative triangle if possible. Figuratively, this tool can then be combined with the PRO-POW strategic framework as in Figure 3.14.

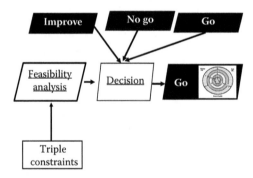

Figure 3.14 Decision-making and the summative triangle.

Timing, cost structure, and norms of quality (with their relationship to the four Ps) are evaluated in parallel to the three scenarios (pessimistic, realistic, and optimistic). This simple method gives the analyst an overview of the project in order to rapidly capture its quintessence.

I also encourage the analyst to prepare a spreadsheet that can transfer the summative triangle into a tabular format (Table 3.7).

The analyst would review this table, of course, and try to pinpoint where POVs would likely find a niche.

Table 3.7 Example of a PRO-POW for the Stradivarius concert

Stage	Task	Calendar *in days*	Costs *in US$*	Norms of quality
Vision	Concert idea			
	Choice of music			
Planning	Grant request			
	Production budgeting			
	Concert hall reservation			
Mobilization	Contracts with musicians			
	Contracts with external resources			
Deployment	Rehearsals			
	Stage conception			
	Posters, ads			
Completion/ evaluation	Concert			
	Media feedback			
Feasible or not?				

3.11   Conclusion to Chapter 3

Projects must be imperatively be defined with precision. Many authors have noted that one of the major causes of project failure is an inadequate project formulation right from inception. A project is formed of the four Ps (the four structural variables) and is expressed by at least two of the three functional variables that are deliverables and/or documentation and/or impacts. Large projects always include these three forms of outputs. A project proposal that does not recognize the role of the four Ps and that does not identify the outputs is not well defined; most likely, unpleasant surprises loom. Structuring a project carefully is indeed part of the four Ps, with respect to Plans. The preliminary stages of a project will influence the entire course of the project.

Projects exist within a contextual framework of risks. We have related risks to vulnerability, a concept that is at the core of the present book. We closely associate risks with calendars, because risks are outside forces that affect more directly a calendar of tasks and activities. The QMA is an excellent example: there was an environmental risk in laying out the foundation of the amphitheater during winter; indeed, the

Shedding light on the project.

project was stalled several weeks due to tough winter conditions. We have linked POVs with cost control management (and we will see that norms of quality are closely associated with an error term ε).

We have also found a mathematical link between calendars, costs, and norms of quality. Setting measurements are crucial to the analyst, and eventually to the project manager; I have offered various tools that, based on my experience, may have some utility. A feasibility analysis template is offered in Appendix 3.1.

In his 30-year search for Tutankhamen's tomb, Howard Carter did not get discouraged by his arch rivals or the lack of results over the years. It is by digging outside the established search area that the clues to the location of the tomb were discovered. It was by pursuing the clues (a stone staircase going deep into the ground) that at last a treasure was found in 1922, which captured humankind's imagination. It is possible, after all, that doing a proper feasibility study may lead to some opportunities to enhance the project beyond its initial expectations.

A useful tip is to begin an analysis by listing all the major tasks and activities of the project, and spreading them along the five stages of the project life cycle. The dissection of each major activity thereafter and the allocation of responsibilities to the four Ps in a detailed way by taking into account the types of links between the various processes, is a mandatory effort, albeit daunting at times. The goal is to shed light on the project from every possible angle.

3.12 What we have learned about POVs: Chapter 3

POVs…

1. Affect the costs more directly.
2. Are minimized by great managers to better handle the triple constraints.
3. Have a tendency to throw the project outside the "football" formed by the three constraints, into an area that we call "chaos".
4. Are actuated by the presence of risks.
5. Can be uncovered by examining everything that can go wrong (PRO system).
6. Are highlighted by the fact that a promoter cannot measure his endeavor or else that he doesn't care enough about finding ways to measure them other than financially or operationally.
7. Develop because there are no measurements/metrics.
8. Are theoretical in nature at the planning stage.
9. Start being active once the [effort | time] curve adopts the linear ascending shape.

10. Represent a source of clear and present danger during the stages of mobilization and deployment.
11. Are sometimes discovered after the project is completed.
12. Are temporally spaced along the calendar of tasks.
13. Are related to proper Description during the planning stage, their Influence during mobilization, their level of Causality during implementation, and their Timeliness during the completion of the project.
14. Can compromise the products, the organization, and/or the work (POW).
15. Are closely associated with cost control management.

3.13   Key managerial considerations: Chapter 3

1. Define the products, organization, and work to be done with attention to detail.
2. Laser focus on what can go wrong in search of POVs.
3. Face risks and reduce vulnerabilities appropriately, and build on strengths while answering adequately to opportunities.
4. Use both a detailed and global picture approach.
5. Evaluate the transformation stages in terms of a calendar of tasks and activities, costs, and norms of quality.
6. Actively search for causal links between tasks.
7. Use the summative triangle to express the concept in a simple format that everyone can embrace.

3.14   Case study Chapter 3: AF Thériault and the Hammerhead military target boats

3.14.1   Introduction

It is by a mere coincidental flow of events that the Hammerhead project came about. AF Thériault (AFT),[36] a manufacturer of boats of all kinds located along the Evangeline Trail in the Acadian part of Northwest Nova Scotia (Canada), had a crew of workers at a trade show when discussions commenced with Meggitt Training System Canada,[37] a military equipment provider well versed in the simulation of law enforcement and military training. The trail is named after an epic story of a young and melancholic woman who falls in love but who, much to her dismay, sees her dreams shattered by the British deportation of the French-speaking families to Louisiana, during the second half of the eighteenth century. Most families have eventually returned to what they called home, which is for the most part stretches of land bordering the Bay of Fundy where seemingly eternal winds have trees whistling night and day, and where the tides are the highest in the world.

It became obvious to the parties involved that AFT was well suited to build target boats that could be used by the navy during its practice mocking the swarming strategy employed by terrorists and thieves along the coast of Somalia, the Straight of Malacca, and the Persian Gulf to cite only these vulnerable areas. The idea of a target boat made available for training against such events was a direct consequence of the U.S. *Cole* incident, which demonstrated to everyone the need for improved protection against small predatory boats. One of the prime customers envisioned to buy the products and services sold by Meggitt were United Nations–friendly countries.

The swarming strategy implies the use of small and fast boats that act in a cohesive manner to attack larger, isolated boats, be they commercial, military, or with a tourist vocation. The name "swarming" comes from the way bees form tight yet highly effective groups with various functions aimed at confusing and attacking the enemy.

3.14.2 A short history

AFT was founded by the current general manager's grandfather, Auguste (Gus) Thériault, in 1938. Vintage photos are selectively stored in the main administrative building on an AFT site, a 10-acre site that comprises a total of 23 buildings, ranging from the storage of advanced composites, cabinets, hydraulic units, paint and spray containers, pipe fitting materials, and propeller machines. There are also hull maintenance and manufacturing shops, where some 175 employees work. Sandblasting, installation, painting, replacing or repairing parts, and finishing are all daily activities. A number of types of boats are built at AFT, including fishing boats, ferries, Marine Corps boats, work boats, and yachts. Table Case 3.A provides pictures of some of the boats.[38]

Throughout the decades, AFT has built nearly a thousand boats of all sorts, sizes, and uses, and has won numerous prizes and achievement awards. Boat building and repair includes work on aluminum, composites, fiberglass, steel, and wood. The maximum boat length is 130 feet (39.6 m).

3.14.3 Hammerhead project

The Hammerhead project developed as Meggitt and AFT realized they could work together to produce a high-speed, autonomous (unmanned) target boats that could be shot during military practice. The life span of a Hammerhead is short yet the boat is instrumental in the training of highly specialized Marine Corps.

Table Case 3.A Some of the boats built by AFT over the years

More specifically, the goal was to build a boat 17 feet long and about 4 feet 3 inches wide, capable of reaching a speed of 30 knots in sea state condition 3 (according to the Beaufort scale) at a maximum cost of C$60,000 (as of the year 2006). The boat would be unmanned, made of solid material, yet its design would not be overly sophisticated, as it would serve as a direct target as opposed to being capable of complex evasive capabilities. Tolerances for the construction were set at ±1/15th of an inch (or ±2 mm) in large part because it was equipped with a set of electronic components given that it was to be unmanned and GPS-driven. No lead would be utilized as this would cause an environmental hazard given that the boat would be shot and would sink into the sea—instead, an iron pig covered with fiberglass was elected where weights had to be installed to compensate for the lack of crew on board. Additionally, the boat was to have a single lifting point so it could be easily moved from one location to the next with an articulating mechanical arm (crane). The initial budget for the project was modest. A timeline for completion was set at 6 months.

Initially, the project had four key stakeholders: AFT, Meggitt, the Canadian Navy, and Mercury Marine Canada (not related to the car industry). It faced a number of challenges. The risk of failure stemmed from the fact that such an unmanned vessel had never been built by AFT. In addition, given that the boat would eventually be shot during military training, a well-executed trade-off had to be reached between the cost and quality of materials and the structural integrity. In short, there was no need to overbuild; the vessel just had to be sufficiently good.

The prototype took four years in the making. Although building boats was the expertise of AFT, several factors had to be examined and tested. Various forces affect such a vessel, which, being unmanned, must be remotely controlled: currents, gravity, the type of water (salt or fresh), waves, and wind. These exert a tremendous pressure onto the hull when the boat travels at speeds of up to 30 knots/h. A special design had to be developed in order to maximize the kinetic energy under the hull in an attempt to achieve a balance between aerodynamics and propulsion. Special (secret) features had to be incorporated into the paint, the shape of the hull had to be carefully drafted, and various weights of up to 100 lbs had to be inserted into the boat to mimic the presence of a crew on board. The Hammerhead hardly capsizes; it can turn at sharp angles without losing its balance and seems, at high speed, to float on a carpet of thin air above the water level.

Over the years, some 346 Hammerheads have been built, with an average annual output of 30. The first year saw the production of three Hammerheads only.

These are boats that can actually save lives, according to AFT. The navy perfects its shooting skills against terrorists at sea; this potentially prevents devastating attacks on civilian and navy targets. As mentioned, such attacks have been made famous in the case of the U.S. *Cole* in 2000, which killed 17 U.S. crew members.

The design of the boat was done by the project leader, David Saulnier, and a colleague, who both played an instrumental role in the development of the boat over the years. Seven of David's staff eventually joined forces in building the prototype, literally starting from scratch. The very first boat was the only one ever built that would be piloted. It was first tested on a local lake, and then reworked to be tested, unmanned, on the sea. David attributes the success of the Hammerhead boat to the hard work and dedication of his team; everyone has been committed and can build a Hammerhead inside out. As of today, some 10 years later, it takes 8 people (not including David) and 5 days to build the boat; reaching the point of autonomy[39] took 5 years in the making, given the numerous changes that were made and requested by the project partners. A manual has been prepared both at Meggitt and AFT that gives the specifications for the Hammerhead components, and a Work Breakdown Structure (WBS) has actually been completed.

3.14.4   What makes an efficient team

David, the project leader, knew well in advance what would make the project successful in terms of the use of authority and power, and in terms of team cohesiveness. "I use a horizontal approach with my staff," comments David. "Status (read: egos) is a work atmosphere killer." Indeed, the mere location of the plant is conducive to such a work culture (psychodynamics): the community of about 9000 people, mostly French-speaking, isolated in a province that has nearly a million people, who mostly speak English, has fostered a close-knit network whereby everyone knows each other and has known their neighbors since childhood. David Saulnier says

> "My staff trusts me because I am easy to connect with; I am not saying easy to deal with – there is a difference. I am the boss, but my approach is convivial. I never hesitate to congratulate my staff for their dedication and hard work. Each boat is scrupulously crafted much like a piece of art. We take a lot of pride in the Hammerhead series. This is something we have developed from scratch."

Communication is by e-mail and telephone with outside partners (such as Meggitt and Mercury) with occasional person-to-person interactions, but within AFT, most of the communication is verbal. Pride resonates among the team members; indeed, it is the glue that unites the various trades, from fiberglass applicators to painters. David Saulnier adds

> "I like to have a warm relationship with my workers. I give them time. I do not use a top down approach. Rather, I am interested in what they can teach me. We realize that we are each a member of a team and that there are no cat and mouse games – I am not scrutinizing them and looking over their shoulders to verify their work. Overall, the trick is to champion a positive atmosphere."

David's approach to project management is, surprisingly enough, based on loss control management. When he planned the project, he took a deep look into areas where losses could occur; vulnerabilities indeed entail costs that eventually plague a project. One of the worst enemies of a team effort, he says, is overtime. Good work is done within the time calendar as scheduled in advance. Overtime is a cost driver that must be avoided. Hence, David makes a special effort to ensure his staff abides by the established Work Breakdown Structure (WBS) and maintains highly cohesive communication patterns. This is not to say that clashes do not occur; conflicts are a natural occurrence. David's technique is to separate the belligerents, cool them down, and try to understand what triggered their emotions at the source of conflicts.

3.14.5   Conclusion

Overall, there is a sense that the project has gone according to the timeline set from the get-go, the budget, and in full adherence to the norms of quality imposed by Meggitt. AFT keeps enough materials at hand to avoid supply crises. All of the skills necessitated to build the boat are transferable, so that the absence or loss of one key team member can be compensated by the coming of a worker drawn from the crew of some 150 employees at AFT.

3.14.6   Appendix A of Case 3: The manufacturing of the Hammerhead[40]

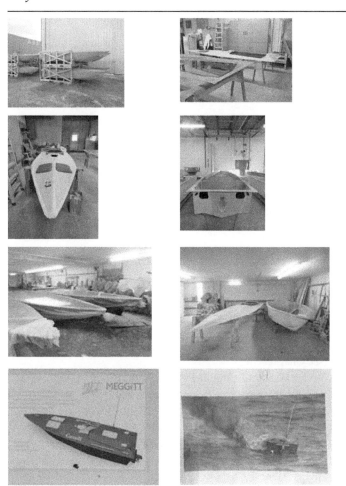

3.14.7   Questions related to Case 3: AFT

1. Discuss how the Hammerhead project responded to an opportunity.
2. List the stakeholders and classify them according to the People's nomenclature: customers, suppliers, Forces of Production, FP, and so on.
3. Give the characteristics of the Hammerhead boat and discuss in what way it corresponds to the characteristics of a project (e.g., innovative).

4. Discuss the Plan.
5. Discuss norms of quality.
6. Discuss the POW.
7. Discuss how trust and collaboration are fostered at AFT.

3.15 Appendix 3.1: Feasibility study template

This template starts with the prefeasibility form, with additional details as follows:

- Project charter
- Detailed analysis of the triple constraints
- Detailed WBS
- Analyses based on the types of analysis (see Chapter 7)
- Six strategic Ps analysis (see Chapter 7)
- Management plan

Conclusion:_____

Signature:_____

3.15.1 Checklist

Yes/No	Did you…
	Assess the six core competencies of management and team members (see Chapter 5)?
	Check the scientific literature background?
	Compare the project with similar projects?
	Create a model for the project (see Chapter 4)?
	Define the project accurately?
	Determine the magic moments (see Chapter 4)?
	Determine the type of feasibility analysis that is required (e.g., environmental)?
	Identify potential Unfits (uncontrolled Forces of Production, FP_{nc})?
	Measure perceived, added, and residual values?
	Propose a summative triangle (see Chapter 3)?
	Receive all the required documentation from the promoter?
	Review the five frames of analysis?
	Review the three scenarios, especially the Pessimistic one?
	Review the triple constraints with walls, ceilings, and floors?
	Examine the line(s) of authority?
	Check the measurement scales?

Of course, the document is to be adapted by each feasibility analyst.

Endnotes

1. O'Shaugnessy (1992, p. 101).
2. Jaafari (1990).
3. Farrell (1995).
4. O'Shaugnessy (1992).
5. The newest version of the PMBOK has abandoned the concept of the triple constraints and juggles instead with a number of concepts that are inputs or external factors. I will review this in more detail further along in this book. For now, I adopt the position that the triple constraints frame a project manager endeavor. Note that the same comment applies to Mulcahy's *PMP Exam Prep* book (p. 28). Customer satisfaction is listed as a constraint. This is not an operational constraint; this is the result of norms of quality, timeline, and costs having been respected after they had been established following a needs analysis (which PMBOK 5 refers to as "collect requirement" in the group of processes 5.2), which indeed is of prime importance (Owusu and Welch, 2007). Risks are also listed as a constraint (p. 28). Risks are external factors that influence the transformation process that a project is; they are certainly not a "guideline" (a constraint) within which to operate. Resources are input, not constraints. Scope is a vision while norms of quality are truly a constraint: a binding component of the project.
6. See Morris and Pinto (2004, p. XV).
7. A number of financial tools are available, of course, but this is not the subject of this book. For a quick overview, see Mulcahy (2013, pp. 267–269) and Schweser (2015, pp. 42–43.)
8. Morris (1989, p. 180).
9. See Morris and Pinto (2004, p. XV).
10. PMBOK 5 states (p. 6) that the constraints are composed of scope, quality, schedule, budget, resources, and risks. I oppose this because scope is a vision, resources are the Means of Production and risks are an external force. The original concept of the triple constraints should have been kept, in my modest opinion.
11. We will see these in more details further along in this book.
12. As a metaphor, we could relate the concept of "mobilization" in targeting POVs to the notion of immune mobilization in the body.
13. My translation.
14. See my study on the link between trust and quality in the SCF case at the end of this book.
15. PMI (2004, p. 1).
16. The absolute standard is, of course, an ascending linear curve with a slope of 1.
17. Note that in Figure 3.3, because transformation is considered an action, it cannot henceforth be treated as an object; therefore, the four Ps that point toward it cannot be structural process elements (because these define an object/element), so that Influence arrows are used (*I*) (see Chapter 4 on modeling Processes).
18. Starting from the second half of eighteenth-century England, this would include: animal power, anthracite, charcoal, coke, human power, hydropower, natural gas, steam, wood, followed by electricity, petroleum, atomic, solar, and wind energy.

19. FP and MP are used in plural form, meaning that FP refers to "Forces of Production" and MP to "Means of Production."
20. PMBOK 5 (2013, p. 169). Equation 3.1 (with its calculated standard deviations and variances) is widely used; of course, my experience tells me that often, however, project managers boost the value of the realistic scenario versus that of the pessimistic scenario in an effort to better represent the reality.
21. This formula is found in countless books on project management dating back decades.
22. See PMBOK (2004, pp. 76–78).
23. See Morris and Pinto (2004, p. 13).
24. See Morris and Pinto (2004, p. 17).
25. Chapter 7 will show that a product tree analysis can be used in this case.
26. Recall Table 2.13.
27. A full strategic analysis would include the six Ps of strategic management (PRO, POW, POE, POV, the four Ps, PWP), see Chapter 7.
28. Note: I use the word "phase" for the input (T) → transformation (T) → output model, which forms the transformation chain, and "stage" for the five steps taking place within the transformation phase.
29. Melville (1851).
30. Pace has been found to be a characteristic of innovation according to Shenhar and Dvir (2007), and innovation is a *sine qua non* condition to the definition of projects.
31. I will expand on the types of variables in Chapters 4 through 6.
32. As seen, this relates to how well the project is defined, that is, how complete its description is as presented by the project promoters.
33. A friend of mine calls the "Project Implementation Groups" the *Pigs* . . . now, that's not nice!
34. For example, PMBOK 5 (2013, p. 38).
35. Variables are descriptive, of influence, longitudinal, or causal.
36. http://www.aftheriault.com/en/.
37. http://meggitttrainingsystems.com/About-Meggitt.
38. Photos generously provided by AFT.
39. See Chapter 4.
40. For obvious reasons, only nonconfidential pictures are made available. These pictures have been taken by the author.
41. Putting the date and the currency on any table that treats costs is highly recommended. Indeed, currency value changes over time and according to country.

chapter four

Processes

So it is, said the Wise Eagle, that the river, like the flow of things, is always the same despite being different every day.

Darloz

4.1   Introduction to Chapter 4

Processes and methods have long been an integral part of project management, the critical path analysis and critical incidents path analysis being two examples among others. The *Project Management Body of Knowledge* (PMBOK) 5 defines processes (p. 85) as "a set of interrelated actions and activities performed to create a prespecified product, service, or result. Each process is characterized by its inputs, the tools and techniques that can be applied, and the resulting outputs."

On pages 11 and 12, it exemplifies processes in the following terms:

> A primary function of a PMO (Project Management Office) is to support project managers in a variety of ways which may include, but are not limited to: managing shared resources across all projects administered by the PMO.

PMBOK 5 also refers to best practices, methodology, policies, procedures standards, and templates.

Processes apply to concrete items such as building materials just as much as to abstract elements or information. In the latter case, data constitute the input. This input goes through a transformation stage, such as collect → sort → analyze → synthesize → use for decision-making → use for action → use to evaluate → archive. The output is the actual document that is being shelved or archived. All processes contain these three fundamental elements: inputs, transformation, and outputs.

This chapter delves into the minutiae of processes, a subject we have briefly touched on in the frames of analysis discussed in the context of prefeasibility studies.

4.2   Transformation

As we have seen in the previous chapters, processes[1] is one of the four Ps (Plans, Processes, People, and Power) of a feasibility study and of project management. Nature has its own processes: two prime examples are parasitism and symbiosis, which are strategies for life survival. Processes, in their most simple expression, consist of leading inputs into a change phase out of which they come as outputs that serve a certain utility, which is embedded in the notion of opportunity. Some sources of reference classify outputs as benefits, ordinary outputs, and results.[2] From our perspective, the results are either products, processes, services, or scientific research, and benefits are most likely tied to financial value or utility.

Transformation is a utilitarian act or else a series of acts. We touched on the idea that utility may be linked to "efficiency" and "efficacy", two terms we will define further along in this chapter. The notion of utility is fundamental; it is vastly used in economics, and we will see in Chapters 5 and 6 that we can resort to it when addressing interpersonal relationships. Put simply, if there were no utility in a project, there would be no successful project. In my initial model, I stated that a key characteristic of projects is that they answer particular needs or, viewed differently, that they respond to opportunities. However, in the end, this response must have a utility. Utility is bound in particular by a calendar of tasks and activities. This is another fundamental aspect of projects. In classical economic science, time is generally not a factor to be reckoned with when discussing utility curves; in project management, time is of the essence. I will come back to this notion of utility and see its relationship to the k constant in the section on efficiency.

We have also seen that some modeling comes in handy when wanting to simplify complex processes and guarantee that stakeholders relate to the project in the same way, thus avoiding confusion once the team members (the Forces of Production, FP) become actively involved. Before we venture too far into the specifics of Processes, we must learn how to create a utility model of the transformation phase. This is what we do next.

4.3   *Modeling processes*

Various authors have commented on the usefulness of modeling. It has been said, for example, that

> Regardless of the kind of reliability and validity checks, models are simplifications of reality. They can be made more or less complicated and may capture all or only a portion of the variance in a given set of data. It is up to the investigator and his or her peers to decide how much a particular model is supposed to describe.[3]

Other scholars specify that modeling be done by grouping concepts while ensuring a sense of coherence, explaining by the same token that modeling is a simplification of reality.[4] Some authors rightfully comment that modeling is meant to make intelligible a complex reality, not to make complexity a simple reality. Overall, the key concern is to achieve simplification. I propose a method that allows feasibility experts to simplify complex processes.

There are, of course, several ways to model processes. Nowadays, the program evaluation and review technique (PERT) is losing ground in favor of the Critical Path Method[5] (CPM). On the other hand, dynamic systems simulation offers opportunities for rich optimizations.

As can be guessed by the reader, I have stuck to a certain number of rules when illustrating the project makeup in the chapters covered so far. Indeed, I use the modeling system that I find most useful because it allows the analyst to detect holes (or POVs) within the model under investigation.[6] When it comes to processes, this method calls for parallelograms and arrows. I detail it in the next subsections. The explanations may seem a bit dry because they rest on a set of rules and procedures, but I find it to be a necessary analytical method in order to pinpoint POVs.

4.3.1   *Straight direct and diagonal flows*

There are two important flows in the transformation phase that need to be taken into account and that are most often ignored. All processes can be divided into two directional flows: straight direct or diagonal. Dominant and Contingency strategies are examples of straight direct processes because things are prepared and evolve according to plan; a Short strategy is an example of a diagonal process because the manager is caught off guard and must resort to innovative ways to get out of trouble.

Production processes that use machinery are generally of the so-called straight direct kind: there is one point of entry, one direction in the production, and one output. There could be various points of entry; for example, in some corrugated carton manufacturing, two corrugated cardboard sheets of corrugated carton of different sizes enter from different entry points to be channeled along some conveyors bordered by rails. Eventually, these two sheets are glued together to produce one single box that can be folded in various ways, much like a piece of origami. This remains a straight direct flow. The bottom line is that the inputs are not diverted; they adopt a linear logic that leads them toward a predict-

Silos—part of a production plant.

able output. To revert to the example of the chess game, straight direct flows can be represented by the rooks. In a good strategic game, rooks are generally deployed once other pieces on the board have been assigned a fair or an advantageous strategic position. From a managerial point of view, this means that the transformation phase, which should be, in theory, completely direct linear, should be deployed once all the other strategic steps have been put in place; that is, once the four Ps have been properly prepared for it. POVs become obvious when the direct linear process starts going awry; if it does, it may be caused by uncontrollable factors that have awakened preexisting POVs lying dormant within the four Ps.

Diagonal processes are typical of human thinking and interaction. Group meetings provide a prime example: a meeting rarely evolves as anticipated. One participant starts to wander, another one diverts to an unrelated topic, and so forth. Diagonal processes seem counterproductive, but in fact, they are an essential part of any project. Each project is unique and contains some level of innovation. Innovation is, by definition, a diagonal process. These occur when one walks off track, that is, when one explores avenues that are not commanded by the routine aspects of life, that have not been planned, that provide new information, and that may, in the end, change the ultimate outcome of the transformation process. Hence, while straight direct processes ensure that the ultimate outcome is what was 100% forecasted, diagonal processes show deviations from the forecasted output. Diagonal processes resemble the way that a bishop moves on a chessboard. Strategically, they are deployed on the board before the rooks and they often play a pinning role: they prevent one piece from moving (e.g., they attack a knight that is in line with the queen so that moving the knight leads to the loss of the queen).

As strange as it seems, straight direct processes present some disadvantages, especially in the area of work culture (psychodynamics; i.e., the way people interact). The groupthink phenomenon is a case in point: members of a group convince themselves that they are correct even when they are wrong. Why? Because they wish to remain as consistent with themselves as possible (a logic exemplified by the k constant). On the other hand, a machine that goes off tangent and throws inputs into different diagonal directions will not produce the anticipated output.

Overall, straight direct and diagonal flows both play a vital role in the transformation process. Typically, most diagonal processes should take place before and at the beginning of the transformation phase, and most straight direct flows should occur during and at the end of the transformation phase. A project is in large part a melting pot of straight direct and diagonal processes. POVs can be detected when this combination starts to be both ineffective and inefficient. A feasibility study should recognize which processes are straight direct and which ones are diagonal, and in fact, it should corroborate the fact that a project plan accounts for both during the entire transformation phase. In other words, some brainstorming sessions during the construction phase are always good, if it were only to discover what goes right (not what goes wrong) in the project.*

An example of such a dynamic between straight direct and diagonal flows is the Apollo 13 mission. All was set in a linear fashion—a straight direct process. However, a problem with an oxygen tank occurred during the flight that forced engineers to think outside the box in order to save the crew and bring it back to earth. It is through a diagonal process that this was achieved—by way of imagining ways of dealing with carbon dioxide and a shortage of oxygen. Technically, the idea of using the moon's gravity to produce a string shot effect is also a diagonal process, given the circumstances. Often, diagonal processes are integrated to become straight direct processes in future ventures. According to Darwin's theory of evolution, it is changes in the normal flow of heredity (diagonal processes) that lead to random improvements which, if they warrant better prospects of survival, will eventually be adopted (become linear processes). Hence, evolution dictates straight direct and diagonal flows.

POVs are brought under control when straight direct and diagonal processes work in tandem to achieve a result that is close enough to

---

* Part of what we do as consultants is hold such meetings to discover POVs and treat them to ensure project success.

the intended output. They become a source of problems when the two processes interfere with each other. Because diagonal flows are mostly intangibles as they often pertain to psychological phenomena, one cannot actually exactly measure them, but we will see in Chapters 5 and 6 on People how we can go around this hurdle.

4.3.2 Parallelograms

Parallelograms apply to any element of a process and arrows are used to illustrate the bonds that exist between each process element, or put differently, between each parallelogram.

Parallelograms have various codes. All models develop from a starting point, the beginning parallelogram is always drawn with a thick outline. The intermediate parallelograms do no use thick outlines and the end parallelogram is filled in black. The best example is that of the basic definition of a project as shown in Figure 4.1.

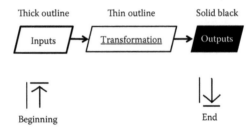

Figure 4.1 Beginning, intermediate, and end parallelograms.

Recall that processes are so-called straight direct, that is, they are linear and without any interruptions; any process that does not respect this assumption should be redesigned in a way that meets this condition. This may require looking at the project in a different way. This is a crucial effort because it eases the detection of POVs. A POV that is hidden in a variety of entry points, end points, or nondirect flows will likely remain undetected until after it causes damage to the project. However, with a straight direct process, the human brain can easily track the problems and their consequences using a back and forth analysis.

The bond that links the parallelograms is, in these cases as in any such project model, coordinated by time (*T*). It is called a "longitudinal" bond and has no polarity (it is neither positive nor negative). However, there is a system where there is a feedback loop to the entry point (or other points along the transformation stages). In this case, the time factor is expressed by a small *t*, or more exactly (*t*). In some software modeling systems (e.g., in some electrical software), a small *r* is sometimes used. I adopt the (*t*) nomenclature in my modeling system.

A typical feedback loop is found in the closed dynamic system that is a hospital. The patient first sees the doctor; then, based on the assessment of his ailment, a bed is reserved for a predetermined period of time, and surgery is scheduled. Costs accompany such a flow of events. The patient will have no choice but to go back (feedback loop) to his/her doctor in order to have his/her recovery assessed and receive permission to leave the hospital.

There are two other kinds of bonds that also contain a time factor: influence (I) and causal (C). I and C bonds have polarity: there may be a positive or negative influence of one parallelogram (project element) on another one (I^+ or I^-), and one parallelogram may cause another parallelogram to exist with positive or negative results (C^+ or C^-). Hence, we arrive at Table 4.1.[7]

Table 4.1 Types of bonds between process elements involving time

Name of bond involving time	Code
Longitudinal	(T)
Longitudinal (loop back)	(t)
Influence	(I^+) or (I^-) or ($I^\pm$)
Causal	(C^+) or (C^-)

All three types of bonds (T, I, and C) are named **consequent arrows**; as indicated, they mandatorily imply a temporal factor.[8] These bonds are consequent in the sense that they lead one parallelogram (project element) to another, either because this will obligatorily happen over time (T), or because one project element influences another over time (I), or else one project element causes the emergence of the other one over time, 100% of the time, given a set of conditions (C). As an example of the latter dynamic, let us take a simple kettle. Placed over fire, it will bring the water it contains to boiling point 100% of the time, given the right heat and atmospheric pressure conditions. There is just no way around it, every single time the conditions are met, the process will take place.

So far, we have what is illustrated in Figure 4.2.

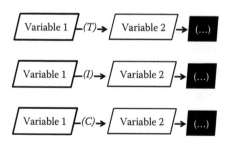

Figure 4.2 Types of consequent arrows.

As a rule of thumb, "time" is assumed to travel left to right; project elements must follow their sequence from left to right.[9] Note that each arrow in any model must be identified, as in Figure 4.2; this is done by way of inserting either (*T*), (*I*), or (*C*) with their respective polarities wherever they apply. This way, anyone can capture the dynamic that links the project elements together.

4.3.3   Corrugated cat litter box example

As an example of a straight direct flow, I will discuss a cat litter corrugated box. Let me first give the background for this innovative product.

As of 2014, U.S. citizens own more than 70 million domestic cats; Canadians own about one-tenth this number. The pet product market is continuously expanding and has the advantage of being countercyclical: when the economy turns sour, people seek companionship and buy more pet products. In England, pet owners spend more on pet food than on their own breakfast.

Cats behave quite differently than dogs. Stray cats can survive; most stray dogs cannot. Cats take great care in maintaining their fur and in using clean waste disposal facilities—cat litter boxes. Cats will not hesitate to boycott dirty cat litter trays. Hence, a cat owner must routinely clean the cat litter tray with a small scoop or else throw the entire litter content away (typically composed of urine- and feces-filled sodium bentonite). This is an unpleasant chore because the litter stinks, it produces dust, and it is generally heavy and hard to handle. Furthermore, cat feces contain a parasite (*Toxoplasma gondii*) that can be harmful to pregnant women. Owners who do not use a new solid plastic cat litter tray every 3 months or so are generally not aware that the uric acid contained in the cat's urine dots the tray with microscopic holes where bacteria find a convenient shelter that promotes their survival and distribution. Plastic trays are rigid and are produced using injection molding technology; other trays are press formed, but are extremely flimsy and hold straight when filled with cat litter. Some cat owners insert a plastic bag inside the plastic tray, keeping the bag handles out. This solution works, but offers two disadvantages: first, the cat's claws can damage the integrity of the bag, puncturing it in multiple locations; and secondly, the cat usually dislikes plastic material and much prefers natural fibers such as corrugated boxes, made out of tree composites.

An innovative cat litter box made out of corrugated cardboard sprayed with a so-called water-resistant Michelman coating offers a solution to the various drawbacks associated with the traditional plastic tray. Corrugated trays exist, but have not yet found favor with the public. Some retail outlets that have tried to sell them have ended up with idle inventories, although more and more pet stores offer cat litter trays made out of recycled tree fibers. Many of the corrugated trays do not fold; if they do, they do not do so in a friendly manner (in particular, the cat litter content interferes with the folding mechanisms) so that the cat owner is left with part of the initial problem.

The innovative cat litter box is for cats, what disposable diapers are for humans. The consumer buys the tray in the flat position—it is thus easy to store and to transport. He then unfolds it in a flip second upon arriving at home. Once he deems the useful life of the corrugated tray has been reached (a maximum of two weeks for an average cat), he simply folds the tray in its ultimate, suitcase-like position and throws it away in a matter of three seconds. The particular benefits of the innovative tray (outlying the opportunity it entails) are that: (1) it is composed of recycled material; (2) it is easily assembled; (3) it is protected by a water-resistant coating; (4) it limits dust generated by such litter as sodium bentonite, a mineral known to absorb eight times its own weight in liquid (e.g., water or urine); (5) it limits the spread of bacteria; (6) it is easy to locate in a retailer's shop; (7) it is easy to store and to transport; and (8) it is ideal for vacationers, condominium owners, and older people.

The machinery needed to manufacture the unique, patented cat litter box was studied by Correx Packaging[10] in the Northern United States during 2000; Correx is a U.S. manufacturer of corrugated products and toilet paper worth approximately US$1.5 billion. The box is manufactured in the following way: Two corrugated sheets are fed at two different entry points, one is large and forms the bottom sheet, and the other one is smaller and forms the top sheet. The two sheets go through two separate conveyors that bring them together, with the top smaller sheet being glued to the large bottom sheet. Once this is accomplished, the unit travels along a straight direct conveyor bordered by various trail and folding mechanisms that force the unit to adopt its final shape, that of a cat litter box. The machine that would accept two entry points for the small and large sheets did not exist among the various Correx production plants. An option was to buy a ready-made machine or else to conceive and manufacture a stand-alone machine that would hook up to the existing so-called Post machines. This option was estimated to cost roughly US$80,000 but had risks, albeit benign: it had never been built before and thus would require some adjustments. Table 4.1 and Figure 4.3 illustrate the entire process. Note that in Figure 4.3, humidity represents an "influence" factor (I) that affects the production process but in controllable ways. Humidity levels would have to be abnormally high or else abnormally low for it to cause defects to the cardboard to the point that the transformation process would be seriously jeopardized. So, we settled in determining that humidity levels are a negative influence factor when either too high or too low and modeled it as (I), that is, as influencing the manufacturing process (Table 4.2).

In this particular setting, there are two entry points (see Figure 4.3).

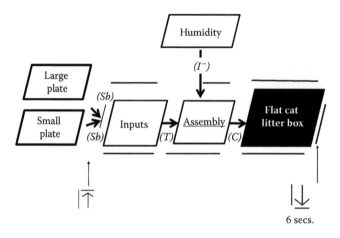

Figure 4.3 An innovative cat litter box manufacturing flow.

Indeed, we specified that two sheets, one small and one large, were introduced into the corrugated folding machine from two separate entry points; they are both essential to produce the output. However, for all intents and purposes, our model should obligatorily start one step *prior* to these two entry points. By adopting this vision of things, we immediately see a POV: what if the two sheets are not fed at the right time, so that they end up not being correctly glued, one on top of the other, thus compromising the folding mechanism that produces the flat product (second column in Table 4.2)? The input starts one step prior to the feeding of the folding machine with the two corrugated sheets. The transformation phase requires the gluing of the two sheets, one on top of the other, and the folding of the resulting unit. The output is the flat box that can be unfolded by the user into two positions: one that can be turned into a cat litter tray and one that resembles a suitcase so that the unit can be thrown into the garbage after the cat has accessed the litter for its useful life.[11] Note here that time is involved. Our model points to one POV (assuming humidity levels can be controlled): the entry point. Figure 4.4 illustrates this fact.

In Figure 4.4, we decided that the process started at the "real entry point" and the process called "input" were to be treated separately, because the entry point is critical. POVs always affect one or more of the three constraints[12]: time, costs, and/or norms of quality (note that each POV has an arrow pointing to the related constraint), which bind the four Ps. In this case, it is not Plans, People, or Power that suffer from POVs, but Processes, and the critical aspect is time, not costs or norms of quality. The cat litter example is meant to show how what could be rendered in a complex way, can actually be illustrated in a simple format that everyone can

Table 4.2 An innovative cat litter box manufacturing

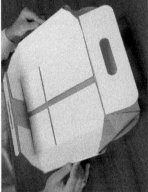

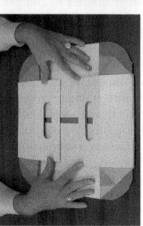

Position 1: The box comes out of the manufacturing process

Position 2: The box is unfolded for use as a cat litter tray in which sodium bentonite is poured

Position 3: The box is folded in its ultimate form for ease of throwing in a garbage bin

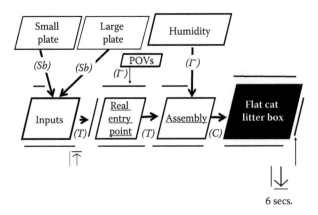

6 secs.

Figure 4.4 POVs in a straight direct flow.

understand, providing the reader learns the codes (or modeling language) that deal with processes.

4.3.4   The Italian Floorlite example

The Italian Floorlite[13] example is chosen to contrast it to the Correx case. Italian Floorlite is an Italian company based in Milan; it has various plants, one of which is located in Genoa. As its name suggests, it manufactures and distributes commercial flooring and industrial rubber on the European continent. Flooring products are mostly sold to architects and designers as well as to educational, health-care, and institutional organizations. Industrial rubber is offered to companies that use it as raw material for further processing. Some components of the manufacturing process include natural rubber and natural fillers, such as clay, limestone, and dolomite.

Early in 2010, Mr. Sergio Valiantino (Ing.) was mandated with supervising the development and implementation of a new machine. This machine had never been conceived before. It had been determined that the existing production process could be cut one step. The normal workflow for the transformation of the rubber paste consisted of feeding it into conveyors where it would be squeezed between stainless steel drums, then directed along different steps to ultimately form the final sheets of rubber having various qualities (with respect to thickness, colors, coating, etc.). One step seemed superfluous; however, the rubber paste was first heated up and then cooled down, then heated up again before being cooled off one last time. What if the two heating–cooling stages were reduced to one stage only, thus speeding up the process and reducing production costs? The project to strip the process of the extra step was estimated to cost US$1.2 million.

The machine was tested in 2013 with positive results—in fact, it was discovered that the new outputs were of better quality than the ones from the old system. The machine and its production settings were achieved on time, and within budget. However, not respecting the temperature needed for the process caused the machine to jam, and cleaning it would consume a great deal of time and effort. The second "discovery" was the difficulty in training the operators. In the past, minute changes in heat levels would not lead to catastrophic results, so that employees did not have to be overly concerned with the precise handling of the various components of the machine. However, the new machine responded differently to human intervention: slight deviations of set parameters caused major problems, 100% of the time, resulting in unwanted waste and the machine locking up. Not only would this generate costs, but it would also nullify Italian Floorlite's competitive advantage that it hoped to achieve with the new machine. The process can be exemplified in Figure 4.5.

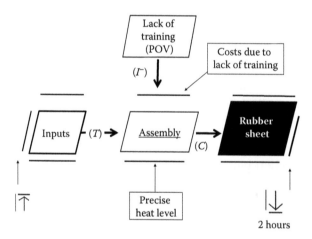

Figure 4.5 Italian Floorlite's improved process for producing rubber-based floor sheets.

The human factor is deemed to be a negative influence. The heating of the machine, however, can be precisely set with a button, much like an oven at home. Hence, this is not a factor, but a set parameter of which management and some team members (the Forces of Production, FP) have full control—it is a minimum norm of quality. There is one POV readily identifiable: a lack of training. The problem here is not that the norms of quality are not met or that they don't exist, but the fact that employees lack the proper training given the fact that they are accustomed to being quite lax with the temperature control as they are habituated to the old machine.

As can be seen from the Correx and Italian Floorlite examples, modeling is crucial. In the first case, humidity is a factor influencing the process—it can negatively affect the process—but the POV is undoubtedly at the entry point: the two sheets—one large and one small—must be fed into the machine at the right time. In the second case, a lack of training invites employees to be careless with the control of the heating and humidity systems, which may eventually lead the machine to jam as it becomes engulfed with overly thick or overly liquid rubber paste. Proper modeling forces the feasibility analyst and the project manager to pinpoint where, exactly, the source of a potential or real problem may be. In all cases, sooner or later, the problems will affect the calendar of activities, costs, and norms of quality or a mixture of these.

Recall that one of the main reasons for project failure is that projects are not well understood by the stakeholders (perhaps because, in part, they have different motivations, different ways of seeing the opportunity). Using a modeling system that palliates for this inconvenience is thus recommended. At a glance, everyone can see the process and figure out what is happening. The manager need not explain the sometimes-intricate working model to his employees; however, the fact that he fully comprehends it facilitates his communication. He can resort to catchy phrases and images that will work just as fine. The goal is to be understood, not to look smart.

Our basic project model is composed of three parallelograms: an input, a transformation, and an output parallelogram. But each parallelogram may be composed of a series of other submodels, with each containing its own sets of project elements (parallelograms).[14] For example, the transformation parallelogram can actually be broken down into a five-project element models, which would include the vision, planning, mobilization, implementation, and the completion/evaluation stages as we have briefly seen in the previous chapter. In all, parallelograms can stand alone, and they can also exist in a series of two, three, four, five, or even more. Of course, the more project elements contained in the model, the more complex the system is and the more potential for misinterpretations to occur between stakeholders. A tip to solve this problem is to indicate that a particular parallelogram (a project element) is actually composed of a submodel by doubling its outline, as in Figure 4.6.

Processes can be examined in two ways: from a descriptive point of view (we did so when we referred to the four Ps being structural components of inputs in a project) and from an active point of view (then the four Ps are labeled Plans', Processes', People', and Power'). In the case of transformation, from a purely descriptive point of view, it is expressed by the well-known life cycle consisting of the vision, planning, mobilization, deployment, and completion/evaluation stages. But we can also look at

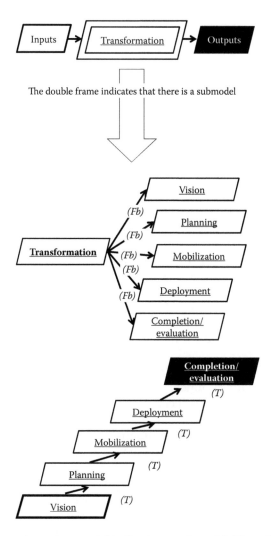

Figure 4.6 The main project model indicating a submodel. The submodel is composed of the five stages of the life cycle of a product, with so-called functional variables expressing the concept of 'Transformation'. The five steps can also be represented in their action, so that each active process element would be underlined; a time factor would unite one parallelogram after the other.

these five elements when they become active and introduce the factor time (*T*), which is a consequent arrow: the consequence, so to speak, of having a vision, is that eventually there will be some planning, and the consequence of planning is that there will eventually be some mobilization, and so forth.

4.3.5   Critical levels of causal bonds

Note that each type of time-related bond has a different critical level: relative time (T) can be stopped. A machine (e.g., the Correx experiment with the innovative cat litter box) can be stopped by pushing the power button or the red alert button positioned along the production line. An influence (I) can be assuaged by adopting certain measures. Suppose, for example, that the corrugated material is sensitive to the humidity level of the production and storage rooms, as can be expected; high humidity or an environment that is too dry may damage the integrity of the corrugated carton, thus making the folding process awkward and susceptible to failure. The humidity level in the rooms can be somewhat controlled with such apparatuses as dehumidifiers or their opposite: humidifiers.[15] In the case of causal bonds (C), the critical linkage level is reached, and hence, carries the highest potential for POVs to cause havoc in the transformation phase. Causal bonds have no way back: once a system is put in place, it cannot be stopped. The consequent arrow will do its magic (leading to the next project element or parallelogram) no matter what. This can have dramatic results. Recall the example of the vessel in the Victor Hugo story: the loose cannon could not be stopped and it ravaged the inner gut of the vessel it was supposed to protect.

I acknowledge the fact that the PMBOK[16] lists some of the consequent bonds, but not with as much detail as in my proposed methodology. Causal bonds are as follows (I complement them with my methodology):

1. The start of an activity is conditional to the end of a previous activity. This translates as the following:

   - End of previous activity (C^+) → start of next activity

   This means that the start of the next activity cannot occur unless the previous activity has ended, 100% of the time. So we can assume that this is a causal bond, because we take for granted that the next activity is mandatorily part of the project, otherwise it would not be listed (this is, here again, a logical statement given that projects are closed dynamic systems). We also assume that we are operating along a critical path. A more precise way of stipulating this causal bond is

   - Not ending the previous activity (C^+) → not starting the next activity

   Which reads: "Not ending the previous activity prevents the (causes the not…) starting of the next activity," which is the most intuitive way of expressing what is happening in the process, because obviously the project planners want the first activity to take place.

In the Québec Multifunctional Amphitheatre (QMA) example, not completing the pouring of the concrete base supporting the amphitheater holds back the rest of the construction project. The same phenomenon can be expressed using C^- instead of C^+:

- Not ending the previous activity (C^-) → start of next activity

Which reads: "Not ending the previous activity has a negative causal effect on the start of the next activity."

2. The end of a current activity is conditional on the end of a previous activity. This translates as

- End of previous activity (C^+) → end of current activity
    Or else, in a more intuitive way:
- Not ending the previous activity (C^+) → not ending the current activity

In the Mervel Farm project, suppose that a parking area is specifically allocated to the visitors of the site. Not closing the parking lot at night would indicate to would-be night adventurers or squatters that the farm site is still operating (albeit this would be a wrong assumption, but one that would fit their desires), so that they would feel free to wander on the site at night, thus possibly bothering the neighbors with loud music and heavy drinking. In other words, in order to consider the site closed for the night, the parking lot must mandatorily be locked, 100% of the time.

3. The start of the current activity is conditional on the start of the previous activity. This translates as

- Start of previous activity (C^+) → Start of current activity

This states that, in a closed dynamic system, the start of the previous activity leads, 100% of the time, to the start of the current activity. In other words, the project manager knows that starting the previous activity goes hand in hand with beginning the current activity. In the Correx box example, the feeding of the large corrugate sheet into the folding machine necessarily indicates that the feeding of the small corrugate sheet should immediately take place, otherwise the project (the realization of the deliverable) is entirely compromised.

4. The end of the current activity is conditional on the start of the previous activity. This translates as

- Start of previous activity (C^+) → End of current activity

In this case, the project manager knows that starting an activity will lead, 100% of the time, to the end of the next activity (put differently, the start of the previous activity leads to the end of the current activity). In other words, the first activity is not techni- cally completed until the end of the second activity has occurred. For example, the mobilization of resources serves no purpose if they are not put to use in the next stage of the production process. Within the transformation phase of the project, the mobilization stage has a causal effect on the completion of the implementation stage (given a closed dynamic system); otherwise, the entire effort is a lamentable loss. The reader can thus see that, under pres- sure to complete the project, the implementation stage will con- tinue affecting the mobilization stage if resources fall short. This can cause delays and increase costs. Thus, this particular setting betrays lurking POVs.

We can estimate the critical level of POVs of these scenarios as shown in Table 4.3—keeping in mind that the more the project runs along a critical path (and gets closer to the due date of delivery), the more critical the level is.

Table 4.3 Critical levels of POVS according to causal relationships

Type of causal link	Critical level of POVs
Not ending the previous activity(C^+) → not starting the next activity	Low because no costs are associated with the next activity
Not ending the previous activity(C^+) → not ending the current activity	Moderate because both activities are assumed to be near their end
Starting the previous activity(C^+) → starting the current activity	Serious because the process will not work unless both activities (which both incur costs) are given the go-ahead
Starting the previous activity(C^+) → ending the current activity	Critical because there is intense pressure once the entire process has begun

The project feasibility expert would be well advised to identify T and I bonds, and to specify the types of causal (C) bonds that exist in the process he is examining, because all of these bonds attract differ- ent critical levels of POVs.

What this effort does is help specify the location of potential POVs. Indeed, the causal link is often an indication of a potential POV. If the activities are relatively inconsequential, then the causal relationship may be dealt with at some point during the project; however, causal links along the critical path are undeniable expressions of potential POVs. If one element of the process drifts, then automatically, the other fails too, altering the course of the critical path ascribed to the project. Hence, assessing causal links in the context of critical paths is one way of unveiling POVs.

Let us take the example of the Montréal Olympic Stadium (MOS). The start of the 1976 Games was conditional on the completion of the main infrastructure; namely, on the functional completion of the stadium (it had not been completed as per plan, but it would be functional). Hence, we have the following:

- End of previous activity (C^+) → start of next activity

Alternatively, we have

- Functional completion of the stadium (C^+) → opening of the Games

This is a causal link in the sense that the Games could not take place until the stadium was functionally completed; there was just no way around it. Putting it in the context of a critical path gives us the following:

- Functional completion of the stadium prior to July 17, 1976 (C^+) → opening of the Games on July 17, 1976.

In fact, to be more precise, we have
- Incomplete functional completion of the stadium prior to July 17, 1976 (C^+) → no opening of the Games on July 17, 1976.

This expresses the conditional aspect of the causal bond. It also shows that there is a POV: the project would not be completed unless the stadium was functional prior to July 17, 1976. The feasibility analyst could then look back and try to identify everything that could prevent such activity (the functional completion of the stadium) from taking place.

We have thus added to the meaning of the PMBOK a proposed list of causal bonds. The reader can see how the methodology suggested in this book goes one step further than the PMBOK and other such books on project management and how it helps uncover POVs. With this last example, the reader can also appreciate how important it is to conceive a proper model of the project.

The influence (*I*) bond should also be understood in all of its peculiarities. There are two kinds of influence: direct and indirect. The direct influence (which can be I^+ or I^-), directly affects a process element (a parallelogram) as in the case of humidity affecting the cardboard in the Correx example. Indirect influences affect process elements in two ways: mediating or moderating. With the mediating way (which can be I^+ or I^-), a process element is introduced between two existing process elements to offer an additional route to the straight line process between these two process elements (parallelograms). With the moderating way

($I^\pm$), a process element can influence the interaction between two existing process elements in one of two ways, depending on the point of view or on circumstances: either positively or negatively, hence the code ($I^\pm$). There exist statistical methods to identify mediating and moderating variables (process elements), which I will briefly discuss in the chapter on People. Table 4.4 illustrates the different types of influence bonds.

Table 4.4 Different types of influence bonds (I)

Some production processes are designed in a way that there is a contingency provision: should one line of production slow down, an alternate route channels the production units that will eventually land on the output deck, where they will be collected for storage. This is a **mediating process**. It occurs in the brain as well: some neuronal paths are networked in such a way that if a flow of information cannot make it, say, from point A in the brain to point B in the brain (e.g., because of delays or damage), then an optional route is offered that sees the flow of information going from point A in the brain to point Z in the brain, and then to point B in the brain. Point Z is a temporary node through which the information flow travels in order to move from point A to point B. Often, point Z presents a certain number of advantages: it allows the route A–B to catch up on delays, for example, or to upgrade the quality of information because point Z contains bits of data that can enrich the information that is traveling.[17]

In a project, it is most useful to identify mediating routes because they help reduce the dangers associated with POVs. If something goes wrong with a particular process, then an alternative path can be used. At times, paths A–B and A–Z–B are engaged concurrently at different levels of usage. This softens the otherwise uncompromising level of the critical path. In short, POVs on critical paths are more dangerous than POVs on paths where mediating options exist. Mediating paths are often designed when preparing contingency plans, by posing the question: "If this does not work, what other option can be put in place?" An example is the emergency staircases in high-rise buildings: if elevators cannot be used to go from, say, floor 12 (point A) to ground level (point B), what alternative can be offered? Well, a staircase going from floor 12 to floor 5 (point Z), where occupants of the building must stop and go to another set of stairs to go from floor 5 to the ground floor. This ameliorates the security of the entire building because it cuts into the possibility of flames engulfing the entire exit path (what would have been a single stairway). Note that mediating paths are diagonal flows in the sense that they take a course off of the straight direct flow; when the unit that is being processed travels along the mediating flow, it is often improved. As such, the deliverable is not exactly what was predicted, an observation that meets the definition of a diagonal flow.

Point Z (floor 5) palliates for the POV that a single staircase represents in such a context. Note that we are still dealing with consequent arrows and parallelograms, that is, with processes. Time is therefore a factor. Indeed, in this case, if it takes too long to go from one set of stairs to another one on floor 5, the entire reason for having set such an emergency pathway is lost: smoke and flames will catch up before the occupants can escape. Time is of the essence in modeling with consequent arrows.

It can be added that POVs weaken any process system: they work against it and thus form a negative force, which the four Ps should theoretically combat. POVs have two ways of exercising this negative force, by way of (1) excesses (e.g., excess procedures dragging the production process) and (2) shortages, acting as a hole in the process (e.g., insufficient information or a shortage of material). In both instances, the process draws to a halt or else moves backward, which is contrary to the objective of the project.

As for **moderating influences**, generally these factors affect a process one way or the other. On some occasions, these factors can be beneficial, on others, detrimental. This is why the symbol $(I^{\pm})$ is used. Often, people (such as scientists) argue because they treat the same process without realizing that the factor affecting it can have an opposite influence depending on circumstances. People then quarrel bitterly over the same thing, occasionally with frowns and menacing stares...uselessly! This aggravates conflicts and misunderstandings. This is why it is so important (I emphasize this again) to resort to proper modeling when preparing the plan for the project.

The use of an external consultant in a project serves as a prime example. Some of the employees will react favorably by recognizing his value, and will therefore cooperate. Others will feel threatened—fearing, for example, that they may lose their jobs—and hence will become defensive, if not unfriendly altogether. This is true in the case of a team of experts in human relations (HR) that is invited to come and hone the skills of an existing project team. Some project members may actually believe that the experts have been hired not to improve their work conditions, but to find a way of laying them off. The experts—characterized here as a process element—have an antagonistic position: they appear good or bad, depending on how one looks at them, and probably on circumstances (the project goes well versus the project experiences difficulties). It is easy to see that moderating process elements can serve as a catalyst to awaking dormant POVs.

Overall, we have what is displayed in Table 4.5.

When the two elements being linked by a bond move in the same direction (when one goes up, the other goes up, or ↑↑), a positive sign is

Table 4.5 Details of the types of bonds between process elements involving time

Name of bond	Code	Impact on POV
Longitudinal	(T)	Moderate impact on POV when not in a critical stage
Longitudinal (loop back)	(t)	Idem
Influence: Direct	(I^+) or (I^-)	Can affect the process element, but not dramatically
Influence: Indirect		
Mediator	(I^+) or (I^-)	Can help deal with POVs
Moderator	$(I^\pm)$	Can cause managerial problems
Causal	(C^+) or (C^-)	Can have a devastating impact on POVs, even when positive, because this may cause excess stress on the system

Not ending the previous activity (C^+ *nens*) → not starting the next activity

Not ending the previous activity (C^+ *nene*) → not ending the current activity

Starting the previous activity (C^+ *ss*) → starting the current activity

Starting the previous activity (C^+ *se*) → ending the current activity employed (e.g., I^+); when they move in the opposite direction ($\uparrow\downarrow$), a negative sign is used (e.g., I^-).

So far, I have examined consequent arrows: what they have in common is that they cannot occur without the passage of time. All project-related processes fall within one of the three types of consequent arrows: longitudinal (*T*), influential (*I*), or causal (*C*). Ideally, the project feasibility analyst should be able to draw up each and every process by identifying the type of arrows that are pertinent between the various process elements.* Recall that a Dominant strategy is achieved when the project manager can control the project, that is, in the present context, when he can verify that all processes (parallelograms and arrows) work according to plan. On the contrary, a Short strategy is implemented in a hurry when the project manager reacts to unforeseen events. Without a doubt, a Dominant strategy has greater chances of being achieved when a proper plan is set, that is, when all processes are captured and analyzed using parallelograms and *T*, *C*, and *I* bonds.

Let us revert to the Mervel Farm example. From a macroscopic point of view, the inputs are Plans, Processes, People, and Power. With time, these inputs enter the next process element (parallelogram): transformation. Within this phase, all four Ps interact with each other (to become Plans′, Processes′, People′, and Power′) and various influences could actually affect this phase. Finally, with time, some deliverables (as well as some form of knowledge and impacts) are generated: the opening of the summer trail for the first Pierreville-based agricenter. This is a longitudinal process. Let's focus now on the input phase and assume that if no government funding is available, then the project will not go ahead despite the fact that it is deemed feasible from a technical point of view. Hence, we can express this by saying that the start of the next stage (the opening day) is conditional to the end of the previous activity (funding). It could also be possible that the project would go ahead without funding being available because the promoters believe that they could eventually collect the money that they need. In the latter case, the start of the next activity (the opening day) is conditional on the start of the previous activity (some funding). We can express the first scenario as the following: if there is no full funding right up front, then there is no opening day. End of previous activity (complete funding) (C^+) → start of next activity (grand opening). Indeed, the government would not grant money if it was not absolutely convinced that the project could materialize. There is a potential risk: what if the money is misused? What if the government postpones granting the money because it is an election year and the dollars could help gain more votes thanks to another project? The presence of an election campaign acts as a moderator for the input phase: it could be good and it

---

* This requires practice or else the assistance of an expert consultant.

could be bad for the project ($I^{\pm}$)—nobody knows the results of the election yet. Assume that the promoters have conducted a marketing study and found that the project is vouched for by most citizens. That's a positive influence (I^+). But it may be the case that citizens living close to the farm fear the flow of traffic, the farm odors, and the noise. Their opinions and actions would exert a negative pressure on the project (I^-). Now, assume the project is set to take place across the entire Mervel Farm land and that a provision is secured to set part of the summer event in a nearby park in case the flow of traffic becomes too big to handle. That's a mediating path: an indirect positive influence. Let's now look at the transformation phase. Its parallelogram would be designed with a double outline, to indicate that there are one or more submodels associated with it. The first submodel would be the five stages of transformation typically found in a project life cycle: vision, planning, mobilization, implementation, and completion/evaluation. Say we take mobilization. This process element (parallelogram) could also be composed of different process elements linked together by T, C, and I bonds.

By applying the same strict logic across the entire project definition, the feasibility analyst minimizes the chances of errors; we could even venture to say that improper modeling is a POV in itself. If it weren't, there would be no issues with team members fretting about not respecting the plan, not understanding the project, not fulfilling the work tasks, and not delivering.

Our analytical foray does not end with consequent arrows. We must now address descriptive arrows, which come in two types: **structural** (they can be "binary" or "continuous") or **functional** (generally measured on a continuous scale).[18] The reader has actually already been exposed to both of them. The four Ps are structural (descriptive) process elements linked to inputs by structural (descriptive) arrows. Deliverables, the book of knowledge (BOK), and impacts are functional process elements linked to outputs by functional arrows. Descriptive (structural or functional) arrows are not related to time. Time is absolutely not to be taken into account. In the field of statistics, some authors[19] set the following conditions for formative variables, which I apply to my concept of structural variables: (1) changing or withdrawing one variable alters the meaning of the core concept and (2) there should be little or no colinearity between the structural variables. Along these lines, it has been said that "omitting an indicator is omitting a part of the construct."[20]

The code for visualizing a structural arrow is as follows: at least two process elements are required to be linked by way of arrows to one single process element (parallelogram). If it weren't the case, the one structural element would simply equal the second structural element, making it superfluous from a modeling point of view. Also, each arrow leaving the structural process element to reach the single main structural element must head for the same single point on the parallelogram that represents it. Let's take the four Ps. The first condition is met: there are at least two elements pointing toward the main process element—inputs. The second condition is that each of the four Ps in the parallelogram points toward the same point on the parallelogram representing the main element—inputs. This is one sure graphical way of differentiating between consequent and descriptive arrows.

There is more. The four Ps represent essential conditions defining the main element—"inputs" in this case. All structural elements forming a main structural element must be *sine qua non* conditions. The main element—inputs—is not fully defined if any of the four Ps are missing. The four Ps form the parallelogram of inputs. One can think of it this way: A bicycle is mandatorily formed of a seat, pedals, a frame, two wheels, and so on. If it consists of one wheel only, it is no longer a bicycle but a unicycle. If it is composed of three wheels, it is a tricycle, not a bicycle. Without a doubt, the two wheels, along with the seat, frame (etc.) form the "bicycle". Again, this is tantamount to formative variables in statistics.[21] The key point for the feasibility analyst is that he must learn to find all the elements that are *sine qua non* conditions for forming the main process element. For example, in the case of the Correx process, because there are two corrugated sheets that need to be brought and glued together, a *sine qua non* condition to the machine that will be built to produce the innovative cat litter is one that obligatorily has two entry feeding points: one for the large corrugated sheet, and one for the small sheet. Hence, everyone can realize that the machine will not produce what it is supposed to produce—the innovative cat litter tray—if it does not have two mechanisms permitting each of the corrugated sheets to reach the main conveyor.

To identify the structural components of a process, two questions must be posed. First, returning to the bicycle example, the feasibility analyst would ask: "Is a seat a necessary element to define a normal bicycle?" If yes, it is a step toward determining that it is a structural element. "Are two wheels necessary to build a *bi*-cycle?" The answer being "Yes," the two wheels are most likely structural elements of the main element—the bicycle. The feasibility analyst has to go through every possible scenario in order to fully define the main process element; failing that, he will open the door to POVs. Indeed, if brakes are missing on the bicycle (thus making the main element—the bicycle—incomplete), the project of riding a bicycle becomes dangerous and the child riding it is in a precarious (vulnerable) position. Not having determined that a normal bike is composed of brakes has led to creating a POV with potentially disastrous effects. The second question that must be posed is the following: "Do the handles exist independently from, say, the seat?" If the answer is "Yes" then that's an indication that it is a structural element. "Do the handles exist independently from the frame?" "Yes." All permutations between each potential structural element must be checked in order to confirm that there is no colinearity (no correlation) between them. If there is, then the potential structural element is not a structural element, but a functional process element.[22]

The reader should be aware of the impact of the absence of colinearity. In a statistical sense, it means that a regression can be run without having to pay attention to the interaction between each of its variables (with functional elements, there is colinearity, thus making regression analyses much more complex). This can be expressed as follows[23]:

$$\text{Main structural element} = \beta_0 + \beta_1 X_1 + \beta_2 X_2 + \beta_3 X_3 + \beta_4 X_4 - \varepsilon \quad (4.1)$$

Where the Xs represent each subelement (a process element that forms the main process element, such as the handles and brakes) and β the level of influence of the respective subelement on the main element. The analyst need not worry about possible interactions between each X because there is none; this is a condition for establishing that the X is a structural element of the main element. In this case, there is no room for diagonal processes; everything is straight direct.

In the examples given so far, each P of the four Ps that is present at the pretransformation phase is completely independent of the other, and each is a *sine qua non* condition for structuring "inputs". Similarly, I have defined People as being composed of customers, suppliers, regulators, and bad apples. I speculate that bad apples are a necessary part of the equation forming People. Customers can exist independently of suppliers (but will not buy anything) and independently of regulators and bad apples, and there can be no People on a project without customers. So "customers" is a structural element defining People. However, it takes at least two structural process elements to form another one, so that if I had only first identified customers as forming People, I would have obligatorily sought to find at least another element (in fact, we outlined three more—suppliers, regulators, and bad apples). This way of thinking ensures that each and every time, the feasibility analyst fully clarifies the parts of the project that is evaluated.

This analytical method is a great way of uncovering POVs because these develop most often when there has been a lack of definition: POVs grow like mushrooms, in dark and humid conditions! Again, a project that is well conceived is a project that can hardly go awry. The advantage of this method is that it helps in the modeling of the project because generally, process elements tend to be symmetrical. If there is an off button, there is also an on button. With respect to the brain, the same process occurs: there are molecules in the brain (neurotransmitters) that facilitate the emergence of certain behaviors and others that inhibit them. In cats, for example, hostile predatory behaviors cannot exist at the same time that defensive behaviors take place. That's because the neurotransmitters that act between three of the brain areas implicated in the mechanics of the behavioral response (the central amygdala, the lateral/medial hypothalamus and the periaqueductal gray area, or PAG) are either activated or inhibited, but never concurrently. When the medial hypothalamus is activated, the lateral hypothalamus (responsible for Instrumentally hostile aggression) is inhibited. We will see this further in the chapters on People (Chapters 5 and 6) and Power (Chapter 7). The point here is that proper modeling tends to generate symmetrical patterns, which greatly helps understanding processes.

One last note on structural process elements: they can be measured in two ways, which I call "binary" or "continuous".[24] A binary structural element is one that is measured by posing the simple question: "Is it present, yes (value = 1) or no (value = 0)?[25]" A so-called continuous structural process element is measured by a scale where a range of values greater than 2 can be used. For example, a scale of 11 alternatives could be used, with 0 not being offensive at all and 10 being critically offensive. Binary scales associated with a structural arrow are marked (*Sb*) and continuous scales associated with a structural arrow are marked (*Sc*). This allows the analyst to quickly establish what kind of measurement is favored for the structural elements pertaining to the project process under review. Refer to[26,27] Figure 4.7.

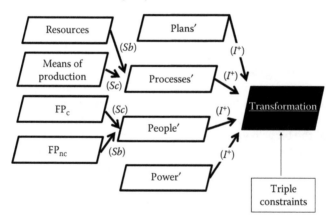

Figure 4.7 Binary and Continuous structural process elements.

In Figure 4.7, I look at the resources and Means of Production from a descriptive point of view: they are *sine qua non* variables elected to define Processes. There can be no process to speak of without the resources and the Means of Production. In the last example, the analyst has chosen to measure resources with a yes/no question (are resources available "Yes" or "No"; he is using a binary measure). However, he has established a continuous scale for Means of Production, so that he may be operating on a question like: "Are all Means of Production prepped to enter the transformation phase?" Also, since the analyst has decided to look at transformation from an active point of view and not from a descriptive point of view, the four Ps exert an influence; they are no longer structural elements (truly, they are structural variables forming "inputs", but "inputs" by itself is not an active concept).

During the vision stage of a project, it may be possible that the promoter is not entirely certain of what characterizes the project in all its details. There will come a point when he will be required to determine all of these details, but in the meantime there is room for flexibility. To account for this, a small *s* can be used: (*sb*) for a binary structural element, and (*sc*) for a continuous structural element. In accordance with the format system adopted in the present methodology, these codes appear graphically on the bond linking two parallelograms. This way, the nature of the bond is specified: it is a structural bond, a tentative one, and it is measured with a binary or a continuous scale. As the reader can appreciate, working this way provides plentiful information that can be used to get a limpid picture of the project processes and to reach prompt and precise decisions.

With structural process elements, we speak of *sine qua non* conditions. This is not the case with functional process elements. Structural elements explain what the main element is mandatorily made of. Missing the inclusion of one of the necessary components (elements, which can be material items, abstract concepts such as "People", or processes in themselves) means missing the opportunity to fully define the main reference element. Imagine sending a man to the moon without reserves of oxygen. Oxygen is an essential input for the space shuttle, 100% of the time. Functional elements do not refer to what the main process element is mandatorily made of, but rather to how it functions. Let us take the example of the bicycle yet again. Suppose the reader does not know what I am referring to. However, I indicate that the object I have in mind is a key element of the Tour de France, that it does not use gas, that the maximum speed it can safely achieve in a downhill is 44 miles/h (70 km/h), and that it uses man (leg) power. At this game of charades, based on pure functionality, most readers would have guessed that I am referring to a bicycle and not a turbojet. A tricycle cannot achieve a speed of 44 miles/h. A car cannot be moved solely by manpower. So the object that I have in mind is, probably, in fact most probably, a bicycle. The more the functionality of the mysterious object is detailed, the more convinced the reader will be that I am talking about a bicycle. So the reader has managed to identify the object (or the process element) without mention of any one of its structural elements (pedals, wheels, etc.). This means that all process elements can and must be established in two ways: structurally (*S*) and functionally (*F*).

Note that the functional aspect of objects is regularly utilized in science. For example, scientists can infer the presence of a planet without ever seeing it simply based on how surrounding planets and suns behave.

Recall that quality was defined as [(Functionality + Design) over Costs]. If we equate functionality with functional variables and design with structural variables, we thus obtain this meliorated version of perceived value definition:

$$\text{Perc'd value} = \frac{\text{Functional process elements} + \text{Structural process elements}}{\text{Costs}}$$

$$(4.2)$$

Logically, if the feasibility analyst wants to determine the quality of a project, he must identify and quantify its structural and functional process elements. Hence, there is no way around properly conceiving a project; most people, anyway, behave in a way as to maximize their value (in financial terms: their wealth). By highlighting the value of the project to the team members (Forces of Production) who participate in it, managers get them even more motivated and dedicated, thus reducing the potential effects of stealthy POVs.

A functional description is coded (F) and can be measured in a binary or, more usually, in a continuous way (Fb or Fc). Functional elements are not *sine qua non* conditions. For example, one could progress without the reference to the Tour de France and eventually guess that the object that is under investigation is a bicycle. It helps to know that it has something to do with the Tour de France, but the fact is that it does not preclude the identification of the object. The more pertinent key functional characteristics that are enumerated, the closer we get to unveiling the mysterious object. Furthermore, functional elements interact with each other—they contain some levels of colinearity. Recall that the output portion of our project definition model has been linked to deliverables, formalized knowledge, and impacts. If the reader goes back to Figure 1.3, he will see that the arrows start from the same point on the parallelogram identifying "outputs" and point toward each of these three items. This is an indication that time is not a factor and that we are dealing here with functional elements. There would be no impacts if there weren't any deliverables; deliverables and impacts have some form of colinearity. There would be no useful formalized knowledge if there were no deliverables. There would be no useful formalized knowledge if impacts, both positive and negative, were not discussed. As can be seen, functional elements have some kind of relationship between them, which makes running multiple linear regressions a challenge because the interactions must be taken into account—doing otherwise is reducing the meaning of the linear regression. Another point is that there can be any useful number of functional elements, yet at least two are necessary, just as was the case with structural elements. But taking

one functional element out of a list of 10 does not diminish the functionality of the main element; *a contrario*, taking one element out of a list of
structural elements is very damaging to the meaning of the main process
element. Functional elements are akin to reflective variables in statistics.[28]

Let us take some examples. The cat litter box can actually have multiple functions: it certainly can be used as a cat litter box, but it can also
serve when fruit picking at the Mervel Farm, or as a gift container at the
gift shop located at the same farm. Should it be used as a cat litter box,
there is little justification for paying extra money to decorate it with fancy
colors—most cat litter boxes are kept in a dark laundry room hoping the
unpleasant odors will not reach the living room or the kitchen. However,
if it were used as a gift basket of sorts, then a colorful print on the otherwise dull brown cardboard could be an asset in order to sell the product.
So, functionality has in some way defined the process element that has
been produced out of two flat, glued-together corrugated sheets. Similarly,
there are projects that have very little impact to speak of, and others that
don't really generate formalized knowledge; yet, the fact that they do
would be somewhat of an indication of the size of the project (and probably of its costs) without ever knowing what projects are referred to.[29] The
same thinking process occurs in a criminal investigation or archeology,
or even geology for that matter. Finding a 2000-year-old skeleton with a
sharp wound on its shoulder blade and a nearby small triangular stone
that is obviously man-made may be a hint that the individual was the
victim of an aggressor while he was running for safety (he was obviously
hit from behind). The functional elements (e.g., the triangular stone is
part of a tool utilized for hunting) help generate a scenario of what likely
happened.[30] Thus, identifying the functional process elements of a main
process is a very handy way of tracking back errors that have disabled a
project, or else of anticipating POVs (the victim's back in this case).

Let us accept the fact that the functional elements of the Québec
Multifunctional Amphitheatre (QMA) are (1) a venue for entertainment
and (2) a sports center (more specifically, an arena built in the hope that
the city will soon be awarded an NHL team franchise, which it once had in
the 1980s with great success[31]). The two main functions of the QMA are set.
Suppose the ice rink does not work. Then the reason for being of the QMA
(the main process element being the QMA) is amputated by 50%. In other
words, failure is knocking at the door. The process of analyzing the fact that
some of the functional elements of the QMA have been amputated can start
by going back along the parallelograms and arrows that have been part of
the entire model, including S and F, T, I, and C arrows. In short, all project
management processes can be pictured by way of two kinds of descriptive
arrows (S and F) and three sorts of functional arrows (T, I, and C). In reality, the S and F arrows fall under one heading: descriptive. S and F are both
used to describe a process element. That's all there is to it. The science of

physics is entitled to its four forces (weak bond, strong bond, electromagnetism, and gravity), so why would project management not have its own four forces as well: descriptive, longitudinal, influence, and causal?!

A chessboard holds 32 pieces—16 white and 16 black. Each piece has its own operating mode. The pieces interact with each other once the clock starts ticking. Billions of patterns exist (an estimate is that 10^{120} games are possible[32]), yet good players can play back their games from memory. Why? Because they have followed a certain logic that can be expressed by S and F, T, I, and C codes. Knowing the S and F of each piece, there are moves that can influence the opponent action (e.g., what is referred to as a "pin" in chess), moves that lead the other player to use time to prepare his response (certain moves in chess are made only to confuse the opponent and force him to exhaust valuable time), and other moves that can cause the opponent's actions (so-called forced moves, with the ultimate forced move being the throwing of the desolate king against the wall angrily, or more diplomatically, accepting the loss). There are just no other kinds of moves given the S and F nature of the chess pieces. When a chess player remembers a game, he usually tries to reason as to why the pieces on the chessboard are located where they stand; he then examines his motivation (to delay the opponent's response, influence a move, or force a move?). The most dramatic situation and the one that the player tries to optimize is one where he forces a move from the opponent. Throughout the game, as previously mentioned, the chess player engages preferably in a Dominant strategy or else in a Contingency strategy, depending on the context; yet whatever the positional strategy, S and F, T, I, and C explain every move he makes. Victory is in sight when the player has defeated all odds of losing (of being attacked on his/her POVs) or of coming to a draw, and similarly, success in a project is ensured when a project manager has defeated all odds of POVs becoming active.

Again, each and every process is fully comprehended from a descriptive point of view by way of identifying structural process elements and from a functional point of view by way of listing relevant functional process elements, nothing less. Both sides (structural and functional) of the object (process) must be assessed, otherwise the object (process) is deemed to be poorly defined. And, as mentioned from the start, a poor plan leads to derailment. Refer to Table 4.6.

Recall that I have resorted to walls, ceilings, and floors when discussing some processes in the introduction of this book. Let us take the example of the humidity factor as a negative influence (I^-) in the Correx cat litter process. Under normal conditions (which have to be assumed in a production plant), humidity levels can reach a certain minimum and a certain maximum level, which will influence the cardboard behavior along the production line, without compromising the process. In the case of the Italian Floorlite's rubber sheets, minute changes in temperature

Table 4.6 Descriptive (structural and functional) parallelograms and
arrows

Type	Avoid	Code
Structural (S)	Doing a regression without independently measuring the main construct Boosting Cronbach's alpha[66]	Binary (Sb) Continuous (Sc) Temporary (sb or sc)
Functional (F)	Not recognizing colinearity	Binary (Fb) or Continuous (Fc)

levels during the heating process can cause havoc. The minimums and maximums are so sensitive that they escape the normality of the production conditions. Hence, in the end, not having respected the quality floor (the norm of quality) causes the production line to peter out, as it becomes jammed with a thick paste or else filled with an excessively liquid paste that quickly penetrates all the mechanical parts of the machine. Establishing walls, ceilings, and floors along the parallelograms (process elements) is just as critical as specifying the types of bonds that exist between them.

To prove my point that modeling is critical to any feasibility study and that levels expressed by walls, ceilings, and floors are a crucial consideration, I follow with a real case involving the use of my modeling technique in a wildlife context.

4.3.6   *Example taken from wildlife*

In 1972, the United Nations published the *Atlas of the Living Resources of the Seas*, which provided a map of the marine areas where fishing was encouraged (I^+).

Over the years, however, the coastal areas of the oceans have literally turned into a gigantic soup of jellyfish. Recently, a Swedish nuclear power plant was put out of action by the presence of these marine animals, which had clogged the cooling filters of one of the reactors.[33] Jellyfish are inexhaustible zooplankton predators, but do not represent much nutritional value for most predators (such as tuna and turtles), being composed of more than 90% water.

The causes of and influences on the proliferation of jellyfish are multiple. Overfishing has left a vacuum by fostering an opportunity for jellyfish (cause C^+) to expand because the competition from fish for the same food sources has been augmented (cause C^+) as resources becomes scarce. Increased ocean temperatures have facilitated (influence, I^+) the population growth of jellyfish. These have now supplanted fish in the food chain in many marine regions. Krill overfishing has also worsened the situation (cause C^+): their relative absence has allowed pelagic animals, known for high rates of growth and multiplication, to occupy a larger portion of the ecosystem (cause C^+). Furthermore, toxic products associated with pesticides employed in modern agriculture that end up in the sea have been found to block the growth of sea crustaceans (cause C^+), so that there is less food for fish and more living space for Salpae and other gelatinous organisms (cause C^+). Also, drug residues discharged into the sea by way of sewers (containing urine) act as endocrine disruptors (influence I^+) that block the reproduction of sexual species (cause C^+), but not that of jellyfish, which use an asexual budding reproduction system (poor guys, they don't know what they're missing!). Finally, jellyfish have developed coping mechanisms (such as the ability for some to regress to an earlier stage of development) so that their level of vulnerability is minimized at the end (influence I^+).

In short, fisheries' policies or the way they were implemented have maximized the vulnerability of certain species of fish through a series of causes and influences (for which we have just made a judgmental evaluation), while it has reinforced the survival capacity of jellyfish, an animal that has gone through 600 million years of evolution.

This process can be exhibited as shown in Figure 4.8.

Recall that there should be only one point of entry, so that the model in Figure 4.8 is only partly true. When we look at it, the problem is not fishing regulation *per se*, but the lack of proper regulations for fishing

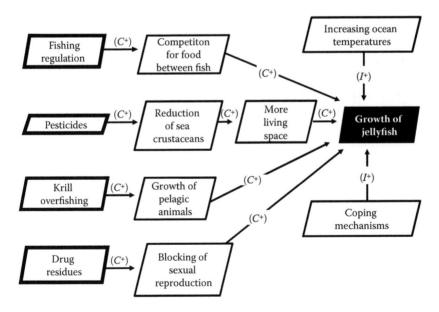

Figure 4.8 An example taken from wildlife.

krill, as well as pesticides and drug disposal. Hence, the entry point is a "lack of proper regulation", which is then expressed by four process elements (functional variables): drug residue disposal, fishing regulation, krill overfishing, and pesticide management. Hence, using this methodology, we get a better picture of what is leading to the explosive population growth of jellyfish.

If we transpose this entire process in a project context, the picture of the situation is crystal clear thanks to efficient modeling. In addition, the reader can appreciate the importance of determining POVs in advance, because the consequences can be catastrophic in a closed dynamic system such as the marine system or any project. In the jellyfish example, nature has it that they have developed coping mechanisms to deal with their own POVs, which has then assisted them with their survival and expansion.

Note how judgmental it can be to draw a model sometimes, such as the aforementioned marine process. Some experts could contend, for example, that overfishing is not a cause, but simply an influence.

The explanatory powers of each of the four fundamental types of bonds (descriptive, S and F; longitudinal, T or t; influential, I; and causal, C) are not equal. A project manager would not be well received by saying that his assignment had turned into a muddy outcome because that's just the way the project was from the get-go (descriptive). I assign a value of zero in terms of the explanatory power of this approach. An expert could claim that the growth of the jellyfish population happens with the passage of time, which is true, but which holds very little explanatory power (let's assume an explanatory power of 0.5). They could claim that there have been factors that, when they increase (↑), are accompanied by a surge in the population of jellyfish (↑); this only goes as far as explaining why the ecosystem is changing, but does not provide the reason for why it is changing in one direction (increasing jellyfish population ↑). We ascribe a value of 0.5 as an explanatory power to this statement. The same scientist could vehemently contend that overfishing is the root cause (C^+) of the jellyfish population explosion, but knowing that there are other variables, such as the presence of endocrine disruptors, makes this statement questionable: if proved wrong, the scientist's reputation would be tarnished. He may be 50% of the time wrong or 50% of the time right; nobody knows for sure yet. We cannot attribute a value of 1 to this presumed causal link, but must rather settle for a value of 0.5 because of the probability that the scientist may be wrong, given that causal links are generally very hard to prove.

If we are project managers who have to explain why our project has not achieved its goals, we'd like to leave all opportunities to justify ourselves while finding the safest stand. By saying that we think there has been a link between what was done and what was achieved, but by leaving the door open to the presence of negative factors, I^-, that may be to blame, or else by claiming that time is the culprit, or else that a possible cause should not be discarded, we are sort of mitigating our damages— we can hardly be accused of not considering all options. Many scientific reports are concocted that way, and even when a causal bond is suspected, a scientist will find a way of questioning it. The example of the tobacco industry speaks volumes: some scientists are convinced that there is a causal link between cigarette smoking and lung cancer, while others say there is no sound evidence of such a causal effect and claim that the causal argument is, well, smoke and mirrors! Most people would agree, however, that there is a link between one and the emergence of the other.

Modeling implies a trade-off between what is certain and what is uncertain. Models are not perfect, they are a simplification of reality. The lesson learned here is that when it comes to project management processes, looking at bonds while acknowledging all of the options that could explain the project difficulties is a sharp way of approaching things, keeping in mind how these options ignite POVs.

4.4   Modeling language

As the reader can judge, I have developed a language specifically for the purpose of conducting project feasibility analysis. The basis of it is that all processes must be brought to their ultimate causal (C) condition, nothing less. It is only by doing so that the expert can see where POVs lie. To achieve this, we avoid completely any routes of escape, such as "if this, then that," "either (…) or," and so forth. Every process must be straight direct. The point here is not necessarily to write an entire production process into this language but rather to use it as a tool to better circumvent a particular process and see what could go wrong.

To show how the modeling language works, let's take another real example, that of *Bleu lavande* ("Blue lavender"). The reader will realize that the language I develop here explains a lot about human behavior as we shall see: most people, consciously or not, bring events to their critical path, that is, to their causal dynamics. It is often because of this critical causal thinking that people tend to disagree with each other or lack proper diligence, resulting in poor processes.

4.4.1   An example: Bleu lavande

Bleu lavande[34] is a medium-sized company based in the Eastern Townships in Québec, Canada, an area known for its kaleidoscope of vibrant colors in the fall, with festive orange, red, and beige hues spreading over acres after acres.

An entrepreneur cautiously chose the location to start a unique business after much research: growing lavender and transforming it into lavender-based products such as essential bath oils, hand sanitizers, and so forth. Lavender was not a traditional plant in Québec, although it had long been a landmark product in France. The fields where the plant grows have to be slightly inclined and there has to be enough sunlight. The business developed nicely, but a strategic miscalculation led it to bankruptcy even though demand was high and the market kept expanding. Instead of using existing specialty stores across the province to sell his products, the entrepreneur established his own stores. But there is a wide margin in the kinds of operations that are entailed in the retail business and farm-based operations.

The business was salvaged and has been successfully operating since 2015. In the summertime, agritourism provides an important source of revenue: the fields as well as the farmhouses and transformation operations (steam distillation) are open to the public, who line up and pay C$10 to access them. The neat thing about the concept is that anyone can actually see how the outputs—condensed lavender oil and its by-product (a near transparent liquid that smells something like lemon, but with a rancid undertone)—are manufactured.

I endeavor here to write the entire process using our modeling language technique to show how POVs can be identified.

First, the seeds are planted in the soil. A concern is to avoid contamination by a plant that emulates almost perfectly the lavender flower and branches, but which produces poor quality oil. Let us start at the point where the plant must be harvested: late in July.

- Lavender plant (T) → <u>harvested</u>

The word "harvested" is underlined because it is an operation. Time is the kind of arrow that takes place because the lavender plant does not influence (I) the transformation process, it does not define $(S$ and $F)$ this transformation process, and certainly the plant itself does not force (C) the business owners to process it. A day late in harvesting would not make a significant difference.

There are three worries during that first step: time, costs, and norms of quality. Only time is a real factor: not harvesting during the critical harvesting season means losing the quality branches and flowers required for transformation. Hence we have

- $|\overline{\uparrow}$ Lavender plant (T) → <u>harvested</u>

Here, we must have an entry wall. Walls refer to time (T). There is no loop (t) as the plant, once picked, cannot be put right back into the soil. Now, we want to turn this into a critical path because that's how we spot

POVs. We assume that the plant loses its essential quality past August 20, and so that is the absolute deadline. We have the following:

- $|\bar{\uparrow}$ July 1, Lavender plant (T) → harvested, August 20
- August 21, Lavender plant (C^+) → poor quality, June 30 $|\underline{\downarrow}$

Therefore, the causal link occurs within these boundaries—August 21 to June 30 of the following year. We assume that all plants have been transformed and sold by June 30 of the following year, so that the end wall is nearly a year after first harvesting. The question that comes next relates to transformation. We know that transformation is the act of management, and where the team members (the Forces of Production, FP, both controlled and uncontrolled) use Means of Production, MP (e.g., land/building, machinery/equipment), to transform inputs into outputs. POVs must therefore be found in FP, MP, or inputs. Note that weather conditions are not a POV, they represent an external risk. The question is: "What are the chances that the FP will miss the entry point of July 1 and the deadline for harvesting of August 20?" For this to happen, there would have to be a strike or a dramatic shortage of labor; in either case, the producer could palliate the problem by hiring local farmers and students. The POV is weak. There could also be a break in the tractors used to collect the plant: this is a more serious predicament. This entails the following:

- $|\bar{\uparrow}$ Poor MP (C^+) → slow (or delayed) harvesting

Slow harvesting would lead to delays and bring the harvest past August 20, so that the process would enter into a critical (POV) zone. So, looking at this process sentence [Poor MP (C^+) → slow harvesting], we immediately seize the importance of checking the machinery before the harvesting season starts.

Finally, the last elements of the inputs are resources (such as electricity, gas) and materials. Resources should not be an issue, and materials, in this case, refer to the lavender plant about to be harvested. Recall that for each line of text, we evaluate which of the calendars, costs, or norms of quality is most sensitive. In the present case, for the process sentence [Poor MP (C^+) → slow harvesting], it is not time anymore that is in question, but quality. The producer could readily harvest the poor quality lavender flowers and branches and produce poor quality oil, and sell it to a dollar store outlet. But his concern is to offer his clientele a high-quality product. In the process sentence [Poor MP (C^+) → slow harvesting], quality is of prime importance. The plant has to be audited and monitored prior to harvesting; failing that, the end product may be below quality. So we have

- Parameters of quality (S) → quality lavender

And

- |↑̄ Poor quality lavender (C⁺) → harvesting July 1 (following year) |↓

This is logical given what we have said before. Poor quality lavender (quality lavender below the norms of quality [(C⁺) → causes...] will force the producer to defer production; the delay is a worst-case scenario. Thus, the producer faces a huge problem. That's a POV to be reckoned with. It all stems from [Parameters of quality (S) → quality lavender]. One can say that the most critical aspect of this business, so far, has been found to be the proper choosing and monitoring of the input material—the lavender seed.

The producer must now establish what the parameters are, knowing that our modeling technique requires at least two process parameters in order to use a structural arrow (S). It could be the origin of the seed, the way they are stored and prepared for transplant into the soil, the quality of the soil, and so on.

When we phrase the process so far as a single paragraph, we have

- ↓ Parameters of quality (S) → quality lavender
- |↑ Quality lavender plant (T) → harvested
- |↑ July 1, Quality lavender plant (T) → harvested, August 20
- August 21, Lavender plant (C⁺) → poor quality, June 30 |↓
- Poor quality lavender (C⁺) → July 1 of the following year |↓
- |↑, ↓ Poor MP (C⁺) → slow harvesting |↓

The underlying most serious POV appears to be "parameters of quality" (S).

In other words, the producer/business owner can probably get by with a disastrous season, or with a temporary strike, or with a break in the machinery, but what makes his reputation, what makes the product sell, what attracts agritourists to the farm is the quality of the lavender.

We can represent the entire process in a simplified format, as shown in Figure 4.9.

This is how the owner thinks based on Figure 4.9: "I am the owner. I am hiring staff to work the field and process the plants. I absolutely need quality lavender seeds. I am planning my year so that I start harvesting July 1 and expect to have all of the products transformed, packed, and sold at the latest by June 30 the following year. I will produce quality oils, learning in the process, and this business is going to have an impact on the agribusiness of the Eastern Townships." This, in a nutshell, represents the business that Bleu lavande is in.

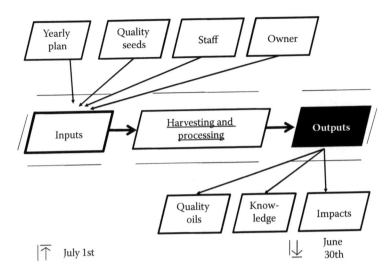

Figure 4.9 Harvesting lavender.

Let us continue with the analysis. Once the lavender bushes are picked up, they are put on a small conveyor and then stacked and compressed into a large stainless steel container. Water is boiled at the bottom of it; as the water evaporates, the droplets filter through the compacted bushes and keep rising, soaked with the oil and by-product that the branches and flowers contain. Droplets reach the top of the sealed container and are channeled down a pipe where they are cooled off; that mist turns back into what's left of the water, and, of course, the oil and its by-product. The by-product is lighter than the oil so that after settling in a container, two layers are formed, with the bottom one being the much desired lavender oil. Without going through all the steps of the analysis, the reader can already realize that there is a causal link inherent to the process: the heating of the container causes the water to boil (C^+) and to evaporate (C^+). With time, the fine water droplets filter through the lavender bushes (C^+). The system operates in such a way that 100% of the time, the fine mist gathers and stores the oil contained in the bushes; this is another causal bond (C^+). There is yet another causal bond, which occurs when the oil-loaded water droplets reach the top of the stainless steel container and are rechanneled through a cooling pipe. The cooling causes the water mist to settle in a liquid form (C^+). Finally, the difference in weight between the oil and its by-product causes the emergence of two well-differentiated layers that can be easily separated; a process that uses time and gravity (C^+). If we target solely causal links, we have

- Boiling water (C⁺) → <u>mist rising</u>
- (Mist | container conditions) (C⁺) → <u>oil extraction</u>
- Cooling pipes (C⁺) → <u>mist in liquid</u>
- Difference in weight of the oil by-product (C⁺) → <u>product separation</u>

Technically, we could associate a POV with every causal link. Realistically, the one that seems most likely to be a concern, should a problem occur, is the oil extraction process. It is this very process that is at the heart of the business; the rest can always be dealt with one way or another. This oil extraction process occurs by way of the forces of nature (water droplets filtering the plant) but is truly operational given the container conditions, as in [(Mist | container conditions) (C⁺) → <u>oil extraction</u>].

The container is an Means of Production, MP (equipment). For all intents and purposes, FP and inputs (resources and materials) can't turn into glitches. However, the container seems critical. A punctured container may cripple the process, a badly screwed lid may cause a leak, and poor quality equipment may produce impurities such as rust. Thus, a POV worth paying attention to is the container itself. It is also probably the most expensive piece of equipment used in the filtering process.

Overall, we have identified two potential POVs: the quality of the lavender (material/input) itself, and the container (equipment/Means of Production). The critical aspect of the lavender is structural (S), whereas the critical aspects of the container are both structural and functional (S and F): it has to be made according to certain specifications and to function as expected. Knowing this, the producer would set his norms of quality for the container in both terms: structural and functional.

This example is very simple. One can imagine what level of complexity a product such as an airplane would entail. The trick is to analyze each step of the entire transformation process and to pose the question whether it is the calendar, costs, or norms of quality that is the most pressing issue. All processes must be brought to their causal form (C), even if this requires scrutinizing every step that composes the process. POVs are most lethal along the critical path.

4.5   *Efficiency and efficacy*

There are two important notions with respect to the measurement of the flow that goes from inputs to transformation to outputs: efficiency and efficacy. "Efficiency" refers to the capacity to maximize the use of resources and eradicate undesirable wastes. Put differently, it refers to the best possible utilization of the Means of Production by management and team members (the Forces of Production).

A project manager wants to maximize the outputs given the inputs while minimizing the error term ε set in the input function.

On the other hand, "efficacy" speaks of the ability to produce what one wants to produce; that is, the ability to reach a goal. In terms of projects, this translates into the capacity to complete the project on time and within budget while meeting the quality standards demanded, as subjected to the interaction of the four Ps during the transformation process. The Québec amphitheater (QMA) is a typical case of high efficacy. The MOS is a painstaking example of poor efficacy. Efficacy is related to outputs: an output of four units given one unit of input is more efficacious than an output of three units given an input of one unit.

The best illustration of such dynamics is at the peak of the (hypothesized) sample curve of Figure 2.6, which we redisplay in Figure 4.10.

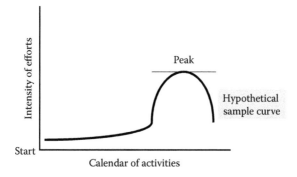

Figure 4.10 A hypothetical sample curve.

Utility is the expression of opportunity; in other words, an opportunity is really worth its name if it provides utility. The opportunity is defined by the utility it brings to the clients who will use the project. I can rewrite my formula from Equation I.5, as follows:

$$k = \text{Risks} \cdot \text{Vulnerabilities} \tag{4.3}$$

We saw that (Equation I.8) (replacing "Opportunities" with "Utility")

$$\text{Vulnerabilities} = \frac{1}{\left(\dfrac{\text{Infrastructures}}{\text{Opportunities}}\right)} = \frac{\text{Utility}}{\text{Infrastructures}} \tag{4.4}$$

Therefore, we have

$$\text{Utility} = \frac{\text{Infrastructures} \cdot k}{\text{Risks}} \qquad (4.5)$$

Indeed, this outlines what makes projects exciting: these elements force the project's initiators to exert prudent planning, to rely on infrastructures, and to take risks.[35] There is no utility achieved even if excellent planning is done (robust management) when infrastructures are poor. A project is useless if it faces high levels of risks and requires many managerial inputs, all the while resting on poor infrastructure. A project that sees, between its beginning and its end, a large positive change in infrastructure, little change in terms of risk, and little need for management implication (read: power games) evidently brings in a lot of utility. This is especially true in the context of some public infrastructures, such as roads/highways and hospitals. Infrastructures are important. Most people are marveled by a bridge that defies natural obstacles such as height or the dangers of extreme weather conditions while providing an important support structure that serves the community. The French bridge Millau Viaduct (world's tallest bridge) is a perfect example.

From this perspective, management has three fundamental preoccupations, managing and minimizing

1. Risks (which ultimately affect calendars)
2. POVs (which we have related more closely to costs)
3. The error term ε (which is a deviation from norms of quality)

Our overall analysis tends to confirm the importance of controlling risks, dealing with POVs, eliminating errors, harmonizing the four Ps, managing the Unfits (uncontrolled Forces of Production, FP_{nc}), and relying on proper infrastructure.

From a process point of view, POVs have two faces: one with respect to the error term ε (linked to efficiency), and one with respect to the final outputs (linked to efficacy). Since ε pertains to inputs allocated at the beginning and during the transformation process, and since outputs arrive at the end of the transformation process, we can state that POVs are temporal: they have to be measured in regard to inputs and outputs. Hence, a classification of POVs can therefore be made, as in Table 4.7.

This is an important statement: POVS are not static. It is wrong to think that POVs can sit there harmlessly and not affect the entire project; it is quite the opposite. Beginning and end POVs are tied by time so that errors at the beginning will affect the outputs; lack of forecasting of

Table 4.7 A timely classification of POVs

Inputs (beginning of and during the transformation process)	Outputs (end of transformation process)
Efficiency	Efficacy
POVs linked to ε	POVs linked to outputs

outputs will empower initial POVs. Hence, the feasibility expert is concerned with POVs as they may exist at the input phase and as they progress toward the closing stage of the process, and his vision goes forward and backward in time (especially by way of using our modeling approach), so that he can anticipate future glitches and reduce the initial difficulties. Because a project is a closed dynamic system, POVs have the potential to create chaos: the system cannot escape its own dynamics. Any overheating during the transformation process (if production is rushed, conflicts erupt, etc.) is self-contained and so both effi-

Working together or not?

ciency and efficacy are compromised. Overall, evaluating them both at the end of a project may be the best measure of success.

4.6   *Dominant strategy and utility*

We have seen that the constant *k* is instrumental in our model of project management. The chapter on People will elaborate on how we arrived at it through years of research with more than 40 groups having participated in different forms of the project (banking, construction, entertainment, manufacturing, real estate, sales, etc.). As mentioned in the Preface, a number of scientific articles have been generated based on my findings.

I have stipulated that a project is a closed dynamic system. It is bound by time, costs, and norms of quality. Such constraints do not exist in classical economic theory but they certainly exist in project management theory. The repercussions of such a state of affairs are worth noting. We need to find a way to quantify the boundaries. To do this, we specify more precisely the context in which k exists. Recall that we had set that [Risks$=k$/Vulnerability] or [Risks$=k$/POVs].

I have explained that risks cannot exist without a sense that one is vulnerable. It may be shortsighted to assess risks without juxtaposing vulnerabilities, because this only gives half of the real picture. As explained, an individual may be a national lifeguard in which case a tumultuous river may not represent a high risk for him/her; yet it is interpreted as a lethal risk for someone who doesn't know how to swim in such dire conditions. The risks are evaluated as a function of one's vulnerability. Risks (external factors) that drag a project are linked to the capacity of management to deal with the project's internal POVs.

The mathematical formula that I have found and that expresses the relationship between risks and vulnerabilities is a rectangular hyperbola. In fact, only a portion of that curve is active in the realm of project management. At any point along the rectangular hyperbola, the surface as delimited by [Risks vs. Vulnerabilities] is the same. This means that the project is the same at any point in time (since the rectangular hyperbola—the function [Risks$=k$/POVs]—is deployed over time, or more precisely, given the triple constraints). In other words, a project can face a high level of risk (be positioned high along the vertical y-axis) and have few POVs (be positioned near the point of origin along the x-axis) or else management may take more time to reposition the project where there are less risks (low along the vertical y-axis) but where there may be more POVs (far away from the point of origin along the x-axis). In reality, not only time should be taken into account but costs and norms of quality as well. We illustrate this particular dynamic in Figure 4.11[36].

In Figure 4.11, k is equal to 1.3. The constant could have other values, but years of research have led us to believe that the ideal value is 1.3 (1.32

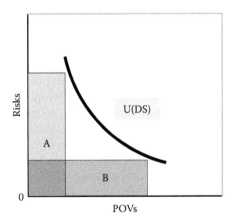

Figure 4.11 The rectangular hyperbola or the DS utility curve U(DS).

to be more precise). Other values, referred to as k', do not provide full efficiency and full efficacy.

Figure 4.11 does not provide the boundaries within which k exists. As I will explain in Chapters 5 and 6 on People, the boundary is 2.3 or [1 + 1.3] for each of the x- and y-axes. Essentially, all there is to remember now is that we set a value that represents an ideal calibration of risks and vulnerabilities at 1.3, and I set an upper limit of the triple constraint at 2.3. Even if we were to change these values, the gist is that we would need to set a value for the ideal combination of risks and vulnerabilities (which are both unavoidable) and a value to express the boundaries set by the triple constraints, which are the calendar (time), costs, and norms of quality, given that we operate in a closed dynamic system. We call the curve [k/POVs] with boundaries set at 2.3 the **Utility of the Dominant strategy**, or U(DS). This curve is similar to utility curves found in economic theory. In short, this curve expresses the fact that given the constraints, the strongest managerial strategy is achieved at [1.3/POVs]; this is where, technically, the project remains equal to itself (does not change or suffer modifications), although other values of k can exist (referred to as k').

The utility curve U(DS) outlines the fact that a set of Dominant strategies is just as good as it allows the manager to face risks while dealing with POVs. This makes intuitive sense: what would be the point of having strong managerial skills if it didn't help, in the end, facing risks that come from outside the project?

A utility curve of this sort has a number of features: (1) it can be associated with a budget constraint, (2) it uses an ordinal measure, (3) more is always preferred to less, (4) no quantities of risk are negative, and (5) utility curves (resulting from different values of k) cannot cross each other.

Various U(DSs) with different values of k are illustrated in Figure 4.12.

The three curves found in Figure 4.12 present a certain utility. We surmise that it is the curve with [k = 1.3] that has the best configuration within the realm of the closed dynamic system that represents a project.

With respect to a budget constraint, this can be explained as follows. On a critical path, a manager may face an absolute choice: to confront all possible risks and assume the project has no POVs whatsoever, or else, to accept all of the POVs but assume there are no risks whatsoever. Both scenarios are irrational and contrary to the nature of projects, since we recognized that there is always an element of risk and some POVs in any project. However, if we were to assume that these two scenarios could exist, in a highly unwanted yet conceivable critical situation,[37] and if we were to join the two points on the y-axis and the x-axis, we would obtain a budget curve. This is illustrated in Figure 4.13.

The budget curve is the name an economist would give to such a straight line with a slope of −1 or near −1.[38] However, in project

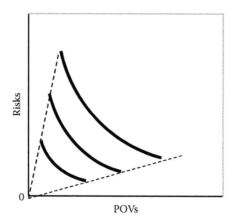

Figure 4.12 Different U(DS) curves.

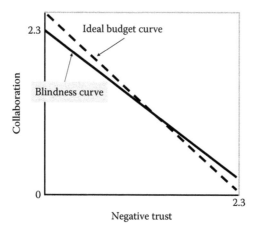

Figure 4.13 The Budget curve or the project Blindness curve.

management, this curve represents management blindness: either the project manager thinks he/she is invulnerable (therefore being able to face any risks) or else they think there are no risks at all (even though they admit to being very vulnerable) given the triple constraints. In either case, the project manager suffers from managerial blindness. Thus, we call this curve the project's "Blindness curve". It is not rare to hear of projects having failed because the promoter underestimated the risks (a typical case when an overly enthusiastic project promoter seeks funding) or else overestimated their own capacity (thus suffering from some form of inflated ego).

In any case, the project Blindness curve touches on the U(DS) curve (Utility of the Dominant strategy curve) as in Table 4.8—note that I stylize the U(DS) curve to ease the reading.

Table 4.8 Project Blindness curve and utility curve of DS

U(DS) Curve k/POV where k=constant, estimated at 1.3	Blindness curve (actually resulting from the Trust–Collaboration function [Collaboration=0.3+0.9 Trust])[67]
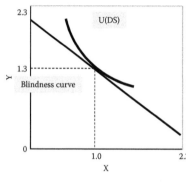	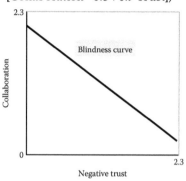
Risks=1.3/POVs	Blindness=2.2–0.9 POVs

The best strategy for the manager is to allocate some element of risk to the project and to admit to some POVs; hence, the project Blindness curve touches the U(DS) curve at its center.

Recall that we set the boundaries of the closed dynamic system to a value of 2.3. Before that point, some time is wasted, some costs are incurred, or quality is somewhat hampered (at least temporarily), or else any combinations of two or three of these elements occur when management is busy dealing, for example, with Unfits (uncontrolled Forces of Production, FP_{nc}). It is part of the game: management accepts to train the Unfits (FP_{nc}). For example, management uses time and incurs costs in order to make sure that the Unfits (FP_{nc}) become normal (controlled) Forces of Production (FP_c), that is, Fits. Efficiency and efficacy are attained by way of the expected return on the investment, hoping that the Unfits (FP_{nc}), once controlled, will become useful Forces of Production, (FP_c). Management will push its managerial skills to the limit in order to sway the Unfits (FP_{nc}) toward the right direction; however, passed a certain point, the Unfits (FP_{nc}) learn new tricks that allow them to clog the project and any new managerial technique that is introduced loses its appeal. This kind of scenario is found in financial crime all the time: regulations are set by governments and abided by the majority of companies, yet soon enough unscrupulous organizations or rogue individuals find ways to cheat the system. We can theoretically assume that management would ideally apply its managerial skills up to the maximum value of the closed dynamic system, set at 2.3, irrespective of the project Blindness curve; in reality, it will move up to the point where it makes sense, that is, up to where it touches on the project Blindness curve, because otherwise it would fail by excess optimism and by underestimating risks or vulnerabilities. Passed the limit of 2.3, the system (the project as a closed dynamic system) does not exist anymore; it is in fact self-destructing. Before that point, though, management has invested time and money, or has somewhat let go of some quality standards.[39] It has done this when in fact it could have expelled the Unfits (the uncontrolled Forces of Production, FP_{nc}), but then there would be other costs and time wasted, with a vulnerability developing in quality control, as the new Fits—the controlled Forces of Production, FP_c—would need training anyway.

To evaluate the shape of the actual DS curve (not the U[DS] curve), I resort to a number of other observations:

1. As shown by Equation 4.7 [$E = z + \beta_0 \, R_n^{\alpha} \, T_h^{(1-\alpha)}$] (and its mirror function E'), which we will discuss, there is an initial allocation with a value[40] of 0.3.
2. We must recall that project managers want to be equal to themselves.
3. We determine that managerial skills gain in momentum over the Unfits (FP_{nc}) at first, they then reach a plateau and then lose momentum, theoretically until the limit of the possible, set at 2.3, is reached. This makes sense: if management did not see results with its improved management techniques, it would promptly abandon them. However, the Unfits (FP_{nc}) eventually learn how to deal with the new managerial techniques and work around them, but Unfits are eventually caught and discharged.
4. Thus, given the particulars of the evolution of the relationship between managerial skills and the Unfits (FP_{nc}), we assume that a sinusoidal function set between 0 and 2.3 best represents its underlying dynamic.
5. We also know that the actual peak of the curve, where managerial skills are maximized versus the Unfits (FP_{nc}), is at [$x = 1$; $y = 1.3$] because this is where the project Blindness curve and the U(DS) meet.

I hypothesize that the equation for DS can be[41]

$$DS = 0.3 + \sin(1.5x)$$

or alternatively

$$DS = (k-1) + \sin\left(\frac{10(k-1)}{2} \cdot POVs\right)$$

(4.6)

I have transformed the function in order to include k in Equation 4.6 because I want to have k appearing in all equations so that they can all be put on the same graph at once. I consider Unfits (FP_{nc}) to be POVs.

Figure 4.14 illustrates the DS curve.

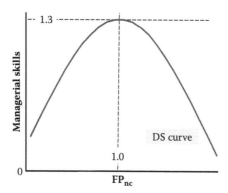

Figure 4.14 Managerial skills/FP_{nc} given the triple constraints.

I must mention that there is what is called in economics an "opportunity cost" associated with the fact that management deploys managerial skills to deal with the Unfits (uncontrolled Forces of Production, FP_{nc}). Recall that we have equated opportunity with utility, as utility is the proof that the opportunity of a given project is worth realizing. Hence, we could say that there is a utility cost prior to the maximum point of 2.3. Let's call this a "utility drawback" so that we don't confuse the use of the word "cost" in this context with its use in the triple constraints, which belongs to project management terminology. It is a drawback because management has to use time, incur costs, and/or sacrifice quality (perhaps only temporarily) in order to achieve its goal. It is also the most secure option because hiring new staff to replace the existing Unfits (uncontrolled Forces of Production, FP_{nc}) may just make things worse, not accounting for the fact that the new staff must be trained anyway. It is a drawback because full efficiency and efficacy cannot be attained instantly without incurring some costs and not without running trials that result in imperfect outputs.

As discussed, a project that sees a compelling change in infrastructure from beginning to end and that improves its clientele's (say a community's) life while having been so well planned that the difference between the

POVs identified at the beginning of the calendar and the POVs left at the end of the calendar amounts to nearly zero, and where there have been no adjustments in the power structure during its course, is a project that presents an outstanding utility. From that perspective, the Roman water system (sets of aqueducts) is an amazing success that certainly helped the Romans to establish their 800-year-long empire. Some portions of that system are still standing today.

A 2000-year-old Roman aqueduct in France.

I can venture to say that a set of optimal Dominant strategies in project management relies on sound infrastructure (the backbone of a project) and superior control of POVs, given a particular utility drawback, that is, given that some adjustments have to be made regarding the calendar of tasks and activities, costs, and adherence to norms of quality. This optimum is realistically reached where the utility of the DS curve meets the project Blindness curve, at $[x=1$ and $y=1.3]$.

The DS curve shows that it reaches a maximum point past which the management's strengths no longer work the way they used to. This situation is found in chess; for example, a pin works up to the point where the adversary's piece being pinned (a king, in the case of a so-called absolute pin) has moved by way of a so-called castle. The pin becomes ineffective and the piece, which was enacting the pin, would probably be better placed somewhere else on the chessboard. As time goes on, the piece that stays in its old pinning position sees its value steadily constrict.

If we graphically combine the U(DS) curve, the project Blindness curve, and the DS curve, we obtain Figure 4.15.

Note that this graph is actually multidimensional. The U(DS) curve reads: [x-axis = POVs; y-axis = Risks], the Blindness curve reads: [x = Negative trust; y = Collaboration], and the DS curve reads: [x = Managerial skills, y = FP_{nc}]. Because all curves touch on the same point of equilibrium at $[x=1, y=1.3]$, I feel entitled to use such a multidimensional system (which will become even more complex as other curves pass through this point). The point where all three curves meet constitutes an ideal scenario. At that point, the utility of the DS shows that the project remains logical with itself, so to speak (the project thus keeps on track), project blindness is recognized as a possibility, but a minimal negative impact on the project is achieved by spreading the risk of excesses equally

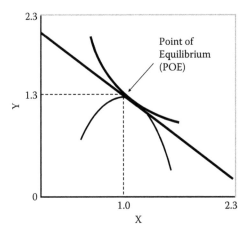

Figure 4.15 U(DS), Project Blindness curve and DS curve.

between complete ignorance of the risks and complete ignorance of the POVs, and the DS curve is at its peak.

We have thus identified a point of equilibrium where the whole dynamic entailed in a project achieves maximum efficiency and maximum efficacy. This **Point of Equilibrium** will be detailed in further sections of this book. If anything, it can be compared to a Point of Equilibrium found in classical economics whereby the demand curve meets the supply curve. In the case of a project, at the Point of Equilibrium, management and the various teams are on track to supply a product that is demanded for by the market and which justifies the investment of time, money, and quality effort.

4.7   Magic moments

Projects have critical moments that can serve as checkpoints in either a feasibility study, if at all advisable, or during their implementation. I call these "magic moments" as they generally have a positive impact on the project. I do not classify POVs as magic moments and prefer to keep them in their own separate category, given their extraordinary influence on projects.

They are three types of magic moments:

1. Benchmarking points, which I represent by a square (because they are, in essence, parameters)
2. Stage gates, represented by a diamond shape
3. Milestones, represented by a black dot

Figure 4.16 is an example of a way of positioning these points along a project life cycle. For sure, the feasibility analyst wants to mark these

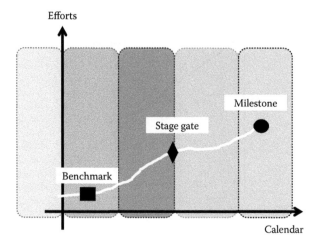

Figure 4.16 Examples of magic moments along the project life cycle.

points because they constitute measurement landmarks, which assist in following up on the project.

Benchmarking points are set when important actions can and should be taken for the benefit of the project development. The point of no return and the point of autonomy defined in Section 4.7.2 are important benchmarking points that are not discussed in PMBOK 5 or previous versions of the PMBOK. Stage gates are present at the beginning of each stage of the project life cycle, but in large projects, they may also exist within each stage that contains substages. Milestones refer to the end of a process; an example is the putting in place of the foundations of a building under construction. They are significant achievements that characterize the project. They are sometimes celebrated by way of a public display (see Table 4.9).

Table 4.9 Different points during the entire process of a project

Name	Type	Symbol	Detail
Benchmark point	Benchmark measure	■	Point of no return Point of autonomy
Stage gate	Stage	◆	Five stages of the life cycle
Milestone	Key achievement	●	Start point Final point of delivery

There are two points that I wish to discuss more particularly because they have a great role to play in a feasibility study: the benchmark points that are the point of no return and the point of autonomy.

4.7.1   *Point of no return*

This is the point in the production process from which the process cannot be stopped or reversed. Going back to the Correx corrugated cardboard production line example, once the two plates of cardboard have been fed into the machine and once they are "swallowed" by it and engaged in the covered conveyors, the process cannot be put in reverse: the two sheets of cardboard have already sustained some level of transformation. A stoppage leads to operational costs as well as to material/time losses and possibly to catastrophic results. A vivid and painful example is the explosion of the Columbia shuttle: the crew at the control center probably suspected that the shuttle had suffered damage during the takeoff period but there was no point in calling it back right away. The best option was to keep on track with the project plan and hope for the best upon return, which unfortunately did not occur. The damage sustained by the shuttle upon takeoff included small holes on its surface that allowed heat to penetrate within it and to eventually cause (C^+) its disintegration upon entry into the atmosphere. In a hospital, once a patient is assigned to a bed, they become a liability for the hospital. The patient cannot leave without letting the hospital know and proceed with due process: they are at a point of no return.

The point of no return presents a particular cost: if a stoppage occurs, all that precedes it is a net loss. More exactly, all that has occurred before it is a net utility drawback, because each element of the triple constraints is affected: time has been wasted, costs incurred, and quality compromised. Everything that ensues is a committed cost, or more exactly a committed utility drawback, that cannot be reasonably avoided. Thus, in that sense, it can become a special point of vulnerability.

In chess, this is similar to so-called *zugzwang*: the player must make a move when in fact they'd rather not play at all because playing will amplify their vulnerability. However, they have no choice, otherwise, they will lose the game on time. They have to wager that the adversary will make a mistake, or that luck will somehow turn around. They are stuck: whatever they do will work against them. In plain English, it is a catch-22. The feasibility expert, of course, wants to eliminate the chance of points of no return transforming into catch-22s; hence, thorough planning will demonstrate control points of no return prior to committing to FP and MP.

4.7.2   *Point of autonomy*

The point of autonomy is reached when the whole transformation process could theoretically operate without supervision. Let's discuss the following example: the Italian Floorlite company acquires a new machine that has been specially designed to reduce the number of transformation steps from four (heating and cooling, then heating and cooling again) to two (heating once and cooling once). The machine operators who work with this new machine need to be trained; in fact, even the chief engineer, Mr. Valiantino, has yet to discover how the machine behaves through extensive pretestings.[42]

There comes a point when all FP (both Fits—controlled—and Unfits— uncontrolled Forces of Production) feel at ease with the machine and know its limits, so that they are confident that they can operate it with their "eyes closed", so to speak. They have gone through their learning curve thoroughly: all of the necessary knowledge or expertise (T_h) has been acquired. At first, the number of worker-hours is high: it has been necessary to contract specialized trainers, consultants, and experts, some of whom will test the machine extensively while others will explain how to operate it. However, the scope of their assignment eventually fades out and soon enough the regular operators no longer need special training. In the end, only a minimal number of operators (work or T[43]) are needed. This is the point of autonomy, it occurs when T_h and T meet.

Put differently, management and team members (Forces of Production) acquire knowledge/know-how expressed by T_h. They also commit to performing a certain number of tasks, which, in project management terms, means that they are part of a given work breakdown structure (WBS).[44]

I have shown in previous writings that T and T_h share a mathematical link, which is easy to explain. The Forces of Production cannot work and learn all at once. There is a trade-off: while they stand in front of the machine learning how to operate it from the machine builder, employees do not actually work, they learn. When they operate the machine at first, they learn how to operate it; they put into practice the lessons they have attended. Sooner rather than later, knowledge will be acquired and they will commit nearly 100% of their time on working with the machine, and 0% on learning. The trade-off can be visualized as shown in Figure 4.17.

We know that management and team members (Forces of Production, FP) utilize the Means of Production, MP. We know that Processes (as one of the four Ps) see the inclusion of FP, MP, and, of course, resources. We have already divided FP into two structural elements: FP_c and FP_{nc}.[45] Means of Production consist of machinery, equipment, as well as building and measuring instruments according to our model. Resources $|R$ are structurally composed of essential resources (e.g., material such as cardboard sheets in the cat litter box as well as electricity to provide energy to the machine) and nonessential resources.

Nonessential resources are excess resources that are not immediately needed for transformation. Recall, however, that any transformation incurs a utility drawback; as such, it is always better to have some reserves (R_n) in order to avoid being caught along a critical path. This strategy helps minimize the project's POVs, a wise management technique (or,

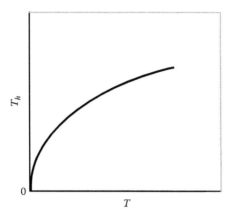

Figure 4.17 Trade-off between Knowledge (T_h) and Work (T).

put differently, a way of adopting a robust strategy). Thus, nonessential resources (R_n) are a structural process element of resources $|R$ just as essential resources (R) are. The same kind of trade-off that exists between T and T_h takes place between R and R_n, as I have shown in previous work. Perhaps the following sentence best expresses the dynamic implied with respect to R and R_n:

> "The trade-offs between goods implies that indi-
> viduals are willing to sacrifice their survival-
> enhancing activities, such as the acquisition of
> nutritious food, of adequate shelter, of health care,
> to acquire goods with zero or negative survival
> value like luxury goods, leisure travel, entertain-
> ment, and so on."[46]

Let us take the example of a cat owner. When this owner goes to a specialty store to buy cat food, he doesn't usually buy cat food for the present day; rather, he buys some for the week(s) to come. He saves cat food—an essential resource—for future use so that it is actually, at that moment, a nonessential resource. The cat will not eat one week's worth of food at once.[47] As time goes on, what was once reserved for future use becomes needed at the present day, so that it is transformed into being an essential resource. Similarly, when the cat owner goes to the specialty pet store, he doesn't necessarily spend his total allocated budget on food; the cat owner may buy his favorite pet a toy, such as a mock mouse. The stuffed mouse is not an essential resource—the cat can easily live without it—but it pleases the cat and makes the owner feel good. Furthermore, while the amount of food the cat will eat will stay relatively constant, the number of toys the cat owner can buy can keep increasing.

Overall, there is a trade-off between present and future use, whether this concerns T, T_h, R, or R_n. In this case, there is a trade-off between R and R_n, portrayed in Figure 4.18.

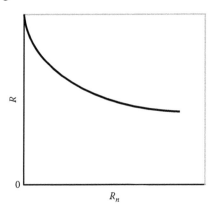

Figure 4.18 Trade-off between R and R_n.

There is more. Technically speaking, people work (T) in order to survive, that is, to pay for the essential resources (R) they cannot afford not to have. At the same time, people engage in advanced education and perfect their skills (T_h) because this improves their market value, allowing them to earn a better salary, which will be spent partly on buying nonessential resources (R_n). People buy Ferraris, but really, who actually needs a Ferrari? (Please don't tell Ferrari I wrote this!)

The point here is that there is an intricate link between T, which permits the acquiring of R, and T_h, which is helpful for possessing R_n. However, as can be guessed, people don't eat and work at the same time, at least not eight hours a day. There is a trade-off between work (T) and acquiring R, just as there is a trade-off between acquiring knowledge T_h and enjoying R_n.

I can summarize my argument by saying that R_n and T_h (nonessential resources combined with technical know-how) increase up to a maximum threshold, and that T and R (work dedicated to obtaining essential resources) subside to a minimum threshold. Where the two curves meet is where the point of autonomy takes place (shown in Figure 4.19).

In theory, efficiency reaches its apex at the point of autonomy. Identifying it helps the feasibility expert to assess the minimum cost of a project once it is operational. If the staff can operate the machine worry-free, it means that POVs have been dealt with, and hence that operating costs are minimal.

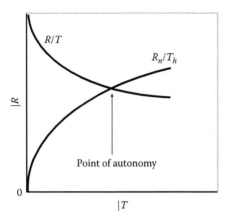

Figure 4.19 Point of autonomy.

We have shown in previous work that the functions underlying the dynamic of T, R, T_h, and R_n are production functions called the "Cobb–Douglas functions", as follows[48,49]:

$$E = z + \beta_0 \, R_n^\alpha \, T_h^{(1-\alpha)} \qquad (4.7)$$

where R_n stands for "nonessential resources" and T_h for "knowledge" or "technology."

We also have

$$E' = z' - \beta_0 \, T^\alpha \, R^{(1-\alpha)} \qquad (4.8)$$

where T is "work" and R is "essential resources."

We need to determine the value of alpha (α) as well as that of z or z'. My past research and database (see the chapter on People) lead us to the answer: when I assume z or z' to equal 0 and set β_0 as having a value of 1, the approximate value I find for alpha (α) is 0.3.

It is useful to hold one variable in both Equations 4.7 and 4.8 as a constant as otherwise the production function requires a three-dimensional approach that is challenging to plot. This treatment allows us to produce the graphs shown in Table 4.10.

The function $[E' = z' - \beta_0 \, T^\alpha \, R^{(1-\alpha)}]$ makes sense. As time goes on (as time is consumed during the project), the team members (Forces of Production) become more and more accustomed to their work, which ends up being routine. Less and less effort is required for the same set of tasks, but essential resources stabilize, much like people's hunger levels reach a satiety point when fed, and yet they keep working: hence, the curve is descending. On the other hand, the function $E = z + \beta_0 \, R_n^\alpha \, T_h^{(1-\alpha)}$

Table 4.10 Production function for $T - R$ and $R_n - T_h$

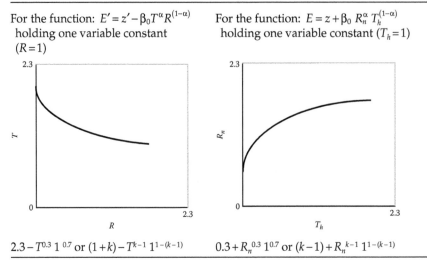

For the function: $E' = z' - \beta_0 T^\alpha R^{(1-\alpha)}$
holding one variable constant
$(R = 1)$

For the function: $E = z + \beta_0 R_n^\alpha T_h^{(1-\alpha)}$
holding one variable constant $(T_h = 1)$

$2.3 - T^{0.3} \ 1^{0.7}$ or $(1 + k) - T^{k-1} \ 1^{1-(k-1)}$ $0.3 + R_n^{0.3} \ 1^{0.7}$ or $(k - 1) + R_n^{k-1} \ 1^{1-(k-1)}$

as set in Table 4.10 expresses the fact that management and team members (the Forces of Production), becoming more efficient, are able to stash larger portions of essential resources (thus turning them into nonessential resources), or else manage to save money to buy more nonessential resources (for the company/project) as the essential needs are answered faster and better than before.

Note again that by having a common variable among all of the key functions we have seen so far (k), we are justified to plot all the curves pertaining to a particular project on the same graph even though the abscissa and coordinate axes differ depending on the process element under review[50] (see the multidimensional Table 4.20).

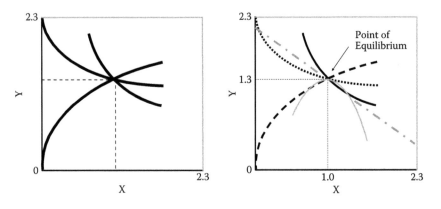

Figure 4.12 Point of Equilibrium (POE)

Note that all curves meet at the Point of Equilibrium: the U(DS), the project Blindness curve, the DS curve, and the two production curves $R - T$ and $R_n - T_h$. They all meet at [$x=1$ and $y=1.3$] within a closed dynamic system bound by the value of [$1+1.3=2.3$]. It is to the best advantage of the project manager to reach this Point of Equilibrium, because at that point, while POVs do exist, they are under maximum control by the team members (through $R - T$ and $R_n - T_h$) and in particular by management (DS curve).

At any point outside this Point of Equilibrium (POE),[51] the following takes place:

1. The project manager's position may be temporarily improved but only to the detriment of the utility drawback (so that the utility drawback worsens)—in the end, the project is worse off.
2. The project manager's position may be temporarily improved but only to the detriment of the Forces of Production (FP), which then retaliate (becoming FP_{nc}; e.g., by way of a strike or sabotage) so that in the end everyone is worse off.

There is no improving the position outside the Point of Equilibrium without alienating the project's nature or the project workforce, or somehow a mixture of both. Put differently, the Point of Equilibrium is where there is the minimal number of POVs that ought to be managed. As project managers get closer and closer to the Point of Equilibrium POE (say at 99% or 0.99) the POVs approach zero. This observation is handy because POVs can hardly be measured directly: one of their core features is that they are often hidden truths, as we have previously discussed. While we do all we can, as feasibility analysts, to uncover them, we cannot be sure that we have uncovered them all. However, the way around this unchartered factor is to measure the Point of Equilibrium. It is perfectly feasible to establish measures for this point, and we are fortunate enough that all curves pass through this point, so that its measurement can rely on a number of angles of analysis. Recall that our initial objective was precisely to shed light on the project from different angles, to leave no shadows (hidden truths) around it. Each curve is a spotlight that illuminates the object—the Point of Equilibrium—so that no POVs can be left unknown. Examples of measures that can be taken include

- For the project Blindness curve: management style, realistic assessments of risks, and POVs.
- For the U(DS): consistency and stability of the project plan as it is conceived and implemented or, put differently, number/nature/salience of the changes, size and relevance of the infrastructures, and so on.

- For the DS curve: sick days, strikes, sabotage, wastes, delays, training, and so on.

As can be seen, the feasibility analyst is not short of measures that can help him assess, ahead of time, whether a project is likely to be a success or not. Otherwise, the expert can set his own list depending on the particularities of the project that he/she is working on.

Let's assume now that all the forces that work in favor of the project, that is, the different curves we have examined so far (the project Blindness curve, U(DS) curve, and DS curve), behave normally and let's set the Point of Equilibrium (POE) as the center point of a normal distribution curve. Refer to Figure 4.21.

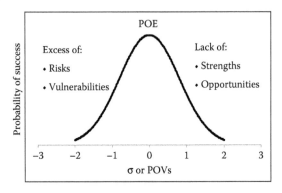

Figure 4.21 Normal distribution for the Point of Equilibrium (POE).

The trick here is to determine what level of variance from the Point of Equilibrium (POE) is acceptable in order for the feasibility analyst to give the go-ahead for a project. Thus, the feasibility analyst has three main concerns:

1. Determining the value and particularities of the POE.
2. Determining the actual kurtosis and skewness[52] of the normal distribution curve.
3. Determining the acceptable levels of variance, with each sigma σ representing a different managerial decision (go—within one σ if we establish that the variance is to be minimized), rectify (between one and two σ's), or no go (two or more σ's).

The nice thing about this overall analysis is that it provides a mathematical framework upon which a feasibility decision can be based by relying not merely on technical, marketing, or financial evaluations alone, but most critically on POVs.

4.7.3 Final point of delivery

A project does not end because the desired output—say, a brand new building—is being delivered to its rightful owner. Measurements must be made that allow project managers to determine whether the triple constraints were respected, and to set formal knowledge that can help in future similar projects. Because we have defined projects as producing outputs—including deliverables, some form of a book of knowledge (BOK), and impacts—a project is not complete until such a BOK is finalized.

A helpful method is to identify the various magic moments of the project to see if measurements can be made at each of these points (especially in an evaluation of the POE) ahead of time, and as they materialize when the project is progressing. This is exemplified in Table 4.11.

Table 4.11 Magic moments up to final point of delivery

Magic moment	Date comments	Assessment during the project	
		Costs associated with the slip-off	Nature of slip-off
Starting point			
Point of autonomy			
Point of no return			
Final point of delivery			

In Table 4.11, the POVs, POE, and other points (stage gates, etc.) are put aside, with the emphasis being solely on the three process points past the starting point, which are regularly ignored or else treated lightly in feasibility studies. I contend that due consideration must be given to these points because they are intimately linked to the success of the project.

Note that the Point of Equilibrium (POE) is not a passage point; it exists within the framework defined by the utility drawback or triple constraints. This means that managers can stay on the equilibrium point at all times, providing they control the level of utility drawback, that is, providing they abide by the calendar of activities, do not exceed anticipated costs, and conform to the preset norms of quality. Hence, the Point of Equilibrium is really a *dynamic* Point of Equilibrium. We will see in Chapters 5 and 6 on People how dynamic it is: very much so, because when two people engage with it, games are played that make the point active (if not shaky) at all times.

4.7.4 *G-rate and g-spread*

We have talked about the variance around the Point of Equilibrium (POE). There is another way of looking at this variance: by expressing it along the different stages of the project. I refer to this variance as the "g-spread", *g* standing for "growth."[53]

Let us set the following scenario while doing a little bit of accounting: A given project is assumed to produce no sale and generate no income (even through investment activities). We set that profits will come in only once the final point of delivery is reached. In the meantime, there is no P&L statement[54] worth the name. The project's balance sheet contains, of course, the list of assets, liabilities, and equity. Applying these accounting terms in the field of project feasibility, we can say that the balance sheet includes the value of the deliverables (assets), the costs incurred or to be incurred (liabilities), and the value of the project (equity). Technically speaking and using an accountant's approach, we want to have sufficient equity to cover the costs (i.e., [Value of project − Value of deliverables = Costs]). Any unplanned increases in costs will affect the project. We know that there is a relationship between each of the three constraints; we have discovered that when costs change, timing and norms of quality (established and controlled by the four Ps) constraints are affected.

Recall that the stages of a project over time are

Vision	Planning	Mobilization	Deployment	Completion/ evaluation

Let's decide that total sales from the project will generate US$5000 once completed, after 5 years. Each of the five stages of the project account for one-fifth of the planned total sales, or US$1000 per stage. We obtain the following result:

Vision	Planning	Mobilization	Deployment	Completion/ evaluation
US$1000	US$1000	US$1000	US$1000	US$1000

However, the time factor affects each stage in a particular way. The investment in inputs (resources, etc.) that is in the vision period is bearing fruit only along each of the subsequent stages. Somehow, we need to account for this; we add a growth factor, which we name "*g*". This *g* (*g*-rate) is similar to an interest rate on a bank loan (sometimes called a "discount rate") or a return on an investment. To facilitate our understanding of the present example, we set this growth factor, *g*, at a value of 10, 10 being the theoretical maximum value and 0 being the minimum value that the factor can theoretically reach. Rounding numbers, we obtain the future value of US$5000 by imputing US$820 per stage given a g of 10[55,56]:

Vision	Planning	Mobilization	Deployment	Completion/ evaluation
US$820 $\|g=10$	US$820 $\|g=10$	US$820 $\|g=10$	US$820 $\|g=10$	US$820 $\|g=10$

Suppose that at the end of a given stage, managers have produced the deliverables (e.g., the foundation of a house) according to the planned schedule, costs, and norms of quality, we can state that the g-rate $= 10$; put differently, 100% of the production step has been completed. If the deliverables, costs, or timing, or a mixture of these elements have not met the expectations set at the end of the first stage, we would say that only 80%, for example, of that particular stage has been completed.

Yet, we want to keep our initial goal of total sales of the final project to US$5000. Therefore, the g-rate must be adjusted upward beyond the failing stage, since we set that sales (the US$820/step initially established) cannot be changed, and so that we have

Vision	Planning	Mobilization	Deployment	Completion/ evaluation
US$820 * 80% = US$660 $\| g=10$	US$820 $\|g=10$ To be adjusted	US$820 $\|g=10$ To be adjusted	US$820 $\|g=10$ To be adjusted	US$820 $\|g=10$ To be adjusted

A simple calculation would show that the g-rate would have to be adjusted upward for the last four stages as follows:

Vision	Planning	Mobilization	Deployment	Completion/ evaluation
US$820 * 80% = US$660 $\| g_1=10$	US$820 $\|g_2=12.5\%$	US$820 $\|g_3=12.5\%$	US$820 $\|g_4=12.5\%$	US$820 $\|g_5=12.5\%$

The reader can see that there is a difference between the maximum set of 10% and the g-rate necessitated by the production problems incurred during the first stage. This gap (called the "g-spread") consists of 2.5 points (2.5%). A negative gap [real g – ideal g] indicates that there is something to be worried about. The team must work more intensely in the last four stages to catch up for the shortcomings encountered in stage 1. Thus, the gap becomes an indicator of shortcomings and a measure of POVs: the project has incurred delays, exceeded expected costs, and/or faced quality challenges that go beyond existing resources allocated to the project. The more difficulties that accumulate over time, the more the *g*-rate increases in value, because the remaining stages of the project must obligatorily compensate for past mistakes.

A negative *g* (a pessimistic scenario being realized) indicates that the project has fallen behind schedule, exceeded costs, or suffered from quality mishaps, or an amalgam of two or three of these problems. A moderate positive *g* (an optimistic scenario being realized), conversely, points to the fact that the project is progressing properly. The shape of the curve of evolution of the *g*-spread can be set as per the example in Figure 4.22.

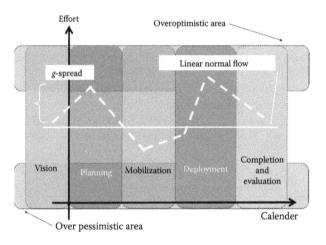

Figure 4.22 The *g*-spreads.

A manager could want to fix the setbacks incurred in stage 1 (vision) only during stage 2 (planning), forcing g_2 to increase to a whopping value of 24%, for a difference of [24−10 =] 14 points above the base *g*-rate. Accomplishing such a deed would probably entail that materials of lesser quality were used, payments of suppliers delayed, and so forth.

An overly positive *g*-spread is also a bad sign: the manager likely condones real problems, or else, tricks the delivery date (e.g., to receive a performance premium), reduces costs unduly (e.g., by cutting quality or fringe benefits granted to workers), and/or circumvents norms of quality (e.g., by ignoring certain standards).[57]

In summary, the *g*-spread is a temporal (longitudinal) representation of the variance σ around the project's dynamic Point of Equilibrium, POE.[58] The *g*-spread can serve as a representation of POVs. When it falls outside a certain upper or lower limit, this means that the project is vulnerable. Too much stress is imposed upon it. The *g*-spread highlights how critical the POVs are: the closer the project is to its deadline, the higher the value of g must be if there have been recent mishaps or cumulative problems. I assume that the total vulnerability is the sum of each g-spread along each of the five stages, so that

$$\text{Total vulnerability} = \sum_{t=1}^{5} g\text{-spreads} \qquad (4.9)$$

The third law of project feasibility is found to be the following:

4.7.4.1 The third law of project feasibility: The law of points of vulnerability

The higher the total vulnerability is (the sums of all g-spreads along each stage of the transformation phase) and the weaker the remedial actions are, the less the project is feasible.

The feasibility analyst is thus equipped with a punctual view of POVs—by looking at the project's dynamic Point of Equilibrium (POE)—and with a longitudinal view of it.

4.8   Not-so-magic moments

We have identified key process points that are sometimes ignored when performing a feasibility analysis. Stage gates, benchmarks, and milestones have a positive connotation; even the points of autonomy, of no return, and the final point of delivery should indicate that the project has been proceeding along. There are, however, points that have negative connotations attached to them, I see them as the "little brothers" of POVs. They require some efforts at troubleshooting, a typical activity for both feasibility experts and project managers.

4.8.1   Triggers, red flags, concerns, bottlenecks, hurdles

Triggers serve as a spark plug for a causal flow that has negative consequences.

Red flags are warnings that something may be about to go wrong. Questions such as the following may be posed by a feasibility analyst:

- Were there any difficulties in meeting deadlines for presenting the proposed project?
- Are interruptions likely to be debilitating?
- Will staff accomplish what they are supposed to do every day?
- Will staff remain productive from start to finish?
- Are the project's priorities likely to be respected?

These questions can be put in perspective when interviewing the project promoter who is seeking funding. His lack of preparedness or his overly ambitious plans are, certainly, red flags.

As I have established in the case of the Québec Multifunctional Amphitheatre (QMA), a list of preoccupations/concerns was made (see Table I.8). These may not be POVs just yet, but they may well develop into a vulnerability given that external risks affect the project in a certain way and with sufficient energy.

Bottlenecks occur during production. However, the feasibility analyst should review the forecasted production flows to verify whether bottlenecks are likely to develop. Surprisingly, bottlenecks are often due to human factors. Those experienced in driving in intense traffic are often witnesses to the fact that there are no obvious justifications for a particular traffic jam or sudden ambling. This phenomenon has been analyzed time and again, to the point that some cities have gotten rid of some sets of lights at specific intersections because city planners realized that the flow of traffic was more efficient when people were left to deal with each other rather than when being regulated.

Hurdles are temporary obstacles that may be anticipated. If so, they should not occur during production; of course, steps should have been taken to weed them out or to negotiate around them.[59] As anticipatory events, they can be included in a feasibility analysis—they are an example of POVs.

4.9   Conclusion to Chapter 4

This chapter was dedicated to the second *P* of project management: Processes. We have seen four different levels of causation, something that no other books, I believe and to the best of my knowledge, have done in the past. We proposed a language to quickly explain project events and to help visualize the *g*-spread. Overall, Processes cannot be treated lightly: in reality, a feasibility study is, by itself, a project, so that it is naturally a reflexive effort. A feasibility study that is a well-processed project is likely to reach conclusions that are far more reliable than one that is vague, inconsistent, and that does not respect the rules inherent to process analysis, with its flow of inputs, transformations, and outputs.

4.10   What we have learned about
POVs: Chapter 4

POVs…

1. Become obvious when the direct linear process starts going astray.
2. Can be awoken by uncontrollable factors.

3. Can be detected when the project starts to be both ineffective and inefficient.
4. Are brought under control when straight direct and diagonal processes work in tandem to achieve a result that is close enough to the intended output.
5. That are hidden in a variety of entry points, end points, or nondirect flows will likely remain undetected until after they cause damage to the project.
6. Always affect one or more of the three constraints: time, costs, and/or norms of quality.
7. Cause havoc in the transformation phase of the project when the most critical linkage level is reached.
8. Have four levels of criticality (low, moderate, serious, critical) depending on the nature of the causal bond in the process.
9. Can be highlighted by the causal links appearing in the process (especially in a critical path assessment).
10. Are more dangerous on the critical path than on paths where mediating options exist.
11. Weaken any process system.
12. Can be awoken by the catalyst effect of a moderating variable.
13. Are sensitive to the types of process bonds (D, T, I, C).
14. Grow like mushrooms (in dark areas).
15. Can be anticipated to a certain degree.
16. Can be identified through our modeling technique.
17. Are to be considered by a manager along with: controlling risks, eliminating errors, managing the Unfits (uncontrolled Forces of Production, FP_{nc}), harmonizing the four Ps, and relying on proper infrastructure.
18. Have two faces: one with respect to the error term ε (linked to efficiency), and one with respect to the final outputs (linked to efficacy).
19. Are temporal: they have to be measured in regard to inputs, outputs, and to the error term ε.
20. Are empowered by the lack of forecasting of outputs.
21. Are linked to the Blindness curve.
22. Are linked to utility (a zero difference between beginning and end POVs suggests a high utility).
23. Are not classified as magic moments.
24. Can be somewhat reduced by having some excesses (R_n).
25. Are at the highest level when far away from the POE.
26. Can hardly be measured directly because they are often cloaked.
27. Can be uncovered by analyzing the POE.
28. Can be somewhat represented by the g-spread.
29. Can be viewed punctually and longitudinally.

4.11   Key managerial considerations: Chapter 4

1. Reduce all process flows to causal relationships.
2. Seek equilibrium.
3. Measure variance punctually and longitudinally.
4. Use modeling and modeling language to extirpate hidden POVs.
5. Be absorbed by three fundamental preoccupations, that is, managing and minimizing: risks, POVs, and ε, the error term.

4.12   Case study Chapter 4: Sea Crest Fisheries

4.12.1   Introduction

Sea Crest Fisheries (SCF) is based just off of Evangeline Trail, in the French-speaking sector of Northwest Nova Scotia, Canada, along the shores of the Bay of Fundy where there is an abundance of herrings, scallops, and lobsters. The company, founded by the father of the present president Brian Saulnier, was traditionally oriented toward herring fishing, with a large part of the business going to Japan—the Japanese are fond of herring roe. Brian has scintillating brown eyes and an abundance of energy, which he expands from 5 o'clock in the morning to late in the afternoon, supervising the plant, dealing with his suppliers, and entertaining his customers.

In the 1990–2000s, a growing business in the area was that of the farming of minks—mink fur was a much-coveted product among European and Russian customers in particular. On the other hand, the fish business in Europe witnessed a decline in sales and reduced profits due to lower market prices. In 1998, Sea Crest Fisheries was bought by four partners who saw an opportunity to control the supply of mink food for their local farms. As a result, Sea Crest's vocation changed and included the manufacturing of mink food, produced with four basic key ingredients: fish, poultry, pork, and grain. The actual transformation process is very simple: it consists of mixing the four core ingredients (plus, in particular, some supplement mixes) with the right ratio to provide maximum feeding to the minks, thousands of them disseminated in a large number of farms located within the vicinity. Production losses are near zero.

The mink food business was not new to the owners of the company. They were clients who had been buying raw materials from it for years, and who made their own mink food on their local farms. However, at the start of 2000, the mink fur business grew exponentially, leaving little room for the farm production of mink food. It became easier and more sensible to buy prepared mink food from an external source. From this, the project of transforming Sea Crest Fisheries into a mink food–producing facility developed. The opportunity came from two opposing trends in the market: a reduction in the fish business and a surge in the demand for mink fur. By having numerous local farmers who were all buying from one source, the purchasing power of this group strengthened, thus leading to lower food costs for mink farmers. Furthermore, there was some form of synergy between mink food and herring: minks eagerly feast on herring.

4.12.2 The project

In 2003, a vision emerged whereby Sea Crest Fisheries would be mostly dedicated to mink food production. Brian Saulnier, president of the company, mentions

> The project came out of a vision to see equipment and building assembled in a functional manner in order to produce mink food for which market demand and sales had previously been anticipated. I knew it had to be framed within a budget constraint and a specific timeline.

This became the goal set in 2005 for Sea Crest Fisheries (SCF): to build a plant devoted to producing mink food.[60] The time line to accomplish this was within 12 months, at a cost of C$2MM (in the end, the project cost an additional C$5000), with norms of quality adapted to fit the requirements of the Canadian Food Inspection Agency (CFIA). Government regulations were, in fact, minimal because the industry was relatively new; however, Sea Crest preferred to be in a position where it could be prepared for anticipated regulations. The time line was somewhat critical because SCF wanted to start the operations in June 2006 so that it would have enough time to prepare and adapt to the peak production season, which occurs in August. However, delays in the installation of the electrical panels pushed the launch production date to the limit of early August. In a sense, this was the most pessimistic scenario happening: excess demand compared with actual production capacity. The original plan had been set in a realistic scenario of preparedness for June 2006, with no real optimistic scenario pushing production ahead of the scheduled launch date.

An informal list of stakeholders was prepared and concentrated on the owners, the clients, and the suppliers. The suppliers represented a risk by themselves because there was no certainty they could supply all the raw material needed for the two formulations that ended up being chosen by the SCF team. SCF had a backup plan though: in case of shortages, inputs such as poultry, herring (which came plentifully and was SCF's original main business) would be used. The project manager and final decision-maker on the evaluation of the project was Brian Saulnier and it is he, also, who prepared an informal project charter, not the actual sponsors of the project. He was given *carte blanche* in about every aspect of the project. Hence, the organization was neither traditional nor a matrix, but rather purely project oriented. Of the three aspects of POW (Product, Organization, and Work) Breakdown Structure, it is the WBS that caused the most difficulty up to the present day. According to Brian:

> This is an area where we learned we should have done things differently. Recall we had a business that was mostly based on temporary work and limited skills; the work load evolved with the fishing season. However, with the mink-producing plant, we moved to full time, year-long jobs. Yet, employees never had a job description. Hence, they acquired rights and privileges that at times were counterproductive and that we could not, as managers, change for lack of proper procedures. We are still experiencing difficulties in this area nowadays, albeit they are being addressed.

An informal survey was done to assess potential customers' needs. The process was rather informal given the nature of the close-knit community. How many minks would be fed and how many different formulations would be needed? As it turned out, each mink grower had developed his own formulation; variations between the different formulations were wide ranging, for example, from 15% to 70% in herring content. In the end, however, two formulations satisfied 100% of the customers, with 80% having chosen one formulation, and the balance, 20%, having chosen another one.

Cost analyses and break-even points were established based on expected prices and volumes. However, new infrastructure and new equipment were needed, with the latter eventually bought from a company in Finland. In 2003–2004, Sea Crest built a freezer out of which items such as chicken and herring were shipped to local farmers. In 2004–2005, the floor layout for the upcoming mink-producing plant was prepared, but the owners soon realized its surface had been understated as the mink fur business grew faster and faster; at some point, the growth rate was about twofold every 6 months. Once landed, the Finnish equipment was installed by local crew and a local company was hired to set up the electrical system and control boards. In 2006, some 40% of the 35 employees marshaled in the new business; the first test, ran in August, took 12 hours to produce 11,000 lbs. As of 2015, some 95% of the 50 employees are bustling producing a batch every 7 min, with a maximum production of 800,000 lbs/day during the busiest month of August. This translates into an average of 13 trucks delivering food to local farmers every day. Between 2007 and 2015, annual production grew by 325%.

Fortunately, not all clients converged toward the newly built plants at first, although demand was certainly very high during the month of August 2006; some farmers kept on producing their own mink food for a while until they realized it was more economically sound to outsource to Sea Crest. Also playing in Sea Crest's favor was the limited level of regulation in the business of mink food production and the mink farming business (a situation that is changing fast due to river and land pollution the likes of which had been seen for decades in many parts of the country with pig farming).

In short, the main milestones and stages of the project were

- September 2005: Ordering the equipment from a Finnish company
- October 2005–February 2006: Building the production, refrigerator, storage, and warehouse areas
- February 2006–July 2006: Installing the electrical wiring and panels
- August 2006: First production batch

The project was innovative in three ways from SCF's perspective: (1) the company had never produced mink food before and its business had mainly been in extracting herring roe from herring fish for shipments overseas; (2) new equipment had to be bought for which SCF had no experience and which needed to be complemented with an adapted design to fit its needs—in fact, the entire control room systems, the electrical panels, and the plumbing were done in house; and (3) the formulations had to be redesigned to meet the majority of the growers' needs, with the help of a research facility attached to a local university.

Using the PMBOK terminology,[61] the project can be seen as a series of groups of processes. The initiating group of processes encompasses the different stakeholders who had a vision for a new plant. Processes *per se* were focused on listening to future clients' needs.

Persons and Power were easily chosen and established, as Brian Saulnier had all the latitude he needed. The planning group of processes (we are still using PMBOK terminology) included a financial plan that was presented to a government-sponsored bank.

A firm calendar was adopted as well as norms of quality (based on government standards). The executing group of processes saw Brian Saulnier ensuring that the plan was being respected and that the processes matched their preset flows (although there were no formal PERT or critical path methods used). Persons that participated in the execution stage included an external contractor for building the actual plant and another one for the electrical system (which proved to be the cause of the delay, starting in February 2006); internally, the mechanics formerly employed to handle machinery related to herring processing were diverted to supervising the new equipment implementation. The mechanic with the most expertise was chosen as an informal second-in-command. Much of the work was done informally using a "learn as we go" rationale.

The monitoring and controlling group of processes was minimal and, again, rather informal. No major modifications were necessary versus the initial plan, except for relocating the ladder that gives access to the control room. No log books were prepared to keep track of modifications, however. Brian and his second-in-command had heartfully intended that the project progressed as initially envisioned.

The closing group of processes experienced some challenges. As mentioned, the closing stage took place toward the end of July rather than at the beginning of June due to difficulties with the installation of the electrical system. A final report was needed by the bank. New employees had to be hired (seven in total) when production started and a new managerial job was devised, that of feed kitchen manager (two people, one each for the day and night shifts).

The sociocultural risk associated with the business stemmed from pressure groups that oppose the killing of animals in fur production. In the United States, a large fur plant had been set ablaze by frantic activists. Internally, two major current POVs come to mind. First, while the plant manager masters the plant's processes and follows up on postproduction testing in its laboratory, many incoming raw materials present some form of contamination (e.g., chicken), which could spread internally. Second, the plant has become the largest and quasi-exclusive producer of mink food (there is another producer nearby—Nova Feeds, of a smaller size); the concern is that it has only one line of production, with few spare parts and backup Means of production (e.g., grinders, mixers) available, so that any major production glitch would virtually shut the plant down and threaten the survival of some of the local mink farms. It would be enough for a risk-factor-turned-POV (an outside activist hiding his intentions and getting hired to work on the production line) to target a vulnerability point (the entry point of raw material or the production line itself) and sabotage it to cause a business disaster.

Environmental concerns arose chiefly from noise levels: local inhabitants complained about trucks driving in and out of the plant at any time of the night. Strict transportation schedules and on and off loading rules were set to minimize noise levels.

One of the biggest baskets of risks and vulnerabilities was the divergence between the various external and internal stakeholders.[62] The tight network of the community helped in solving the issues at hand and in ensuring an overall positive business climate.

No doubt, Brian Saulnier's personality played an instrumental role in keeping all parties content. He is a laid back leader who trusts his staff and who dislikes confrontation, favoring instead mediation and compromise. His realistic and pragmatic views add to his skills and reinforce the respect his staff displays toward him. His management approach is not about status games, but about being reasonable, believing in people, and encouraging them to do their best. Face-to-face exchanges instead of e-mails are a preferred method of communication. Meetings are kept to a minimum; walking through the plant as it is being built is worth more than sending directives without knowing exactly what happens on the ground.

Business has been good for Sea Crest Fisheries. As stated by Brian Saulnier:

> We are a small part of the country. Here, business is done with handshakes, not with lengthy, overly complex contracts. We have close bonds with the employees and our suppliers and everyone out there seeks to help one another. Management shows gratitude by way of flexibility, for example, with vacation allowance and bonus pay. If I have to summarize our project, it was all about loyalty.

China, however, has entered into the equation. It has long been a substantial buyer, but it now has become a provider of mink fur, with its own methods; add to this its control of its exchange rates on the international scene to facilitate its exports.

4.12.3   Quality audit

A quality audit was performed in 2015–2016 to verify whether improvements could be made to reduce the possibility of contamination. The plant was examined in two different conditions: the first one during idle hours, when no production was taking place and the second condition, discussed in the next section, took place during production.

A tour of the plant was done, observations made, pictures taken, and questions posed. Two categories of observations were listed in the final report: one relating to quality control, the other one relating to safety and security (i.e., relating to costs). A third look at the plant focused on time, that is, on efficiency, maximizing the various flows of personnel movements.

With respect to quality control and the risk of contamination, the following notes were prepared, which are completed with recommendations.

4.12.4   Inputs: Quality audit

Major product inputs include poultry, pork, grains, and fish; antibiotics and vitamins are also added. Poultry is the main source of contamination, especially in the summertime. Two options are readily available: to put all arrivals in quarantine until tests (salmonella and *Escherichia coli*) are done and the product is released, or else to incite the supplier to arrive on the deck with a certificate of conformity. The first option does not solve the problem of what to do with the rejected product (return it or destroy it). This may cause tensions with the supplier, just as the second option would do. Poultry coming in is currently tested in the plant's laboratory, but it happens that the poultry has already entered the production line before the results are out. This practice is not recommended.

One of the Quality control desks at SCF.

As mentioned, there are a number of options to deal with poultry contamination, which can reach elevated levels during the summer.

First, there could be a monitoring system at the shipper's end; however, little control or enforcement can be exercised. Control upon arrival is standard, of course, but at times production rush requires to use poultry right away. At other times, a poultry truck may sit for quite some time outside, in the sun, because of holidays, weekends, or timing related to production.

An option to reduce contamination is to use dried poultry meat (supplied by Nestlé Purina); however, this entails changing the formulation (which is at the heart of the operation and which at last was agreed upon with the mink farmers). In addition, the dried ingredients tend to jam some pipes and screws in the facility, where the mink food is so-called wet food rather than dried (in kibbles).

Acid is put into the formula in an effort to lessen the potential for contamination, but its use is limited for obvious reasons.

Another option (currently partly used) is to freeze all chicken meats—however, to destroy bacteria effectively, the freezing has to be intense. This can be costly and may delay production at times.

Fortunately, to date, contamination has not proved to cause harm to the minks. However, stronger controls are required.

The other entry point that causes concern is the trap set to channel incoming fish shipments.

This trap is located along the side of the main production building. It has no cover; an errant or curious rodent attracted by the smell could climb it and fall down the pipe that ends in the hopper. This trap should have a cover when not in use.

The open chute.

The room where additives (antibiotics, vitamins, etc.) are kept is also of concern. The area is hardly sealed and the product sits on pallets, which sit on the floor.

Typically, such pallets should be shelved, with the bottom shelf roughly 2 feet above the floor (mice can jump about that high). In addition, there should be a space along all the walls of about 2 feet to deter rodents from traveling in dark alleys and to facilitate inspection.

Dry components room.

Sea Crest Fisheries should check whether the antibiotics require special storage conditions such as room temperature and humidity levels.

Entry materials—poultry, pork, and fish—should pass through a metal detector, which has not yet been installed. Pallets must also be inspected—a procedure must be set for such quality control concern. Old, damaged pallets may include wood splinters and nails that can get into the food or else that could allow mold to grow.

Hidden storage space: There is a space that is difficult to access where rodents and bird droppings have been noticed, with no signs of pest control.

Employee's kitchen area: This room is filthy. Employees must learn to keep the place impeccably clean. There are many entry points for mice but one single mouse trap is located beneath the sink.

4.12.5   Transformation: Quality audit

During the transformation phase, a number of areas present a potential for contamination.

Hopper: There is a hopper beside the main processing units and mixers where access is obviously difficult. As a result, the hopper and the pipe do not appear to be perfectly clean; mold can likely develop, bacteria can easily grow given that the ambient temperature, and humidity levels are high.

Large hoppers: There is a room besides the main processing room that contains large hoppers, some of which do not appear to be completely clean.

Ring: There is a ring that is not made of stainless steel and that is painted—this can cause metal/paint particles to enter the food. The paint should be shaved off the rings.

One of the hoppers.

Sitting products: Some pallets of products sit before the entry leading to the freezer where they are being thawed to room temperature. As the product defrosts, it tends to spill on the floor. A space of about 2 feet should separate the pallets from the walls and the pallets should be encased by rails so

Ring with flaking paint.

that circulating individuals and forklifts don't tend to touch them. The pallets should preferably be put above the floor so that contaminants carried on the boots of employees do not readily access the food. The time and day at which the pallets are left to defrost should be indicated on the pallets and recorded in a registry to guarantee traceability.

Pipe: One of the pipes has been hammered several times in order to let the content slide through gravitationally.

This suggests that the formulation is too thick and that some of it may actually be stuck inside, potentially allowing bacteria growth.

Vitamin input: The information on the formulation is written manually on a piece of paper. There should be a software program that manages all aspects of formulation in order to secure traceability.

Control room: The control room is filthy and it was noticed that the

Thawing area.

operator smokes in it (because a concealed ashtray was found after smelling the cigarette smoke), despite having been advised not to do so for two years. There is a dead spot where he cannot see part of the production (especially on top of the stairs near the mixers)—mirrors should be installed to cover dead spots. Even though there are cameras, this may skip the attention of the controller; should the person emptying his bucket of vitamin content inside the mixer be in trouble, there are no alarm buttons on the upper floor.

Hammered tunnel.

Cleaning during production: A couple of employees kept the floors clean during production. Overall, the plant is tidy.

4.12.6   Outputs: Quality audit

Output products are tested in two ways: (1) at the Sea Crest laboratory and at the independent Nova West laboratory, located in a remote location, where measurements include ash, fat, moisture, protein, and as *E. coli* counts. This double-check attests to the fact that the data are accurate. Sample batches are well identified.

4.12.7   Relationship between the appreciation of management and quality

SCF performed a quality audit in March 2016 based on an approved questionnaire aimed at measuring the relationship between the appreciation of management and quality. This effort was deemed useful in order to statistically measure whether the general appreciation of management (in essence, the project and operation leader—Mr. Brian Saulnier) meant higher concerns for quality work and control on the part of employees. Based on preliminary exploratory results,[63] a simple statistic was computed and the data shown in Table Case 4.A were obtained.

"Dependence" relates to the participant's (employee's) need for his job, "Trust" relates to how much the participant feels he that can count on the top manager, "Fairness" relates to his perception of being treated

Table Case 4.A SCF Core behavioral constructs ($n = 29$)

	Dependence	Trust	Fairness	Collab.	Quality
Mean	75	81	77	71	91
Std. Error	19	12	18	18	13
Minimum	14	54	36	31	57
Maximum	100	98	100	96	100

fairly, "Collaboration" refers to the collaboration that exists between the participant and top management, and "Quality" expresses the effort made by the employee toward ensuring quality standards are met or exceeded. Levels of Dependence, Trust, Fairness, and Collaboration are normal and are typical of so-called relational interactions. Both higher and lower numbers would indicate potential conflicts. The level for the construction of Quality is positively high, which speaks strongly in favor of SCF.

An initial multiple linear regression analysis shows that the so-called psychological core ([Trust + Fairness + Collaboration]/3) is significantly related to quality (at $p > 0.05$), with an adjusted R^2 of 0.425.[64]

Hence, it can be said that the plant is well managed and that this is an important factor in the achievement of quality standards, within the limits of this exploratory research. Indeed, an initial factorial analysis (25 iterations, varimax rotation) shows that when Conflicts are near zero, quality improves while when Conflicts approach 100%, quality deteriorates (refer to Table Case 4.B).

Table Case 4.B SCF Brief factorial analysis

Construct	Group 1	Group 2
Fairness	1	0
Conflicts	0	1
Trust	1	0
Collaboration	1	0
Quality	1	0

Let's now move to a simple cost-based audit, looking for areas where costs would be incurred in case of an accident or where they are currently incurred because of certain managerial practices.

4.12.8   *Inputs/transformation/ouputs: Simple cost audit*

Untidiness: In many areas, including in the electronic board control area, the maintenance area, and receiving area, various items are left sitting idle, such as buckets, coffee cups, hammers, rollers, and so forth. This is cumbersome and distracting and may actually cause an accident. This requires employee training.

Loose hoses were noticed, especially in the final output hopper area where the trucks wait. An employee could potentially trip and either fall into the hopper, with no chance of escaping alive, or fall two storeys down onto the concrete floor.

Maintenance room.

Language: Language is not uniform. Some employees speak English only while the Francophones are all bilingual. In case of emergencies, it is not certain whether Anglophones would understand the instructions or not, if expressed in French. Employee training is warranted.

Forklifts: Another safety concern is with respect to the loading areas for the forklift that are near to the robotic equipment. A stopper should be installed on the floor to prevent a forklift from hitting the conveyor.

Control board.

The forklifts do not emit a loud enough noise when backing up, this may lead to an accident.

Emergency: There are a number of emergency devices that are well positioned, especially near the acid bin.

One of the receiving areas.

Documentation: The material safety data sheets (MSDS) all seem to be readily available in case of acid spillage. The pallets are identified with a ticket that is sent for entry into the computer system once the pallet is being moved to be unloaded. The formulations are entered by hand; as discussed, this should be done through a software program.

Follow-up forms are readily available and information is entered into a computer tracking system. However, the forms should include the time of the operations for better traceability. Units of measurement must be neatly identified on all production sheets (e.g., lbs).

The robot sits at the back of the conveyor.

4.12.9   Conclusion

The project of implementing a brand new plan for producing mink food has gone remarkably well, despite not having respected all of the project guidelines provided by the PMBOK. This is due in large part to the kind of close, friendly, and mutually trusting relationships that existed and still exist among all internal and external stakeholders. The operation is well run. The point of autonomy (the point at which the operation could run by itself without additional training or efforts) was achieved in October 2006 as far as machinery handling goes, and somewhere around June 2007 as far as formulations are concerned, for the mere fact that some formulations could not be tested before due to the seasonal nature of the mink farming business.

4.12.10   *Appendix A of Case 4: Floor plan and construction*[65]

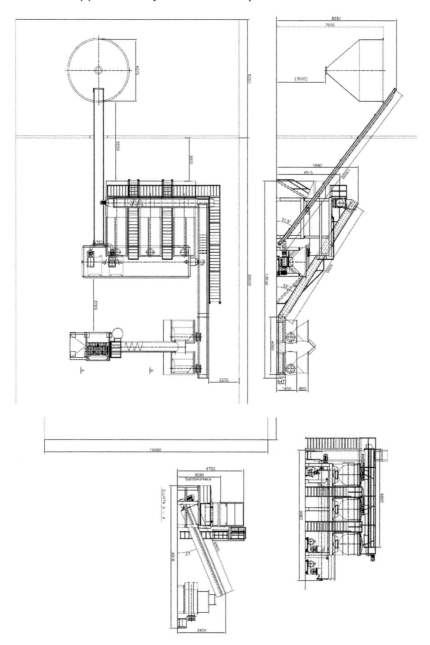

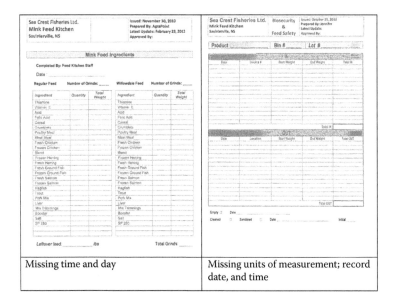

| Missing time and day | Missing units of measurement; record date, and time |

4.12.11   *Appendix B of Case 4: Production*

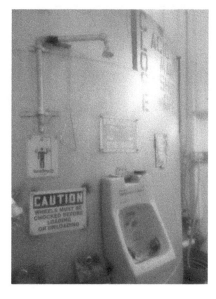

4.12.12   Questions related to Case 4—SCF

1. Describe the opportunity that arose and that led to the establishment of the mink food business.
2. Draw the production process using the methodology taught in this book.
3. Write the production process as explained in this book.
4. Describe the five stages of the mink project life cycle, including the vision stage.
5. List the People involved, including the regulators and the bad apples.
6. Discuss the triple constraints, including the calendar.
7. Refer back to the teachings of the PMBOK and identify the core elements such as the project charter and the management process groups.
8. Discuss the POW.
9. What is your take on the needs analysis?
10. Identify and discuss magic moments.
11. In what way was the project innovative?
12. Discuss the four Ps.
13. List the POVs that you can identify in the initial project and discuss their characteristics, presence, and importance.
14. Position the POVs along the entire production chain.
15. Identify and discuss the risks.
16. Discuss the norms of quality and determine what steps can be taken to improve quality.

17. Discuss the relationship between quality and management appreciation.
18. Discuss the role of infrastructure, risk, and robust management.

Endnotes

1. The PMBOK refers to five groups of processes (integrating, planning, executing, surveillance and control, and closing); however, as I shall discuss further along in this book, these are not processes *per se*, at least not the way they are explained. This being said, PMBOK 5 also seems to make some additions from a modeling point of view on p. 59: it introduces the process of "overall project control" which does not appear in its main matrix. This highlights the importance of using simplified models and of carefully integrating constructs or processes that are a vital part of the set project. Similarly, on p. 61, I could argue that knowledge area "9. Project HR Management" is in fact a subcategory of knowledge area "13. Project Stakeholder Management" since among the project stakeholders are those who benefit from a HR Management plan. Note that the word "direct" in Section 4.2 of p. 61 refers to a controlling activity (a Power activity: to direct) and thus does not really belong in the so-called executing process group. A better word given the context and the four Ps structure that I propose would be "implement." I feel that "Initiating process group" and "Planning process group" belong to Planning, "Executing process group" and "Closing process group" belong to Process, and "Monitoring and controlling process group" belongs to Power. From my modeling point of view, it would be more attuned to production logic to have a functional activity group separated under: Plans, Process, People, and Power.
2. For example, TSO, p. 21.
3. Ryan and Bernard (1994, p. 782).
4. Olivier and Payette (2010, p. 18).
5. The goal of this book, however, is not to delve into PERT, the CPM, activity-on-node (AON) or activity-on-arrow (AOA), and precedence diagram methods. Plenty of books are available on the subject. See, for example, Devaux (2015). The French version of PMBOK 5e edition makes an error on p. 156 when defining finish-to-start and start-to-finish, using exactly the same definition for two different concepts.
6. I developed this method and it has been the subject of various articles and of three books, one published by Springer Psychology.
7. Note: The codes are in parentheses because that's the way they appear in a modeling diagram, where they are positioned on the arrow linking two constructs.
8. For example, see Goldfried and Davison (1994, p. 26).
9. Creswell (1994, p. 85) states: "Position the dependent variable on the right in the diagram and the independent variables on the left."
10. Names and location have been altered to ensure confidentiality.
11. The logic behind the group of processes promoted by PMBOK 5 is not necessarily truly functional in real life. PMBOK attempts to separate the five groups of processes it has identified from the life cycle stages of a project, when in fact they make a whole lot more sense, from my modeling point of

view, when integrated into this life cycle. Additionally, the name "group of processes" is troublesome because the model is somewhat erroneous: there are inputs for sure, but there are no actions that follow—only tools and techniques are listed, which then presumably lead to outputs. However, tools and techniques are not a transformation action that move inputs into a category called outputs; from a modeling point of view, they are a Means of Production and hence, an input. Truly, the groups of processes are simply core functional activities. The complexity of the links between the groups of processes, their inputs, actions, and so-called outputs point to the fact that they must instead be considered as managerial activities that certainly have some logical flow, but that include a vast array of iterations and connections. All process diagrams have inputs, an action phase called transformation, and outputs.

12. In the strictest of terms, POVs have affected first costs (risks affect time and errors affect norms of quality).
13. I changed the names and locations to preserve anonymity.
14. Each model and submodel must be kept to a minimal level of complexity. Many studies suggest that most people rely on a maximum of three, and at times four pieces of critical information to decide, act, and live.
15. In some settings such as museums, humidity is a truly set parameter rather than simply a factor of influence.
16. PMBOK 5 (2013, p. 157).
17. In some network systems, Z would be called a "node".
18. When it comes to neuroscience (on which I will draw substantially in Chapter 7), one refers to structural connectivity as a "set of physical or structural (anatomical) connections linking neural elements" and to functional connectivity as "patterns based on time series data from neural recordings" (Sporns, 2011, pp. 36–37) where interdependence between neurons is recognized. An element of time (in milliseconds) is included in the definition because of the connectivity component of the term.
19. Jarvis et al. (2003).
20. Bollen and Lennox (1991, p. 308).
21. See Diamantopoulos et al. (2008).
22. Collier and Bienstock (2009, p. 284) mention that formative variables in statistics are theoretically uncorrelated (or sometimes negatively correlated, a statement I disagree with in the context of my methodology).
23. Recall we use an error term with a negative sign for reasons we will see further along in this book.
24. One must be careful; the terminology here is different than the one found in statistics for continuous data. An example of a continuous data in statistics is weather: it can be of 12 degrees, 12.5, 12.3, or 15, and so forth to an infinity of possibilities. In my example, continuous means a scale composed of more than two measurement units; in other words, any scale that is not binary (yes/no).
25. Do not use present = 2 and absent = 1, that is, use only 1 and 0 to express the binary nature of a scale. Doing differently could cause difficulties when running statistical analyses or when creating mathematical formulae.
26. The term "structural" is not related to structural equation modeling (SEM).
27. One of the reasons why the analyst needs at least two process elements to form the main process element stems from the fact that he may want to define a process element by what it is and what it is not (black and white). Therefore, a main process element is necessarily formed by *at least* two subprocess

elements. In the context of POVs, this is especially important. White cells in the body, for example, attack invaders by recognizing which bacteria are part of self (and our body contains billions of bacteria) and which are not. The immune system (which ensures there are no POVs left to the open) is dependent upon this differentiation between "self" and "not self."

28. See Diamantopoulos and Winklhofer (2001); Diamantopoulos and Siguaw (2006).

29. Large projects normally, however, produce deliverables, accumulate some formal knowledge, and generate impacts.

30. A useful method to identify the functional variables defining a process element or any object of investigation (e.g., a construct) is to churn out a list of questions that the analyst would want to pose to five different experts on said theme. This forces the analyst to envision what the said theme could do; by actually collecting answers from experts, functional variables have an even better chance of being fully identified.

31. The Québec's *Nordiques*, a fierce rival to the legendary Montréal's *Canadiens*.

32. Delahaye (2015, p. 78).

33. *Guardian* (2013).

34. Real name and location.

35. If we equate infrastructures to 1 (or 100%), we obtain the formula for the Utility of the Dominant strategy U(DS).

36. I show a stylized version of the rectangular hyperbola, and only a portion of it: the one that is significant for my study.

37. For example, the project manager would think of himself very highly and consider that there are no POVs in his project or else he chooses to be oblivious to them.

38. We will see in the chapter on People why it is near −1 and not exactly −1.

39. For example, it has run some trials with inexperienced or uncommitted staff knowing the output will not be to the level of quality that is ultimately demanded.

40. This initial value was found out from a longitudinal study I did and which we examine in Chapters 5 and 6 on People.

41. This function is hypothetical: it is an attempt to portray how behaviors relating to a Dominant strategy look like. It was developed through a series of transformations operated on my core functions.

42. From this perspective, we could refer to the new machine as an MP_{nc}, that is, a Means of Production that is not yet controlled/mastered.

43. *T* is for "Travail" which means "work" in French.

44. As an example, the WBS was set by using the computer-based project planning and scheduling (CBPPS) system software during the 1988 Olympic Games in Calgary.

45. Note that FP_{nc} being considered a structural process element of FP, means that there cannot be a project without FP_{nc}; we have seen in the section on innovation why that is.

46. De Fraja (2009, p. 51).

47. Cats are wiser than dogs in this respect; they only eat what they actually need, and they are very fussy on their choice of food. In fact, cats, unlike dogs, need taurine in their diet, because they don't produce it naturally. Thus, taurine is an essential resource, without which they die.

48. I do not include an error term in order to facilitate my presentation.

49. We have seen that R and R_n entertain a contrasting relationship. When an individual accumulates essential resources R, he cannot gather nonessential resources R_n at exactly the same time. For this reason, in the output functions, R and R_n are presented in a different order: $[E = z' + \beta_0\ R_n^{\alpha}\ T_h^{(1-\alpha)}]$ where R_n comes first, and $[E' = z' - \beta_0\ T^{\alpha}\ R^{(1-\alpha)}]$, where R comes second.
50. Otherwise, the graph would have to be multidimensional. For simplification, I do not name the axes directly on the graph.
51. This will become clearer in the chapter on People.
52. I suspect they change according to the type and size of projects.
53. The literature on project management is not short of valuable concepts with respect to quality concerns, such as the notion of schedule variance, cost variance, and schedule performance index.
54. Profit and loss statement.
55. There is no initial investment and measurements are taken at the end of each stage, not at the beginning of the stage.
56. Read, for example: US\$820 | $g = 10 \rightarrow$ US\$820 given $g = 10$.
57. Structures such as bridges that collapse because of preventable construction defects provide an example. Often, construction materials have been deliberately chosen for their price regardless of their low quality, or else normal construction procedures have been short cut.
58. In typical project management literature, one speaks of upper and lower control limits (according to quality expectations) and specification limits (according to needs) in a control chart. An accepted heuristic is the *rule of seven* by which any set of seven or more points appearing consecutively in either the lower or upper range (on one side of the mean) is considered not to be a random occurrence, suggesting the process is actually out of control.
59. See Harrison and Lock (2004).
60. This paragraph is in line with the concept of a "project charter" used in the PMBOK 4 (2008). *A Guide to the Project Management Body of Knowledge—Fourth Edition*. USA: Project Management Institute.
61. *The Project Management Book of Knowledge* (2013), 5th edition, proposes five groups of processes, which I subdivide into Plan, Processes, People, and Power (authority).
62. Indeed, a project feasibility analyst must look for divergence of interests as an important source of points of vulnerability.
63. I intend to perform analyses that are more complex in the near future.
64. $F = 21.666$ and *sig.* $= 0.000$. Residuals are normal. Durbin–Watson is acceptable at 2.580.
65. Pictures in Appendix A of this case have been graciously provided by Mr. Brian Saulnier, president, Sea Crest Fisheries.
66. Many psychometric or marketing questionnaires are built in order to boost the Cronbach's alpha, regardless of sound psychometric rules. See work by the authors on the subject in the bibliography.
67. Interestingly enough, when I take the standard network formula for calculating communication channels, $[n(n-1)/2]$ and find the value of n to achieve the k constant of 1.3 (more precisely 1.32), I obtain $n = 2.2$, which is the upper limit of the Blindness curve. The average k for many groups I studied is actually often 1.32 or near this value. For fun, I found that $1.32 = 1 + 1/\pi$, and had a drink afterward!

chapter five

People (Main constructs)

So it is, said the Wise Eagle, that sometimes Man is
an unmanned error of Nature.

Darloz

5.1 Introduction to Chapter 5

We have seen that "People" is part of the four Ps of project management,
and in particular of project feasibility analyses. Indeed, interpersonal
skills are fundamental in any project.[1] Various authors have identified
some competencies as being the ability to communicate, to manage stress,
to show empathy, and to solve problems. Some authors make a very valuable point that one should scan organizations in search of competing
interests;[2] they certainly represent a point of vulnerability (POV).

In our discussion, when people are looked at as external agents, I refer to the concept as "sociocultural risks"; when they are examined as an integral part of the project, I recognize their potential for creating vulnerable points. This is especially true when People engage in the transformation process, where I divide them as Forces of Production (FP), which can be controlled (FP_c or Fits) or uncontrolled (FP_{nc} or Unfits)—the latter representing the largest potential for fostering POVs. On the outside of the project, People have been classified as clients, suppliers, regulators, and bad apples—the latter representing the highest risk. A deadly combination, as we hinted before, is the mix of poor planning and Unfits (uncontrolled Forces of Production, FP_{nc}), or even worse, poor planning, (uncontrolled Forces of Production, FP_{nc}), and bad apples.

I can express my viewpoint figuratively in Figure 5.1.

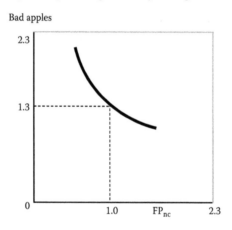

Figure 5.1 Bad apples (as external risks) versus FP_{nc} (as POVs).

Of course, we cannot claim that the utility curve that is hence generated represents a Dominant strategy (DS)—far from it. Figure 5.1 expresses a doom scenario; the present chapter shows how this can be integrated with the U(DS), model we have discussed so far. It expands significantly on the role of people in project management; it digs into psychological phenomena, resorts to the large database that I have accumulated over the years, spans across many more teams of various cultures on two continents, and presents key concepts such as "Trust" and "Collaboration," which should not be ignored in a feasibility study.

Any project manager will have experienced challenges such as interpersonal rivalries, jealousy, or the situation of a group leader desperately trying to establish legitimacy, for example. I integrate all behaviors into my basic model whereby the four Ps are deemed separate inputs that enter into the transformation phase and then interact with each other, as shown in Figure 5.2.

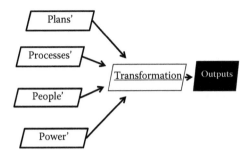

Figure 5.2 People and the basic process model.

I pretend, perhaps unlike the core of the *Project Management Body of Knowledge* (PMBOK), that it is not enough to rely solely on standard human resource management techniques such as human resources management philosophy, individual roles, and job descriptions.[3] Human nature is simply more labyrinthine than that! On the other hand, my research has shown that humans adopt certain behavioral patterns that are bound by predetermined values, the constant k being a prime example.

It is at times very difficult, if not nearly impossible, to get people to admit to their mistakes, to conduct interviews, or to distribute surveys that gauge people while they are in the midst of a crisis. In those circumstances, people feel threatened and find any excuse not to participate in the research. I have been fortunate enough to witness situations whereby outright or abstruse problems plagued a particular project. I have seen POVs in action. The present chapter reports some of my findings.

5.2 Persons

The PMBOK covers general human behaviors from a very technical point of view and on the surface only. I endeavor to do more. The various PMBOKs[4] provide an overview of stakeholders that are in some way involved in projects. These are the project promoter; the project leader; the management team, the team, and other stakeholders. The categorization favored by PMBOK is different from mine, and in fact, perhaps not as systematic.

The four groups that I have identified on the outside of the project can be classified according to their degree of unpredictability (of risk). Going from lower to higher risks, we have

1. Regulators (government, professional associations, etc.);
2. Suppliers (e.g., investors, partners, contractors);
3. Clients;
4. Bad apples (e.g., unethical pressure groups, illegal groups, etc.).

Bad apples are those that act and react under the radar, so to speak; they are on the margin of legality or, in some cases, they are bluntly illegal. However, one must be careful because the definition of bad apples depends on the politico-legal system that tolerates or bans them. In a dictatorship-like company, as can still be found in many parts of the world (some of which I have visited and seen in action), an individual who simply expresses their opinion and is a bit disgruntled may be treated as a bad apple. However, the external groups that benefited illegally from the Montréal Olympic Stadium (MOS) project can be labeled bad apples. In many parts of the world, collusion and abuse of funds generously provided by international organizations are still very much present and are truly risk factors; those who use these techniques are bad apples. Some bad apples may have at times a beneficial influence, however; for example, environmental groups who force a company to respect a natural habitat or else to preserve the way of life of some of the tribes in the remote Amazonian forest may actually help to put the project back on a normal humanity-oriented track. They remain, however, highly unpredictable and hence, from a project management point of view, are risk factors.

According to the PMBOK, the influence of the various stakeholders during the project life cycle diminishes with time, being relatively steady at first, but then retreating substantially midway through the deployment stage. Let's assume that this applies to external groups, that is, to sociocultural forces. Indeed, as the project nears completion, there are fewer reasons to object to it, disrupt it, or wish its discomfiture; in any case, the full responsibility of many projects is transferred to their owners or to their buyers. Hence, from a managerial point of view, risks vanish with time.

We can use the example of the music concert discussed in Chapters 1 and 3 to list the different stakeholders, as shown in Table 5.1.

Table 5.1 Examples of risks and POVs for the concert

External (risks)			Potential POVs
Clients	Regulator	Bad apples	Forces of production
Music lovers	SOCAN	A competitor who plans to disrupt the concert	Musicians
Media	License (alcohol)	Media that have a biased, negative view of the concert	Musical director

Of course, these actors intervene at different moments during the project's life cycle (shown in Table 5.2).

Tables 5.1 and 5.2 are only given to illustrate the role of external and internal People and to show that People, in all forms as we have described them, cannot be omitted from a feasibility study. I derive the fourth law of project feasibility as follows:

Table 5.2 Risks and POVs by transformation stage for the concert

| | External (risks) | | | Potential POVs |
| | | | | Forces of |
Stage	Clients	Regulator	Bad apples	production
Vision	Music lovers	—	—	Musicians
Planning	Media	SOCAN + License (alcohol)	—	—
Mobilization	—	—	—	Musicians Musical director
Deployment	—	—	Unfair, competitive concert	
Completion/ evaluation	Sales	—	Media's biased, negative coverage	—

5.2.1 Fourth law of project feasibility: Law on the Forces of Production

> With respect to POVs, the more Unfits (FP_{nc}) > Fits (FP_c), the more the probability of failure increases.

Let's put ourselves in a project environment where countless pressures are exercised upon us from the outside (political changes and upheavals, pressure groups, unhappy investors, etc.). At the same time, the workforce starts a strike while acts of sabotage multiply. As the deadline for project completion nears, managers (FP_c) are under more and more pressure to handle and control the situation. However, the more they try to deal with the increasing amount of grief, the more time elapses and the more the negative forces (e.g., employees on strike) feel empowered. The situation becomes desperate; disintegration is fast approaching.

Note here that I do not refer to the probability of success, but merely to the probability of failure. This distinction is important. Having excellent controlled Forces of Production, FP_c, is not a guarantee for success—external risks could be overwhelming and disrupt the project. However, having more Unfits (uncontrollable FP) than Fits (controlled FP) leads to a high probability of disruption—this is undoubtedly a failure 'factor'. This is an example of the difference between Key Success Factors (KSF) and Key Failure Factors (KFF), which we have discussed in Chapter 2.

A feasibility study should investigate the role of People in a given project ahead of time; relying on an analysis of curriculum vitae to evaluate core technical competencies is far from enough. Recall that we are chasing POVs just as some meteorologists are chasing hurricanes.

5.3   Observables and modeling psychological constructs

One way for the project feasibility analyst to gather valuable information is to observe. Many experts that I know and whose work it is to decide whether government funds should be granted to a particular project or not explicitly say that they rarely rely solely on written documents and on the application forms they receive. It is when meeting with the project promoter, when visiting his office or plant, or when talking with some suppliers and customers of his/hers that they feel they get the true picture. It has been said:

> Years of experience have taught me to detect possibilities of failure, which I can often judge by testing the trust level that exists between the key stakeholders, including us, the funders. We check the curriculum of the applicant, his credit and criminal (if any) background, past bankruptcies and verify whether the project is overly optimistic. We look for camouflaged details or motivations by way of the documentation, interviews and visits to the facilities.[5]

A large portion of this effort relies on observation. What the feasibility analyst does is collect so-called observables that, when analyzed, guide his/her decision. **Observables** are parameters as they generally apply to human behavior (but also to objects). An observable is by definition a change or a movement; for example, the person being observed has an abrupt gesture when challenged on the financial data he/she puts forth, or else he/she claims loudly that they can conquer the world in an instant with what they believe is a revolutionary idea. Observables are not chosen lightly. Just like metrics in processes are precisely set (e.g., the humidity level surrounding a cardboard-producing line), observables need to measure what the analyst is intending to evaluate, that is, the related construct. Appendix 5.1 offers the reader a review of some 100 scientific articles taken from various sources related to project management. The constructs that come out most often are Trust and Collaboration, with a whopping 60% share of the discussion compared with other psychological constructs.

A dotted view of reality!

Observables fall into a **general** spectrum (habitually, happy people laugh or smile) or into a **contextual** framework (e.g., according to Japanese tradition, one doesn't shake hands but bow). All observables are a reflection of what is going on inside the participant's mind; they are truly functional, but because they are a parameter, they are represented graphically by a rectangle, as shown in Figure 5.3.

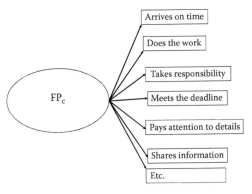

Figure 5.3 Examples of observables in the case of FP_c.

When it comes down to deciding on a psychological construct such as Trust, at least three observables are necessary. Some authors mention that "(...) at least four measures of a construct are needed for an assessment,"[6] but I contend that this causes more difficulties than it solves problems. This is why: Should the feasibility analyst find two observables that describe a behavior in a certain way—say, two behavioral patterns that suggest anger, and two in an opposite way, say, patterns that indicate happiness—he/she will not be able to reach a conclusion. An odd number of observables will do the trick. As a rule of thumb, psychological observables come in multiples of three just like judges, in a boxing match, amount to three, so that no decision deadlock situation occurs.

When too many observables are garnered as a result of observations made by the feasibility analyst regarding the behaviors of People—whether external or internal—there are four ways of reducing the complexity of the situation.

First, a term commonly used by those being observed in their day-to-day operation may turn out to regroup behaviors that seem otherwise disparate. In my many interviews conducted with project participants, I found that the term 'win-win' was extensively mentioned and that it could include such psychological constructs as Fairness and mental games. I thus summarized all the different names under one simple construct: Fairness. A project leader told us that

As a project manager, I endeavor to respect knowledge, ideas and experience team members bring
in. Each team member has a voice on each of the
ten projects that I head concurrently. Managers,
lead trades people and trades people are assembled
according to the project needs and bond based on
fairness. This is how we work here, and our company keeps growing.[7]

Second, the analyst must decide if some observables are worth keeping or else if he/she should delete them after due consideration. Third,
the analyst can find an existing model in which he/she can make them
fit (e.g., the AIDA model—Attention, Interest, Desire, and Action). Lastly,
he/she can divide their main construct—say, Trust—into subconstructs
(such as Affinities, Benevolence, Abilities, and Integrity as we shall see)
and elect at least three observables to each subconstruct. To facilitate this
effort, I provide in Appendix 5.2 a questionnaire that gauges the key psychological constructs often discussed when interviewing project managers or project team members by way of surveys or face-to-face meetings.[8]

It must be noted that constructs are the equivalent of process elements
found in process modeling and that exactly the same types of descriptive
(S and F) and consequent arrows exist (T, C, and I—including media-
tors and moderators).[9] In the case of causal bonds, however, the burden of
proof is much heavier with psychological phenomena than with the laws
of nature that apply to different processes (such as water boiling given
enough heat and a certain atmospheric pressure).[10] Many authors have
recognized over the decades that causality in the psychological domain is
most difficult to prove.[11]

I strongly recommend that the feasibility analyst collect information
from various sources, not merely from the documents handed in by the
project's promoter. Some authors[12] point to the fact that the use of a single
source of information reduces the confirmatory power of analyses and
could lead to systematic or random errors—something we want to avoid.
Even relying on a single expert to judge the feasibility of a project is shaky;
one cannot possibly be an expert in all domains. It has been stated, along
those lines that "(...) obtaining data from multiple informants has been
recommended as superior to such an approach."[13] Observables should
produce data that are exhaustive (near or at saturation), discriminatory
(that are pertinent and nonredundant), and cohesive.

To be more precise with respect to modeling, there are actually three
ways of generating models. This can be done according to

1. Numbers, as in the case of financial models (parametric models
   using tables and rectangles)

2. Processes; as we have seen in the previous chapter, using parallelograms and arrows
3. Concepts, using bubbles and arrows

All three forms of modeling can work in tandem. In fact, I do not consider a feasibility analysis to be complete unless all three approaches have been used. Most often, as previously indicated, it is the People side of a project that is lacking (the psychological or sociocultural modeling) in feasibility studies. In the end, it is the most important one, given that POVs are to be detected and dealt with by People. Table 5.3 offers a view of the three modeling efforts that I deem essential.

Table 5.3 Types of analysis and codes

Type	Code	Type of analysis
Parametric	Tables and rectangles	Data
Processes	Parallelograms and arrows	Machine
Conceptual	Bubbles and arrows	Human

The various risks we have identified in Chapter 2 can be better evaluated using the modeling techniques shown in Table 5.4.

Table 5.4 Modeling preferences by type of risk

| Frame of analysis | Type of model | | |
	Parametric	Processes	Conceptual
Financial	•	—	—
Organizational	•	•	—
Environmental	•	•	—
Technological	•	•	—
Marketing	•	—	•
Sociocultural	•	—	•
Legal	•	•	—
Political	—	•	•

All these modeling techniques should serve to identify causality as best as possible, that is, to assess the critical chains that exist between data, between parallelograms, between bubbles, or between any of these three angles of analysis. As mentioned, this is important because it helps in identifying POVs. The level of causality will partly determine the level of vulnerability. Let us take, for example, an F1 driver. Assume his/her project, so to speak, is to make it to the finish line of the Monaco Grand Prix and rank in the top three drivers. To do so, his/her car must be filled with gasoline. No gasoline means that the car sits idle on the starting grid. Hence, the procedure to feed gasoline ahead of time into the car speaks of a point of vulnerability on two fronts: (1) gasoline allows the car to move and (2) the driver cannot even hope to win without gasoline in his/her car. If the left mirror breaks due to a benign collision during the race, this will lead to seconds being wasted when checking the oncoming competitors, but it may not compromise his chances of attaining a top position. That particular point of vulnerability has a null value. In a feasibility study, the analyst must understand the major tasks that are to be performed, identify the nature of their linkages (as per the four scenarios: S and F, T, I, and C), and decide which are critical to the success of the project, both in terms of processes and in terms of psychological efforts. Once this is done, they can highlight POVs. If, as a purely hypothetical example, it were possible for the F1 driver to repair his/her mirror during a six-second pit stop, then this would be a corrective measure that would alleviate the problem. Hence, it can be said that points of vulnerability augment in function of

1. The strength of the causal links that tie different activities together during the transformation phase
2. The value of these activities to the successful completion of the project
3. The capacity to implement corrective measures should things go awry

These three options imply some psychological inputs, that is, the presence and action of people.

5.4   Teams

PMBOK 5 often refers to teams as groups. Teams are formed by individuals, of course, each with their own way of interpreting and acting upon psychological constructs such as Trust and Collaboration. Teams (and more generally stakeholders) have been defined in various ways:[14] people whom the organization (the project) depends on; people who have an interest in the organization; people who can have an impact on the organization; people who bring necessary[15] skills and resources to the organization; people who produce value for the organization; as well as people characterized by power (capacity to get others to do what needs to be done), legitimacy (perception held that the behavior is desirable and appropriate), and a sense of urgency (or "criticality"). In short, we could state that teams maintain a symbiotic relationship with the project organization, whereby each one provides resources in order to achieve a value-creation goal that serves (in the best of scenarios) their mutual interests.[16]

More precisely, stakeholders (not only team members) have been defined[17] as people making and/or being affected by decisions as well as performing project-related activities. The PMBOK[18] states that "project stakeholders are individuals, groups, or organizations who may affect, be affected by, or perceive themselves to be affected by a decision, activity, or outcome of a project."

Team members are bound by all four types of links that exist between processes and between constructs: S and F, I, as well as T and C. That is, team members can be described by their structural variables (e.g., their competencies) and by their functional variables (what tasks they perform), by how they change with time (e.g., promotion), by how they influence each other (directly, indirectly), and by how one member exercises so much power that they could force a behavior onto another team member (causal), whether by means of authority, charisma, or other characteristics. This can make for a very complex system by itself; in typical process modeling, only one type of bond links two parallelograms, yet at first glance, it looks like all four bonds can exist between two team members. This is true; I will provide a model in Figures 5.4 and 5.5 that results from years of research in the field and that simplifies this state of affairs in a very logical way. Complexity must be acknowledged; indeed, there is an exponential demand for elaborate, global projects that requires specialized knowledge.[19]

Groups vary and evolve over the course of a project's life cycle.[20] The start-up group is active, of course, during the planning stage: it is responsible for preparing the project charter. As the name indicates, the planning group is mostly preoccupied with the planning stage. The execution group looks after mobilizing the resources, Forces, and Means of Production. It expends a variety of efforts: among many other tasks, it looks after corrective measures, information flows, plan modifications, and Work Breakdown Structures (WBS). It is by far the busiest of all the groups. By comparison, the surveillance group does not accomplish as many tasks; it spreads its efforts equally along the five stages of transformation. It approves the changes and proposed corrective measures, makes recommendations, prepares forecasts, produces follow-up reports, updates the plan when necessary, and verifies the deliverables. Finally, the control group (which can be the same as the start-up group as it has been inactive since the mobilization and deployment stages) makes sure that the outputs are completed; including the deliverables, the realization of the BOK, and the measurement of impacts.

This goes to show that team efforts during the project are unequal, which implies another set of complexities. From a processing point of view, tasks have to be lined up in an efficient way; from a psychological point of view, collaboration is incremental.

It must be noted that People on the inside (Forces of Production, FP) are inevitably in touch, directly or indirectly at times, with People on the outside (clients, suppliers, regulators, and even bad apples). This raises the level of complexity one notch. Indeed, a feasibility analysis cannot escape the requirement to look into the **work culture (psychodynamics)** of projects.[21]

A project's work culture (psychodynamics—PWP) refers to the psychological processes that are implied when dealing with projects given the triple constraints, and which are guarded by a particular work ethic.[22] It is useless to examine psychological constructs that do not specifically relate to the work implied in the realization of a project, such as, say, attachment. It is also pointless not to frame a construct such as Trust within the parameters of a particular work ethic: one that emphasizes discipline and rigor, for example, ought to command trust toward other team members.

From this perspective, the feasibility analyst does not merely wish to determine how people that participate in projects are going to aim for the same goal, to commit to their tasks, to trust each other, and to work together. He/she is rather keenly interested in identifying the work culture (psychodynamics) of projects, that is, how people assume their robust (Dominant and Contingency) and Short strategies. While I have provided the reader with some clues as to how to go about measuring Dominant, Contingency, and Short strategies (see the introductory chapter), there are also measurements that can be thought of for evaluating a project's work culture (psychodynamics). Appendix 5.2 offers one way of going about it, which can be adapted according to the analyst's needs. To give the reader a more concrete example, work psychodynamics can be illustrated by looking at people's general beliefs concerning the Germans: they are seen to work diligently, with discipline, and in conjunction with their coworkers. This contrasts with cultures where respect for time (calendar), rules, and procedures is deficient altogether. Poor psychodynamics contains the germs of failure.

5.4.1   Contemporary teams

The previous comment takes on a more powerful sense when realizing that contemporary teams are increasingly multidisciplinary, multicultural, and dispersed, relying more than ever on remote communication tools. An example of multidisciplinarity is the first heart implant: the device itself required expertise in battery, epoxy, silicone materials, simulators, surgical techniques, and transistors to name just these elements.

These three features of contemporary teams may signify a lack of synergy, encourage the appearance of dead spots within or between the stages of the project, and foster gaps and ambiguities,[23] all of which are fertile ground for POVs.

5.4.2   Small teams

To think that people are not conscious of their own vulnerabilities may be shortsighted. Many times indeed, people feel threatened by a newcomer, by someone who is more active or who seems to know more than them. Their routine is challenged and a sense of discomfort is induced, unwillingly and at times unbeknownst to the newcomer.

At the same time, as some authors comment, "people trust others, even when there is no guarantee that the trustee will respond benevolently."[24] The settings of projects whereby small teams are created and members are brought together in order to achieve a common goal calls for both trust in others and trust in oneself. This means accepting that one's own vulnerabilities do not impede on the project's success.

Interestingly enough, people are more prompt to trust when money is part of the equation: after all, management and team members (Forces of Production) are paid to accomplish certain tasks with the understanding that they must rely on others to fulfill them. However, even the scent of money has an emotional component;[25] psychodynamic forces such as trust and distrust are at play and influence decision-making, for example, when the time comes to collaborate with others in a small team environment[26] or under stressful conditions.[27] In fact, team members may put aside the best interests of the project in favor of noneconomic, socioemotional objectives such as nurturing an exaggerated sense of acceptance on the part of others[28] or holding a grudge.

The emotional side of decision-making is at rest when the environment is predictable and safe, such as is the case with functional small teams; indeed, the human brain seeks regularity even amid troubled times and limited information.[29]

Cohesive small teams present a number of advantages. First, they form a "bundle of resources, skills, and capabilities."[30] This allows for a faster response time in the case of adversity (when Short strategies are needed). Second, they generally offer excellent utilitarian value because efforts are not wasted: what one does is promptly checked by what the other person does. There is little room for escape, for free riding[31], or for other forms of abuse.[32] Thirdly, trust is often a given in a small team environment; this generates a higher propensity for commitment toward the project. Indeed, when it comes to financial decisions, for example, studies show that people invest money more readily in companies that they feel they share common interests with, than in other companies.[33] Finally, small teams encourage the building of powerful bonds: people simply relate more strongly to other human beings than to organizations.[34]

However, small teams also have downsides, some of which are well documented. These include a potentially dysfunctional ability to operate due to the tightness of bonds between team members, tunnel vision, and even egoistic interests, thus leading to inconsistent performance. In particular, the comfort that small teams provide prevents members from examining original solutions (diagonal processes) to particular concerns, especially those offered by external advisors, or to admit to mistakes.[35]

Overall, it can be observed that there are upsides and downsides with respect to small teams. Often, a project is composed of a handful of key people; the Québec Multifunctional Amphitheatre (QMA), for example, was managed by a team of six people only, even though it was a 400 million dollar undertaking. The feasibility analyst may be well guided to seek information on such core teams if at all doable. Elements that may be taken into account include the absence of malevolent intentions; an impetus for win-win situations; a history of working together on other projects; a proven ability to learn and to collaborate; a sense of attachment to the team due to common ambitions, desires, interests, and/or tastes; high levels of integrity; and the manager's personality. These are intrinsic behavioral components that cannot be assessed by merely looking at a curriculum; instead, observation and background inquiry may adduce reasonable information to the analyst.

One can guess that the Unfits (uncontrolled Forces of Production, FP_{nc}) fall short on one or every one of those characteristics. For example, they will not be able or willing to collaborate, they will act selfishly, or they will lack integrity. Detecting malevolent intentions—a prime source of team vulnerability—is a hard chore, yet hidden agendas[36] infest many projects. Unfortunately, even honest people may at times fall for potential ulterior, negative motives[37] because of stress factors or due to the influence of others. I will discuss corruption in more detail in Chapter 7, dedicated to Power.

5.5   *A model of interpersonal competencies*

As the reader can sense, there are a limited number of key psychological constructs that come into play in any project, especially when People become People' during the transformation phase, that is, when all People interact with each other[38] to produce the deliverables. These constructs are Control, Transparency, Trust, Fairness, Collaboration, and Commitment. Other constructs are permutations of these core constructs, such as Dependence and Distance.[39] Table 5.5 gives a succinct summary of the main themes found in a representative sample of scientific articles addressing project management.

I have conducted a large number of studies, both by examining current and past literature and by surveying and interviewing over a

Table 5.5 Most common themes found in project management literature

Theme	% of authors
Trust, including acceptance of others, affinities, technical competencies, conformity, team spirit, integrity, reliability, loyalty, open "mindedness," security, sensitivity to other cultures, transparency	29
Collaboration, including adaptation, collaboration, common goal, communication, flexibility, proactivity, reciprocity sharing, and team training	25
Control, including feedback and follow-up, leader commitment, leadership, performance, project definition, resource allocation, and task planning	21
Commitment	15
Fairness, including impartiality, neutrality, and win-win.	8.5
Cultural *distance*	2

thousand participants.[40] This effort has allowed me to propose a model that has been published many times over and which I present in the following sections.

Various authors recognize the importance of competencies[41] and have had their opinions on core competencies with respect to projects; they probably all have valid points. For decades, many others have conjured that competencies are what makes an enterprise competitive.[42] Some scholars[43] identify the following competencies: the ability to establish and to maintain appropriate contacts within and outside of the organization; the ability to tolerate ambiguity, that is, to function effectively when facing risks; the ability to train, to direct, and to motivate colleagues and subordinates; the ability to negotiate and to resolve conflicts; the capacity to identify the concerns of others; the capacity to take a broad view of the priorities and the project itself; leadership by way of the ability to bring together stakeholders and form effective working groups, as well as by way of defining and delegating responsibilities; and finally, by showing personal enthusiasm. PMBOK 5 (pp. 53–54) proposes that "decision-making (and...) trust building" are skills indeed.

I'd like to think that our six core competencies model offers a useful and workable summary of the different interpersonal competencies that various authors and field experts have acknowledged over years of research and experience. I believe it may assist in the hiring process, for example, and even speed it up.[44] The six core competencies model appears in Figure 5.4.

The model reads as follows: project managers must balance their Control of the project with a sense of Transparency, which is defined as the

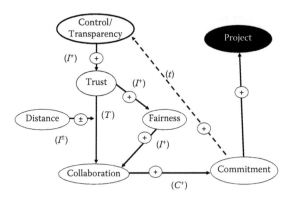

Figure 5.4 Six core competencies model.

sharing of useful information with others to facilitate their work and fire up their motivation. This balancing act between Control and Transparency has a positive influence on the team's willingness to Trust each other (I^+). Enhanced Trust helps build a spirit of collaboration that develops with time (T). This link between Trust and Collaboration is influenced by a Distance factor ($I^\pm$) that can play for or against the project: for some team members, physical distance is a way of avoiding being constantly interrupted, thus improving performance, while for others, it is a means of delaying the completion of preset project tasks. A sense of Fairness within the team and with management strengthens the spirit of Collaboration (I^+, I^+). As team members and more generally stakeholders learn to Trust each other and to work together, this encourages them to commit more and more to the project over time[45] (T). Commitment thus serves as a manifestation of an intrinsic motivation, which many authors recognize as a mandatory element of team building and project success.[46] The consequence of this process is a behavioral adherence to the project (C^+) within its three critical axes: time, costs, and norms of quality. There is a feedback loop (t) that departs from the Commitment construct and ends at the construct 'Control/Transparency'.[47] This corresponds to research by some authors[48] who have noticed that lack of commitment reduces the capacity to build relationships among team members. Some experts rightfully comment that "The involvement of stakeholders from all levels both within and outside the organization within the framework of the communication ecosystem is required to make the project successful."[49,50] Says one project leader:

> People like to have a good career and to work for a company where there is a future and a sense of leadership. People like to see a company building its own future. This spawns a lot of self-confidence.

> When our team members see growth, they are
> inclined to jump in and to embrace challenges.[51]

Note that the model I have presented will vary somewhat depending on the group that is being analyzed (as I demonstrate further on). This is normal; a model is a simplification of reality and is not meant to represent all conceivable real-life situations.

According to my model, the project manager displays three core competencies: (1) the capacity to exercise control; (2) the willingness to be reasonably transparent (e.g., by sharing critical information with staff when needed); and (3) the ability to be fair. Indeed, most people long for reciprocity and fairness in the way they are treated.[52] Of note, a sense of Fairness works hand in hand with Trust[53] and Collaboration.

Ideally, staff have the following interpersonal competencies: they are (1) trustworthy and self-confident, (2) willing and able to collaborate, and (3) committed.

I would venture to say that managers and staff that present such well-focused characteristics could surmount many obstacles and achieve a substantial amount of work. Together, these six competencies that can be shared between managers and staff instead of existing in silos between the two groups, imply a sound work culture (psychodynamic). Faithful employees are likely to complete their tasks, to show up on time, and to take pride in their work, all of which can be observables that reveal the existence of the construct of Commitment. I conducted hundreds of structural equation modeling (SEM) analyses to arrive at this model and I feel confident it can be of use to feasibility analysts.

Going back to Table 2.13 from Chapter 2, one can recognize that the six core competencies model is in line with past findings on Dominant strategies (refer to Table 5.6).

What my six core competencies model does is simply highlight the key competencies and their dynamic relationships, rather than listing them in a passive way. Another conceivable representation of the model is shown in Figure 5.5.[54]

The maximization of the utility of the Dominant strategy is achieved, in terms of work psychodynamics, by the optimization of the ratio Control/Transparency, and of Trust, Fairness, Collaboration, and Commitment. This makes intuitive sense. Unfits (uncontrolled Forces of Production) and bad apples (people who often act at the limit of legality or else who act illegally) will generate distrust, and will cause delays and additional costs, and/or product defects (g-spreads). As delays, costs, and defects mount, distrust grows tremendously. As the situation worsens, the probability of failure inflates. I bet that each one of us has experienced a similar dynamic in a given project or work environment. This seems to reflect what happens in the "real world."

Table 5.6 Main themes and underlying constructs

Characteristics (or KSF)	Underlying construct
Nonbureaucratic	Control/Transparency and Agility
Sound decision-making	Control/Efficiency in decision-making
Leadership	Control
Control measures improvements	
Adequate control procedures	
Excellent conflict management	Control/Transparency and Sound conflict management
Good management–employee relationships	
Trust	Trust
Management support	Trust/Supportive
Skilled staff	Trust/Competencies
Experienced project management	
Team work	Collaboration
Frequent feedback from parent organization	Collaboration/Communication
Good communication	
Empowerment	Collaboration/Sharing
Commitment	Commitment

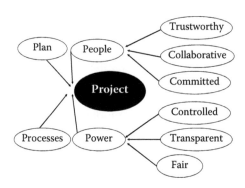

Figure 5.5 Stylized project competencies.

We now turn our attention to improve our understanding of some of the individual psychological constructs contained in the six core competencies model.

5.6   Trust

Outside risks twined with an inherent tendency to feel vulnerable results in being less trustful of others. Trust is intimately tied to vulnerability;

when I discussed this in terms of People in our four Ps model, we saw that this means that Trust and POVs go hand in hand. Multiple studies have defined Trust as the willingness or tendency to be vulnerable in the belief that the *vis-à-vis* (the person facing us, such as a team member) is well intentioned.[55] To that point, some authors have linked trust to vulnerability as follows:

> The definition of trust proposed in this research is the willingness of a party to be vulnerable to the actions of another party based on the expectation that the other will perform a particular action important to the trustor, irrespective of the ability to monitor or control that other party.[56]

For their part, other scholars state the following: "In our analysis, we define trust in terms of confident, positive expectations regarding another's conduct, and distrust in terms of confident negative expectations regarding another's conduct."[57] Some project leaders have said: "Projects drive trust, not the reverse."[58] Some others comment as follows:

> We like to make team members part of the decision process. We display a high level of flexibility. In this region of Clare (Nova Scotia, Canada), most of the population is French-speaking; many have been ordered to go to English schools as children, and were at some point in history segregated against, and even thrown out of stores for speaking French. Most Anglophones have not had the opportunity to learn French. Yet, we all communicate and understand each other despite the language differences; nobody is left out. This is the making of a great work team.[59]

Some authors[60] have dug into various sources, including documents from the United Nations, and stipulate that vulnerability refers to an inability (rather than a lack of willingness) to protect oneself or else refers to a susceptibility, incapacity, or even defenselessness when facing stressful events or pressures that can cause harm or generate a loss, whether emotional, material, physical, or social. Various scholars[61] elaborate on so-called affect-based trust whereby affect is an integral part of human behavior;[62] yet, other authors expand on so-called cognitive trust (trust based not on emotion but on reasoning).

Through my multiple studies, I have identified four structural sub-constructs of Trust, which are Affinity (or how closely team members fit with each other), Benevolence (how team members actually care for each other),[63] Ability (technical competencies characteristic of each team member), and Integrity.[64–66] Figure 5.6 illustrates the psychological construct of Trust.

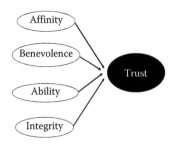

Figure 5.6 Trust.

The trick for the feasibility analyst is to measure Affinities, Benevolence, Abilities, and Integrity, and do this ahead of time. In most instances, this is not feasible, because trust is always directed toward someone (oneself or, in this case, others), so that the analyst would need to have seen the team members in action before he could commit to giving his opinion about the underlying work psychodynamics. Often, the best that they can do is to authenticate the curriculum of the candidates, check their backgrounds, distribute some psychometric tests, and guess. This opens a wide door for POVs to root themselves in a team setup. Thus, it is most advisable to prepare accordingly by way of thorough job description and superior human resource management techniques. If anything, any signs of social dysfunction, of egoistic tendencies, of shortages of skills, and of dishonesty indicate a real or latent Unfit (uncontrolled Force of Production, FP_{nc})—these must be dealt with before the project is launched.

My various studies have shown that Trust can be posted, in terms of behavioral patterns, on a scale that goes from Blind trust to aggression.[67,68] This is what I call the **barometer of Trust** (refer to Table 5.7).

As can be guessed, with lagging levels of Trust come an increasing number and higher salience of POVs. Note that Blind trust is just as dangerous as predatory behaviors (e.g., the likes of financial predation—recall the U.S. predatory mortgages of 2006–2008). I discuss this in the section on blind trust. Essentially, the ideal zone is the one of conditional trust: here, team members trust each other without falling into excesses while maintaining control over their feelings.

Table 5.7 The barometer of Trust

Behavioral pattern	Zone	POVs
Predation	Out of control	Dangerous
Aggression	Zone of conflict	Very high
Provocation		High
Frustration		Moderately high
Indifference	Indifference zone	Average
Random trust	Zone of comfort	Somewhat low
Conditional trust		Low
Blind trust		Dangerous

5.7   Pretrust

I conducted various longitudinal studies in line with my core model and found that Trust exists on the condition of some predisposition to Trust, or, put differently, to some level of Pretrust. This corresponds to observations made by some scholars, who state: "Successful behavioral exchanges are accompanied by positive moods and emotions, which help to cement the experience of trust and set the scene for the continuing exchange and building of greater trust."[69] One of my studies consisted of four small groups, which were measured longitudinally, at $time_0$ and at $time_1$. I provide some of the main results I obtained in the following subsections. Results were computed once the questionnaires were filled, before ($time_0$) and after ($time_1$), the completion of the project's tasks was by groups of approximately 20 individuals each. I acknowledge that the overall sample is too small to reach firm conclusions; the results are part of an exploratory effort.

5.7.1   Regression analysis

In Table 5.8, I combine the results of $time_0$ and $time_1$ for a four-person team's psychodynamics study that I performed a couple of years ago in order to get a larger sample.

A number of observations can be made with respect to the data I obtained looking at both $time_0$ and $time_1$ results taken individually or amalgamated (as per the graphs in Table 5.8). First, there is an initial endowment of Trust, which I call 'Pretrust'. Second, there is a strong correlation between Trust and Collaboration for both $time_0$ and $time_1$, and of course, for $time_0$ and $time_1$ combined. Third, the regression line forms a near perfect 'football' shape, which is a good indicator of a normally behaving population.[2] Fourth, there is a significant improvement in the correlation coefficient (Pearson coefficient or R^2) from $time_0$ to $time_1$ (Group 1: 0.615 to 0.713: +19%; Group 3: 0.652 to 0.825: +27%; Group 4: 0.549 to 0.796:

Table 5.8 The combined *time*₀ and *time*₁ groups (*n* = 103)

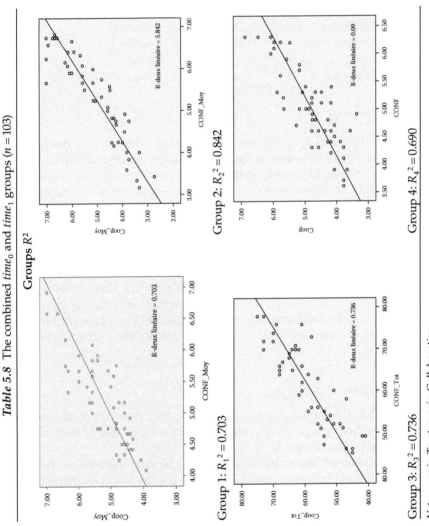

Note: x-axis: Trust; y-axis: Collaboration.

+45%) except for Group 2 (0.890 to 0.796: −11%). For the team members, this means that working together over time has fortified the correlation between Trust and Collaboration: in other words, team members became more trusting of each other and collaborated more as time passed. Indeed, trust takes time to build (*T*). This is in line with past findings.[71] This observation reinforces the comment made by some authors[72] who notice that for any team, trust-damaging occurrences generally weigh more heavily than positive events; this explains the difficulty of building and maintaining effective teams.

Another observation is that the correlation coefficient meets that of previous similar studies performed in suppliers–clients' contexts.[73] A mega-study indicated that 90% of all studies that were examined confirmed the bond Trust => Collaboration.[74,75]

5.7.2   *Factorial analysis on apprehension*

Throughout this chapter, I will perform a number of statistical analyses,[76] including factorial ones, as they are considered a key measure in a psychological investigation.[77] I use a varimax rotation with 25 iterations on all constructs except where indicated. For all research I did, except where specified, I resorted to the psychological construct of **Apprehension**[78]: in my view, it is formed by the ratio of a 'Defensive position' (one's sense of vulnerability—labeled as 'DP') over an 'Instrumentally hostile position' (one's ability to attack—labeled as 'IP'). Apprehension is not the opposite of Trust: Trust is an emotional time-building experience, whereas Apprehension is a momentary perception of present dangers. Apprehension does not encourage Trust, of course. Figuratively, one can view Apprehension as the ratio between a lonely sheep position and a hungry tiger position. Most likely, the more one feels like a lonely sheep when facing a hungry tiger, the more apprehension builds up. Apprehension is defined in this book as the sense that an individual has that he/she may be the target of the action of others, especially of their negative intentions. In the present book, the human threat itself is the act of taking advantage of another person's vulnerability for one's own benefit, causing the other harm (e.g., loss), *by surprise.*

My definition of Apprehension differs from that given by other authors on the subject, or on the subject of a perceived threat. For example, some scholars see a perceived threat as being caused by a trigger (and maybe by a hungry tiger as well!). This generates emotions, which then influence the individual in his attempt to avoid the source of said perceived threat, which then leads to an increased perceived threat when escape is not possible.[79]

Note that my focus is on the internal mechanism of Apprehension for which it is the ratio of Defensive over Hostile positions that counts, and whereby the surprise element cannot be ignored. Indeed, in wildlife (including in universities!), a perceived threat is all about the creatures assessing their vulnerabilities when facing danger and ensuring that they are not caught by surprise, hence their heightened level of vigilance. "Lonely sheep versus hungry tiger" is the title of that movie!

Various authors propose their own models, with one in particular that suggests that Apprehension (perceived threat) is a force to contend with when discussing human behaviors.[80] The latter estimates that risk perception is skewed by information (e.g., quality of feedback), personal (e.g., education, experience, and proclivity to assess the risks), the project (e.g., availability of alternatives), and indirectly by organizational factors (e.g., slack resources). Risk perception would then lead to allegiance to the project. This model emphasizes the role of perception as a focal point where various elements play a role. With reference to so-called personal factors, other authors[81] recognize that they influence a project outcome. We can generalize by stating that the perception of risks and POVs has an impact on a project's outcome. To be more precise, it is the perception (and not the reality) that People have about risks and their own vulnerabilities that influences their behaviors and that dictates the levels of trust and collaboration that they are willing to allocate to a team.

In particular, the construct of Apprehension expresses how a team member feels when management drives the project with exaggerated control or not enough transparency:[82] the two constructs (Control and Transparency) are intimately linked with one pertaining to the psychological aspect of projects (Apprehension or the ratio of Defensive/Hostile positions. Apprehension is organized around a feeling of fear—fear of others, fear that the project may go astray, and so forth—with fear being the sentiment that accompanies a sense of vulnerability. Thus, Apprehension is, from a psychological perspective, an internalization of personal POVs and their accompanying urgent need for a response; failing that, the integrity of the individual will dissolve (C^+). From a work psychodynamics perspective, a Dominant strategy takes place when the project team member or manager arrives at an appropriate ratio of Hostile/Defensive positions (the reverse of Defensive/Hostile positions), which I have found to be the constant value of $k = 1.3$ as I will show as I progress through this chapter. Note that the constant k is found given some small variances in each of the teams that I approached, and that passed two completely different questionnaires that tested for the construct of Apprehension. For the sample just discussed, I obtained the results shown in Table 5.9 when running the factorial analysis; two subgroups (A and B) were created, which stand in sharp contrast, from a behavioral point of view.

Table 5.9 The four combined groups: Factorial analysis (n = 103)

Construct subgroup	Group 1		Group 2		Group 3		Group 4	
	A	B	A	B	A	B	A	B
Trust	0.909	0.200	0.916	0.256	0.945	0.012	0.902	0.026
Fairness[136]	0.901	0.016	0.876	0.083	0.895	-0.040	0.921	0.059
Collabor.	0.946	-0.006	0.874	0.400	0.917	0.260	0.884	0.269
Hostile (IP)	0.167	0.914	0.428	0.806	0.123	0.871	0.215	0.821
Defen. (DP)	-0.613	0.540	0.017	0.952	-0.111	0.897	-0.112	0.886
IP versus DP	IP> DP	IP> DP	IP> DP	IP< DP	IP> RP	IP< DP	IP> DP	IP< DP

If we round up the numbers (to the nearest of 0 or 1) to get a clearer view, we obtain Table 5.10.

Table 5.10 The four *combined* groups: Factorial analysis ($n = 103$)

Construct	Group 1		Group 2		Group 3		Group 4	
Trust	1	0	1	0	1	0	1	0
Fairness	1	0	1	0	1	0	1	0
Collabor.	1	0	1	0	1	0	1	0
Hostile (IP)	1	1	1	1	1	1	1	1
Defen. (DP)	0	1	0	1	0	1	0	1

I note that there is an inverse relationship between one's lack of self-confidence (the impression of being a victim; i.e., Apprehension) and the constructs of Trust, Fairness, and Collaboration. I refer to this as the **law of Apprehension:**[83] the more one feels that he/she is a victim, the less likely they are to trust, reciprocate, and collaborate. Most notably, every single time one team member feels that he/she is in a position of control (Instrumentally hostile position > Defensive position), the values of Trust, Fairness, and Collaboration near the rounded number 1. As soon as the individual feels that they are at a disadvantage (he/she feels more like prey, he/she is more defensive—i.e., more vulnerable—than hostile; in other words, as soon as [Hostile position < Defensive position]), the values of Trust, Fairness, and Collaboration tend obligatorily to diminish radically, toward the rounded number of 0. Even when [Hostile position = Defensive position], those values near zero. Hence, people are very cognizant of their own POVs and this drives their behavior.

Thus, POVs evidently not only exist with respect to Plans and Processes, but also with respect to People, as a psychological phenomenon deeply rooted in a person's identity.

5.7.3   *Multiple linear regressions on a small group study*

I ran multiple linear regressions in order to measure the four structural variables that form Trust (Affinity, AFF; Benevolence, BV; Abilities, AB; and Integrity, IN) with their relation to the general work psychodynamics (how well people got along) and obtained the results shown in Table 5.11.

I observe that the majority of the explanatory variables of Trust are emotional in nature (Ability and Integrity are judged cognitively as they can be measured, somewhat; e.g., Ability can be assessed with a university degree). The emotional variables—Affinity and Benevolence—cannot be as easily measured. This suggests that a large part of a team's effort is based on emotions rather than on purely cognitive appraisals of situations and of others.

Table 5.11 The *combined* groups: Multiple linear regression analysis of the structural variables ($n = 103$)

	Standardized coefficients		Standardized coefficients		
	B	Std Error	Beta	t	Sig.
Group 1					
(Constants)	7.376	2.001	—	3.686	0.001
Avg AFF	0.621	0.352	0.246	1.763	0.084
Avg BV	1.048	0.474	0.407	2.212	0.032
Avg AB	0.208	0.416	0.073	0.501	0.619
Avg IN	0.044	0.388	0.016	0.112	0.911
Group 2					
(Constants)	6.034	1.820	—	3.315	0.002
Avg AFF	−0.037	0.182	−0.035	−0.204	0.839
Avg BV	−0.208	0.152	−0.234	−1.364	0.180
Avg AB	0.596	0.184	0.611	3.241	0.002
Avg IN	0.314	0.141	0.363	2.231	0.031
Group 3					
(Constants)	3.596	1.238	—	2.906	0.006
Avg AFF	0.250	0.010	0.277	2.464	0.018
Avg BV	0.393	0.120	0.455	3.267	0.002
Avg AB	0.058	0.130	0.068	0.447	0.657
Avg IN	0.152	0.113	0.170	1.344	0.186
Group 4					
(Constants)	2.707	0.589	—	4.599	0.000
Avg AFF	0.370	0.102	0.510	3.642	0.001
Avg BV	0.190	0.158	0.188	1.207	0.233
Avg AB	−0.121	0.133	−0.130	−0.913	0.366
Avg IN	0.112	0.123	0.133	0.911	0.367

Note: Dependent variable: General work psychodynamics.

5.7.4 Values of the main constructs

I transformed the values found on the questionnaire used to assess the psychodynamic structure of the four groups from a seven-point Likert scale to a percentage equivalent, a strategy that makes the reading of data easier for most of the main constructs (Table 5.12).

I present the following observations: first, for the groups at $time_0$, at $time_1$, and the two groups combined, the level attained by each construct is

Table 5.12 Average values of main constructs ($n = 103$)

Construct	$Time_0$	$Time_1$	Difference % (rounded)
Trust	72	75	4 (n.s.)
Fairness	70	72	3 (n.s.)
Collaboration	69	78	12

Note: n.s. = not significant.

around 70%. This suggests a so-called relational dynamic. For more on this, see Appendix 5.3, which provides the standards based on years of research. It is enough to mention here that a relational work culture (psychodynamics) implies polite, yet not overly formal interactions, a limited disclosure of personal life and sentiments, and a work-conducive atmosphere.

Second, there has been a positive change between $time_0$ and $time_1$, with the construct of collaboration having much improved (from 69% to 78%).

When we compare the construct of Apprehension (the feeling of being vulnerable) between the two time periods, we obtain Table 5.13.

I make two observations: first, as can be seen from Table 5.13, k' or the ratio between Hostile and Defensive positions (the reverse of Apprehension) is near 1.3. It is when $k = 1.3$—as we have found over the course of many studies—that the levels of Trust, a sense of a win-win relationship (Fairness), and Collaboration are most functional within the realm of project management. Second, the value of k' (Hostile/Defensive positions) has gone up over time (from $time_0$ to $time_1$): this suggests that team members have gotten accustomed to one another and that they experience a lesser sense of vulnerability.

5.7.5   Cluster analysis

I ran a cluster analysis to see if so-called classes could be inferred from the four groups. I did indeed obtain two classes, as shown in Table 5.14.

I note that the two classes show that when the value of the Hostile position/Defensive position ratio augments (from $k' = 1.41$ to $k' = 1.65$), the values of Affinity, Trust, and Collaboration increase. In fact, the reverse dynamic occurs as well: as team members find common affinities, develop trust, and collaborate over time ($time_0$ to $time_1$), they gain in self-confidence, they feel less vulnerable and more in control of the work psychodynamics. In this sample, there are two classes of people, with one being slightly more confident than the other.

5.7.6   Structural equation modeling

I ran a tentative SEM, the results[84] of which are shown in Figure 5.7.

Note that the indices for the SEM are shown in Table 5.15; they show that the model provides a reasonable fit (especially with respect to the CFI and IFI) compared with what was expected.

Table 5.13 The four groups *before* and *after* the
negotiation ($n = 103$)

K' (the reverse of apprehension)	$Time_0$	$Time_1$	Difference % (rounded)
Group 1			
Value (k_1')	1.27	1.82	43
Maximum	2.83	4.00	41
Minimum	0.71	2.00	182
Standard deviation	0.52	0.77	48
Group 2			
Value (k_2')	1.22	1.43	17
Maximum	2.29	2.78	21
Minimum	0.86	0.92	7
Standard deviation	0.31	0.45	45
Group 3			
Value (k_3')	1.40	1.44	3
Maximum	3.40	2.00	−40
Minimum	0.92	0.90	−2
Standard deviation	0.56	0.35	−38
Group 4			
Value (k_4')	1.24	1.33	7
Maximum	1.89	2.45	30
Minimum	0.77	0.77	0
Standard deviation	0.30	0.43	43
Average			
Value (k_{avg}')	1.28	1.50	17
Maximum	2.60	2.81	8
Minimum	0.82	1.15	40
Standard deviation	0.42	0.50	19

I make the following observations: First, all four structural variables of Trust (Affinity, Benevolence, Abilities, and Integrity) play a role at some point or another ($time_0$ or $time_1$) in the work psychodynamics. Secondly, Affinity's role emerges during $time_1$, once team members have had a chance to work together. Thirdly, initial Trust (at $time_0$) (with a β value in a linear regression that I found to be $\beta_0 = 0.3$) is a key influence on current Trust (at $time_1$). Trust plays an enduring role in work psychodynamics and is built over time.[85] Lastly, Integrity is the only structural variable that

Table 5.14 Cluster analysis ($n = 103$)

Construct	Class 1	2	Diff. in %
Time_0			
Affinity	64	81	27
Trust	64	75	17
Collaboration	64	72	13
Time_1			
K' or actual Hostile position/Defensive position; (the reverse of Apprehension)	1.41	1.65	17
Affinity	13.12	17.61	34
Collaboration	4.71	5.75	22
Number of valid participants (25 are missing)	41	37	—

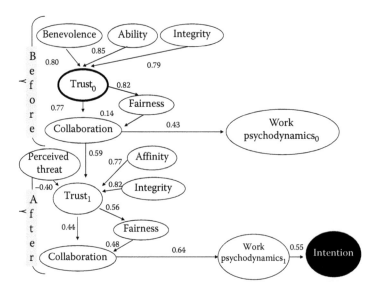

Figure 5.7 The revised working model with SEM ($n = 103$).

appears in both $time_0$ and $time_1$, pinpointing the importance of this cognitive variable.

5.7.7 Discussion

My research on four small groups tested at $time_0$ and $time_1$ shows that Pretrust affects Present Trust. In fact, it is Pretrust that most influences present work psychodynamics (PWP). Collaboration is the key direct indicator

Table 5.15 SEM metrics and actual values (*n* = 103)

Adjustment indices (*fit*)	Key value	Actual values
Absolute Index		
GFI	>0.9	0.833
Incremental Index		
CFI	>0.9	0.900
IFI	>0.9	0.902
Parsimony Index		
PCFI	Lowest comparative value	0.681

of the general psychodynamics prevailing among the team members. Trust is hard to measure; *a contrario*, acts/gestures of collaboration can be easily assessed with such measurements as the flow of communication, the respect of time lines for meetings, task completion, and so forth. Thus, for a feasibility analysis expert, key measurements of a work culture (psychodynamics) that present the most objective evaluation are based on the way that tasks are completed and whether they are indeed completed—this can even be assessed through psychometric testing before beginning the project. Collaboration is the most essential valuation tool of work psychodynamics within a project environment, because it is the easiest to measure. Regarding the Trust construct, the measurement is most easily done in a work psychodynamics context by way of an assessment of abilities, and by checking the background of employees and that of management.

Overall, my observations point to the fact that emotions play a large part in a work culture (psychodynamics). This is something we will see time and again in the remainder of this book.

5.8   Interdependence

I have discussed the fact that People must trust each other when working together on a project. Essential components of PWP (work psychodynamics) include Trust and Collaboration. In fact, people develop a feeling of attachment to their colleagues, to their jobs, and to their work. They build an emotional aura around their daily tasks that nurtures a sense of accomplishment. From a processing point of view, People that work along the different transformation stages of a project have no choice but to rely on others to accomplish their tasks correctly; this includes completing tasks at the right time. There is a level of interdependence (most dramatically expressed by causative bonds [C]) between people, and between Forces and Means of Production.

My research as well as that of others[86] has shown that dependence serves as a moderating variable ($I^\pm$) between Trust and Collaboration. That is, dependence may slow down the formation of trust and collaboration, or the reverse: it may promote it. Contexts and individuals are ultimately the ones that decide what form the bond of influence takes: positive or negative. I speculate that together with trust, dependence participates in building the feeling of attachment.[87,88]

Interdependence increases the stakes of any chain of tasks. The fact that Forces of Production depend on others makes them vulnerable; as they become vulnerable, they generally tend to trust less (or else the opposite, to trust more in order to obtain favors). An interdependent bond between one team member (Force of Production) and another who is unreliable shelters conflicts;[89] this is especially true when this concerns access to resources. One project manager says:[90] "Conflicts happen when resources are rare." Rifts, in turn, tend to make people even more unreliable.[91]

5.9   Blind trust

Some authors[92] convincingly explain that trust has a downside: it may actually impair the perception of what is often called 'moral hazard' (e.g., in the financial industry), or what are often called 'threats', thus increasing one's vulnerability. This means that trust can sometime go "too far."[93] One of the ways by which people tend to go too far, according to other authors,[94] is that they trust either in full or else not at all: there's hardly half measures when it comes to trust because a half stance is a counterproductive one. It is a little bit like being half pregnant! Constantly being overly vigilant obliges management and the team members (the Forces of Production) to spend energy that could be directed more profitably to deal with real problems rather than with their apprehensions; by the same token, not being vigilant opens the door to abuse and to potential negative outcomes. Trust that is pushed to the limit, that is, to its blind spot, offers artificial relief in the face of danger and signals excessive dependence. In all cases, this generates difficulties somewhere along the transformation process, as the phenomenon of groupthink quite accurately exemplifies. As seen before, a groupthink phenomenon occurs when members of a group narrow down their decision-making options with the feeling that they hold the ultimate truth: they have Blind trust in themselves. This situation is very dangerous, of course.

To illustrate Blind trust, we need to revert back to the U(DS) curve. Recall that this curve is a utility curve, which is the result of balancing external threats and internal POVs given a utility drawback, that is, given the triple constraints. This curve describes a function of utility because it serves to smooth out the pressure exercised by risks and POVs upon the system that is a project. Let's transfer this scenario and replace Process with People, so that the U(DS) expresses the ability of managers (Forces of Production) to assume a Dominant strategy through the interplay of risks and POVs. We have seen that a manager can be blind in the sense that he could discard completely any personal vulnerabilities or else that he chooses to ignore all external risks. We have also seen that the entire project system is contained within boundaries—the triple constraints—and we have expressed these boundaries along the x- and y-axes by the value of 2.3 (or $1 + k$). Recall that we represented this as shown in Figure 5.8.

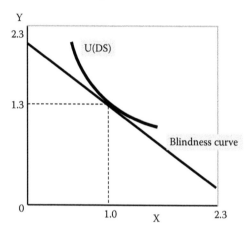

Figure 5.8 U(DS) and the Blindness curve.

As mentioned, Blind trust happens when Trust is pushed to a value equal to or superior to 2.2 on a scale that has a maximum limit of 2.3—this is 96%, and we know that we have an allowance of 4% POVs at the minimum in each project. We will discuss the construct of Trust in the next section and see how it relates to the construct of Collaboration.

5.10 Collaboration

My six core competencies model posits that Trust is intimately linked to Collaboration. Collaboration is, in a sense, the behavioral expression of Trust. In fact, tough times may even reinforce the bond between Trust and Collaboration by bringing people closer together. It has been said that cooperation (or Collaboration in our context of project management) "captures the level of coordinated and complementary actions between exchange partners in their efforts to achieve mutual goals."[95] I adopt this definition.

My research has shown that Collaboration is expressed by four functional variables, as displayed in Figure 5.9.

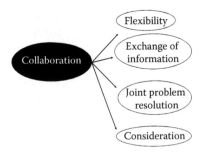

Figure 5.9 The four functional variables of Collaboration.

Recall that psychological constructs are illustrated by bubbles and not by parallelograms. Functional variables are expressions of a construct, generally have a high level of colinearity, and are not *sine qua non* conditions for the meaning or the action of the construct, although obviously having all four variables in this case only makes for stronger collaborative efforts.

Flexibility (FL) alludes to the capacity to adapt to change, knowing change is part of the definition of any project, since projects entail transformation. Exchange of information (EI) refers to communication in any form that is required, joint problem resolution (PR) speaks of a consensus and a capacity to find solutions to dire situations, while so-called Consideration (as in 'being considerate', CD) is the willingness to listen to and acknowledge the opinions of others (to support them).

I know from my past research that Collaboration and Trust are linked in the following manner:

$$\text{Collaboration} = 0.3 + 0.9\,\text{Trust} - \varepsilon \tag{5.1}$$

Or, in k terms:

$$\text{Collaboration}_t = (k-1) + 3(k-1)\,\text{Trust}_t - \varepsilon \tag{5.2}$$

Where $\beta_0 = 0.3$ corresponds to the initial endowment of Trust, that is Trust_{t-1} or, to use the terminology favored before, 'Pretrust'. Note that Trust_t can never reach the maximum value of 2.3, because that would force Collaboration to exceed its maximum value of 2.3 (the upper limit in our model). This is why, at a value of [$\text{Trust}_t = 2.3$], I refer to Blind trust. In fact, Blind trust is located in the area of [$2.2 \leq \text{Trust}_t \leq 2.3$].

This means that present collaboration is dependent on previous experience of trust (or pretrust). This seems normal—rarely will people start collaborating with each other without getting a sense as to whether these other people can be trusted or are deemed to be reasonably reliable. Thus, there is an initial capital of goodwill[96] that is required before collaboration takes place, whether this goodwill results from past experience or from positive hearsay about the *vis-à-vis* (the person one is collaborating with) is a matter of circumstances.

The Blindness curve is simply Equation 5.2, which has been reversed within the boundaries of the closed dynamic system of the project, with maximum boundaries being valued at 2.3 (or $1 + k$). I hypothesize that people trust and collaborate in order to confront risks and POVs.

Given that [Collaboration $= 0.3 + 0.9$ Trust] and knowing the boundaries of the system, I obtain the following:[97]

$$\text{Blind trust} = 2.2 - 0.9 \text{ Negative trust} \qquad (5.3)$$

Or

$$\text{Blind trust} = \left(k + 3\left(k - 1\right)\right) - 3\left(k - 1\right) \text{ Negative trust} \qquad (5.4)$$

With "3," of course, being actually [10 (k − 1)].

The reverse of the Trust $=>$ Collaboration curve[98] is simply stating that Collaboration abates with the reverse of Trust, that is (for simplification purposes), with negative trust (trust for which I have flipped the x-axis). As negative trust amplifies (the feeling that one is exposing his vulnerabilities to a menacing *vis-à-vis* situation), it is much better to limit one's exposure to risks (not to collaborate with those about to abuse one's vulnerabilities). Another way of looking at this is that the manager who completely ignores risks (or else all POVs), that is, who is at the extreme of the y- or x- axes, one way or the other, is not likely to generate trust or else to foster a good collaborative spirit on the part of the team members. He/she may be seen as a fanatic that is best not obeyed. He/she becomes an Unfit (an uncontrolled Force of Production, FP_{nc}). Note that because I have set boundaries (the project being a closed dynamic system), I am able to produce curves that all relate to each other (as we shall see in the next section—see Figure 5.13, for example) and that can all be plotted on the same graph, around the constant k. My model is indeed multidimensional.

5.11   Distance

Nowadays, many projects include teams that are physically far apart from each other.[99] At times, physical distance is compounded by time zone differences[100] as well as by cultural and technological factors. Distance may dilute a group effort. Yet, ultimately, a solid group effort produces, in general, good performance,[101] which is what the project manager is after.

Through my numerous studies, I have found the construct of Distance to be a moderating variable between Trust and Collaboration. For some people, it makes communication easier because they do not like face-to-face encounters or are too busy to pay attention to them, while for others it is a hindrance in their capacity to perform.[102] In general, though, distance plays against group cohesion.[103] Recall that the QMA chief manager had requested that his team members work out of the same office—the building actually facing the construction site—precisely to promote more efficiency.

To test Distance and its role in project management, especially given that more and more projects are spread around the world and include people from different backgrounds, I conducted a study with some 120 participants from 13 different nationalities.

5.11.1   Discriminant analysis

I did not have enough participants representing each country for the sake of statistical analysis, but for cases where I did, I ran a discriminant analysis, the results of which are given in Table 5.16.

Table 5.16 Discriminant analysis and equality of means ($n = 120$)

Number of respondents per nationality[137]	38	10	10
Control	84	79	80
Transparency	65	60	62
Trust	83	78	76
Fairness	82	79	79
Collaboration	85	78	79
Commitment	83	85	77
Distance	72	73	57

Note: Equality of means (ddf1 = 12; ddf2 = 71).

	Wilks lambda	F	Sig.
Control	0.835	1.168	0.322
Transparency	0.770	1.767	0.071
Trust	0.810	1.389	0.192
Fairness	0.887	0.752	0.696
Collaboration	0.839	1.136	0.346
Commitment	0.822	1.278	0.251
Distance	0.691	2.649	0.005

Note: (ddf1 = 12; ddf2 = 71).

Interestingly, only Distance generates a significant difference (at $p < 0.05$) between the constructs.

5.11.2 Factorial analysis

Table 5.17 provides the factorial analysis (varimax rotation, three iterations) that shows that Distance helps in separating groups of constructs.

Table 5.17 Factorial analysis with respect to Distance ($n = 120$)

	Without distance		With distance			
			Component			
	No distance factor	Tends toward	Component with distance factor			
			1	Tends toward	2	Tends toward
Control	0.918	1	0.928	1	−0.064	0
Transparency	0.770	1	0.792	1	−0.172	0
Trust	0.790	1	0.765	1	0.295	0
Fairness	0.868	1	0.855	1	0.183	0
Collaboration	0.932	1	0.937	1	−0.019	0
Commitment	0.804	1	0.784	1	0.237	0
Distance	—	—	0.037	0	0.963	1

I observe the following: First, when Distance is not introduced, only one component is formed, and when Distance is introduced, two components appear, with one being radically different from the other. Secondly, when the value of Distance tends toward 0, all other construct values tend toward 1, but when it tends toward 1, all other construct values approach 0. Distance makes trust, fairness, collaboration, and commitment difficult. Third, there is an intricate relationship between the construct Control/Transparency (or put differently, the reverse of Apprehension[104]) and Distance. A high value of Distance sees low values of Control and Transparency taken individually, and vice versa.

Note that the construct Control/Transparency is to management what the construct Hostile position/Defensive position is to subordinates. Indeed, when employees face a manager who exercises too much control and who is not willing to be transparent, the latter triggers some distrust, just as when an employee faces another employee who seems to want to abuse his/her own vulnerabilities will raise suspicion. Both the ratio Transparency/Control (the reverse of Control/Transparency) and Defensive/Hostile behaviors express somehow the construct experienced by the individual at the receiving end of the interaction: Apprehension.

5.11.3   Cluster analysis

I performed a cluster analysis (25 iterations) with the construct of Distance not being introduced. This reveals three classes, as shown in Table 5.18.

Table 5.18 Cluster analysis without the construct of Distance

Class		1	2	3
Number of valid observations		47	25	30
Control	128.225 (0.000)	79.92	72.38	89.30
Transparency	64.557 (0.000)	61.11	53.73	69.37
Trust	42.390 (0.000)	80.25	73.41	88.41
Fairness	88.923 (0.000)	80.26	70.66	87.09
Collaboration	123.196 (0.000)	81.06	71.72	90.95
Commitment	50.780 (0.000)	79.21	71.88	87.35
Distance	—	—	—	—

Note: F value, (*p* value); for class (df = 3) and error (df = 98) means squares.

I observe that three classes are obtained. They fall within 'relational' individuals (class 1), 'transactional' individuals (class 2), and people who favor close relationships or 'interpersonal' class 3 individuals (see Appendix 5.3).[105]

With Distance being introduced, four classes are formed (refer to Table 5.19).

Table 5.19 Cluster analysis with the construct of Distance

Class		1	2	3	4
Number of valid observations (respondents)		36	21	20	25
Construct		**Values**			
Control	52.169 (0.000)	78.10	72.94	83.99	88.87
Transparency	31.103 (0.000)	59.30	53.97	66.27	68.10
Trust	24.059 (0.000)	80.56	72.77	80.83	88.57
Fairness	66.218 (0.000)	79.27	69.46	82.78	87.32
Collaboration	65.669 (0.000)	79.50	71.10	85.93	90.33
Commitment	41.451 (0.000)	78.93	71.12	79.27	88.80
Distance	48.323 (0.000)	79.76	65.53	55.66	77.95

Note: F value, *p* value (<0.05). For class (df = 3) and error (df = 98) means squares.

I observe that class 1 and class 3 are both relational; however, they differ significantly on the value of Distance (79.76 vs. 55.66). Distance appears to be a key ingredient in the formation of classes.

5.11.4 Regression analyses

I checked for the normality of populations and residuals; I find that many constructs show significant linear regressions between the variables as determined by their R^2 (Table 5.20).[106]

Table 5.20 Various regressions

Regression	R^2	ANOVA F (p value) $\alpha = 0.05$
Control => Transparency	0.515	106.084 (0.000)
Control => Trust	0.429	75.220 (0.000)
Transparency => Trust	0.244	32.348 (0.000)
(Transparency/Control) => Trust	Not significant	—
Distance => Control	Not significant	—
Distance => Transparency	Not significant	—
Distance => (Transparency/Cost)	Not significant	—
Distance => Trust	Not significant, but shows an inverse triangular distribution, suggesting a possible moderating role	
Trust => Collaboration	0.449	81.403 (0.000)
Trust => Fairness[138]	0.388	63.523 (0.000)
Fairness => Collaboration[139,140]	0.626	167.481 (0.000)
Distance => Collaboration	Not significant	—
Distance => Fairness	Not significant	—
Collaboration => Commitment	0.493	97.202 (0.000)
Trust => Commitment	0.361	56.509 (0.000)
Fairness => Commitment	0.440	78.723 (0.000)
Distance => Commitment	Not significant	—
Commitment => Control	0.498	99.051 (0.000)
Commitment => Transparency	0.175	21.281 (0.000)
Commitment => (Transparency/Control)	Not significant	—

I observe the following: First, single linear regressions adjust well with the model that is displayed in Figure 5.10, resulting from a structural equation modeling (SEM) analysis. Secondly, an interesting relationship occurs between Control and Transparency. The ratio of Control over Transparency provides the following average value: 1.32, which is the ideal *k*-value. Values for the ratio Control/Transparency are overall within a functional range (mean = 1.32, standard deviation = 0.12, minimum = 1.10, and maximum = 1.59)—see Appendix 5.3. These findings suggest that managers generally wish to maintain a higher level of Control over the project than that of Transparency. On average, stakeholders wish to have 30% more

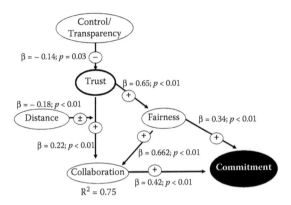

Figure 5.10 Emerging model of international core competencies versus physical distance.

Control versus Transparency (as k-value $\approx$ 1.3). Finally, my analyses tend to show that Distance acts as a moderating variable, a comment that matches some past research; indeed, some authors[107] allude to the fact that physical Distance serves as a moderator between leadership and performance.

5.11.5 *Structural equation modeling*

I performed an SEM using WarpPLS4 because of the relatively small number of participants and the need to assess the moderating role of the Distance construct. I obtained the emerging model shown in Figure 5.10.[108]

This model suggests that too much control and not enough transparency on the part of project management leads to a reduced sense of trust. However, with a well-balanced share of control and transparency, trust builds up. As trust develops between team members and between them and management, collaboration efforts step up. A sense of fairness reinforces positive feelings. Physical distance serves as a moderator between Trust and Collaboration. Commitment (a proof of motivation) is a natural outcome of a positive environment imbued with trust, fairness, and collaborative efforts. Of course, culture is a major force: it affects all of the aforementioned constructs; indeed, some authors conclude that the success of an international project is linked to the abilities of the project manager, their level of imagination and flexibility, and their capacity to deal with the constraints or peculiarities of foreign culture.[109]

5.11.6 *Discussion*

I believe that the results of my investigation make sense: generally speaking, it is harder to instill trust when one cannot see the other person, that is, when there is distance separating the two parties. For some employees, however, distance offers a shelter: they can then use it as an excuse for not performing or for delaying responses (FP_{nc}).[110] Distance is indeed a moderator, serving opposing ends depending on context. This may explain why some authors do not agree on its actual role.[111] For the most part, my results indicate that Distance exercises a negative influence on the ability to collaborate given a certain level of trust.

Key project-related competencies are part of a dynamic system. These competencies are linked together by specific bonds, be they direct or indirect, moderating ($I^{\pm}$) or mediating (I^{+}, or I^{-}). Logically, distance requires better information/communication systems;[112] this becomes more and more evident as teams are increasingly dispersed, multicultural, and multidisciplinary.

5.12   Fairness and DS revisited

My database indicates that the participants who filled out my questionnaire share a number of characteristics: (1) their Instrumentally hostile (IP) level is on average higher than their Defensive level (which means that they don't feel overly vulnerable and hence that they are most probably functional); (2) their Instrumentally hostile (IP) level is not too high—I know that at a level of or above 4.6, individuals tend to become dysfunctional (a little bit too hungry as hungry tigers!); (3) their Instrumentally hostile/ Defensive ratio (a proxy of self-confidence) approximates the constant k with a value of 1.3 or thereabouts; (4) their levels of Trust, sense of a win-win relationship (a sense of Fairness), and Collaboration take place in a relational[113] as opposed to an interpersonal[114] or a transactional mode.[115] This also points to a functional group of respondents who are not overly emotionally involved and yet not bluntly cold toward the project in which they participate.

Recall that I resorted to Cobb–Douglas functions in Chapter 4 on Processes to find the point of autonomy. I can find an actual approximation of these functions by looking at my database. Let us set the following equation:

$$\text{Fairness} = z' - \beta_0 \, \text{Collaboration}^{\alpha} \, \text{Trust}^{(1-\alpha)} \qquad (5.5)$$

I turn this into log functions and annihilate z' and β_0 in the equation, to allow us to guess the value of α. I picked 16 groups out of my database and plotted their levels of Trust, Fairness (F), and Collaboration (Coll). The approximate value I find for α is 0.3, which is $[k-1]$.[116] In the present case, I have found a way to relate all three constructs—Trust, Fairness, and Collaboration—by way of the constant k. Knowing the maximum boundary of the closed dynamic system that a project is, we have the following for $[F = z' - \beta_0, \text{Coll}^\alpha \text{Trust}^{(1-\alpha)}]$:

$$\text{Fairness} = 2.3 - x^{0.3} \cdot 1^{0.7} \tag{5.6}$$

or

$$\text{Fairness} = (1+k) - \text{Collaboration}^{(k-1)} \cdot 1^{(z-k)} \tag{5.7}$$

Figure 5.11 illustrates the hypothesized Fairness function that I have found, holding Trust constant.

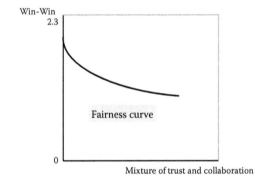

Figure 5.11 Hypothesized Fairness function.

As can be seen from Figure 5.11, this is a descending curve: the sense of Fairness decreases with the interaction of Trust and Collaboration. Trust is something that must be constantly built and fed, nothing less. Most people will eventually develop the impression that they are not being treated fairly if they keep working hard given a steady level of trust, even more so if they are obliged to collaborate. Many kids' favorite response to a problematic situation is: "It's not fair!" This, of course, applies even more eloquently in an environment where risks are potent, and where POVs (which are intimately related to Trust) are present and invigorated. Knowing that trust means accepting to impart one's vulnerabilities (or one's POVs) in the expectation that the *vis-à-vis* has positive intentions (as opposed to malevolent intentions), we can see how the psychological core is related to POVs. Earlier we set a barometer of Trust (Table 5.7). I do not venture at this time to firm up a mathematical formula between Trust and vulnerabilities because I feel this requires further research. Certainly, when one feels that their vulnerabilities are not threatened (they do not face external risks), they tends to trust more.

The psychological core of Trust–Fairness–Collaboration assumes no hidden agendas and no guile (Apprehension or perceived threats = 0). However, from the moment this assumption is betrayed, the psychological core becomes infused with apprehension. This is a very fragile system, as opposed to a robust system, hence the necessity of developing a Contingency strategy and of preparing for Short strategies as discussed in the first chapter.

Indeed, from a People's perspective, the success of any project hinges on the proper handling of POVs, hence the importance of recognizing them as early as possible, including when doing a feasibility study. In fact, we can probably set forth from the assumption that if team members perceive any malevolent intentions, then most likely there will be no commitment at all, or else very little.

In the absence of POVs, there is no apprehension—in fact, I assume that Apprehension is close to the limit of −2.3 (if such a limit existed). As POVs develop, apprehension takes on a peculiar evolution that forces managers to constantly stay on guard. Figure 5.12) illustrates the evolution of apprehension as POVs grow in intensity.

I hypothesize the Apprehension function to be

$$\text{Apprehension} = -2.3 + \sin\left(1.5\,\text{POV}_{t+1}\right) + 1.6\,\text{POV}_{t+1} - \text{POV}_{t+1}^{0.3}$$

or

$$\text{Appr.} = -(1+k) + \sin\left(10\left(\frac{k-1}{2}\right)\cdot\text{POV}_{t+1}\right) + (2k-1)\,\text{POV}_{t+1} - \text{POV}_{t+1}^{(k-1)} \quad (5.8)$$

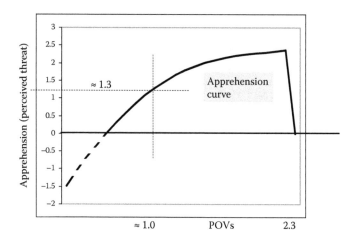

Figure 5.12 Hypothesized Apprehension function and POVs.

In this function, POVs are measured as what they are going to be in the future (hence POV_{t+1}).[117] Note that the function is nearly ascending linear; very near to the critical boundary point of 2.3, the curve crashes. The curve starts at a negative value (which is not a realistic value because the project system is bound by 0 and 2.3), then engages in a positive range of apprehension at approximately $POV = 0.3$. The curve shows that there is a maximum threshold (very close to 2.3), past which the system collapses; this is very much how the brain (and especially the hypothalamus and the senses) functions. There has to be a minimum level of threat to trigger an alarm response, and the system cannot pass its maximum coping capacity otherwise it fails completely.

I do not pretend that my functions are all complete and that they are the actual mathematical functions that govern behavior. However, I reckon that they approximate human behavior and tend to explain how People act and react. My formulas are based on my own research and certainly further research will add valuable insight to my proposed model.

5.13 $|R$ and $|T$

Let me now continue with a logic flow that is in line with what I have presented so far. I postulated in Chapter 4 that team members (Forces of Production) commit to work (expressed by T for *travail* in French) given some essential resources (R). I have shown in past literature that there is an intricate relationship between the work the team members perform and the use of resources that are absolutely essential for the survival of the project (R). These team members (Forces of Production) need to have a certain amount of knowledge, T_h (T_h stands for "technology"), and to keep training and learning on the job while they also accumulate nonessential resources (R_n) in the form of, say, a building, equipment, excess inventories, or materials. These surplus items are not immediately needed and will not cause the project to spin out of control or to stop if they are not immediately put to use. Again, my past research has shown the mathematical link between these two process elements: T_h and R_n. I plot these relationships in[118] Table 5.21.[119]

These curves can be read as follows: on the left side, holding R (essential resources) constant, the amount of effort tends to shrink. This is normal; employees become accustomed to their tasks and need to spend less energy to perform. They don't need to accumulate essential resources (hence the fact that I assume R to be constant) just as a person does not need to constantly eat during the day—a satiety point is eventually reached. On the other hand, the time saved on reducing the amount of effort can be spent on accumulating nonessential resources R_n: put in layman's terms, an individual is likely to relish his/her pay after a hard day's work, buying themselves a little something extra

Table 5.21 $|R$ and $|T$

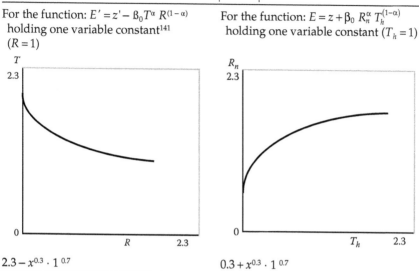

For the function: $E' = z' - \beta_0 T^\alpha R^{(1-\alpha)}$ holding one variable constant[141] $(R = 1)$	For the function: $E = z + \beta_0 R_n^\alpha T_h^{(1-\alpha)}$ holding one variable constant $(T_h = 1)$
$2.3 - x^{0.3} \cdot 1^{0.7}$	$0.3 + x^{0.3} \cdot 1^{0.7}$

besides food, such as a ticket to the local theater, which is not an activity essential for survival. As knowledge stabilizes, teams can pile up nonessential resources R_n that they may want to use later on. Hence, there is a trade-off between R and T, between R_n and T_h, between R and R_n, and between T and T_h.

Let's make a summary of the various curves we have seen so far (multidimensional Figure 5.13).

Trust is important. There would be no use for T and no accumulation of R_n if there was no trust. Team members must trust each other in order to collaborate (in most cases, and unless coerced, of course) and they would not care about compounding their wealth (in the form of R_n) if they were certain others would steal it (would have malevolent intentions).

At the dynamic Point of Equilibrium, POE, (with $x = 1$ and $y = 1.3$), all active psychological forces that drive the project are aligned and, presumably, productivity is optimized. This is a much desired point: team members and management all trust each other (although trust is fragile) and are robust. They balance risks and POVs, collaborate, commit to their tasks, manage POVs in the best achievable way, maintain a high level of utility, maximize their total strategy, and work diligently while preparing reserves for the future. As previously stated, we are able to plot all these scenarios, which provide many viewpoints for the same object—the project—because we have found a constant k by analyzing my database, which we have assumed plays an integral part in every mathematical function I have discussed. As we shall see when I discuss a neurobiological study

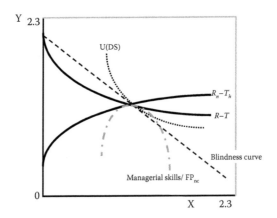

Figure 5.13 Summary of the curves.

I did in 2013, this should come as no surprise: the body's homeostasis (or equilibrium—referring to heat, hunger, thrust, sex drive, and so forth) is controlled by the hypothalamus (which has its own sets of parameters or critical levels that, once reached, trigger the search for food, for example) and which from its lateral portion (the lateral hypothalamus), is responsible for predatory behaviors, generally described as instrumental aggression, and from its medial portion (the medial hypothalamus) for defensive behaviors, generally described as defensive hostility (escaping, freezing, resisting). Thus, the fact that I have identified a hub—the Point of Equilibrium—is simply logical and in line with the way the brain is organized…and people are exactly that: brains!

5.14 Satisfaction

Stakeholders are proud of the projects they participate in. They gladly attend a ceremony when the project is brought to full completion, especially when the project has met the triple constraints of calendars, costs, and norms of quality. We can measure this level of satisfaction by theorizing that the U(DS) curve combines geometrically with the Blindness curve. This combination forms the Satisfaction curve, $S(q)$, where q is simply the independent variable, the achievement of a goal: to reduce POVs down to zero.

This function can be easily calculated since we know the U(DS) and the Blindness functions. It is

$$S(q) = q + \frac{2.6}{q} - 2.3$$

Alternatively

$$S(q) = q + \frac{2k}{q} - (1+k) \qquad (5.9)$$

The multidimensional Figure 5.14 illustrates the concept of 'Satisfaction' and shows how it tends to move the Point of Equilibrium to the right: higher satisfaction is achieved by moving in that direction.

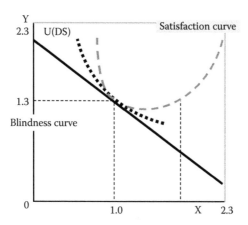

Figure 5.14 The Satisfaction curve.

At its optimum point or near it [$x = 1$ and $y = 1.3$], satisfaction is actually at or near a minimum; the minimum is actually past the Point of Equilibrium. Note that past the minimal point, team members feel more and more satisfied, in an increasing and quasi-linear ascending fashion. They hence have fewer reasons to be vigilant; they lower their guard, thus providing a fertile ground for POVs to gain in activity and power. It is interesting to note that team members cannot sit on their laurels because POVs constantly loom behind the scene. The Point of Equilibrium is tossed between two opposing trends: Trust and Satisfaction exercise a complementary yet opposing force upon the Point of Equilibrium.

There is one more reason why satisfaction cannot be taken for granted: People' interact with each other during the transformation phase. Team members at times perceive others as potential threats. They may fear for their jobs, or be concerned that their team members are not willing to cooperate, or else that a particular member is so sloppy that the project is jeopardized. A work culture (psychodynamics) cannot ignore the fact that team members are composed of individuals trying to gain satisfaction by marching away (to the right) from the Point of Equilibrium. This causes tension.

The best way to illustrate these work psychodynamics is to look at the smaller team unit, that is, two people, having to work together. An Edgeworth box best represents how each team member is forced to stay at their Point of Equilibrium despite their desire to gain more satisfaction, because doing so would compromise the satisfaction of the other team member, who would react by wanting to push their own agenda. Figure 5.15 expresses the reality that every Force of Production experiences when working, even on the simplest project.[120]

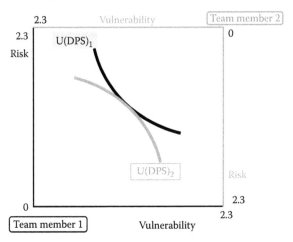

Figure 5.15 Two team members.

As can be seen from Figure 5.15, a team member cannot improve their position (move their U(DS) curve further along its growth curve, which has a slope of 1), which would move the Satisfaction curve further up and to the right for team member 1 (TM_1), without entering into the territory of team member 2 (TM_2) and pushing their own Point of Equilibrium backward from their perspective. I feel that the Point of Equilibrium shared by team member 1 and team member 2 represents a Cournot–Nash equilibrium: neither team member can improve their strategic position without making it worse for the other, because the other team member will retaliate. I posit that it is also Pareto-efficient: each team member's position cannot be reinforced without damaging the other team member's position. Note that because the Blindness curve has a function of [2.2 − 0.9 x] for TM_1, it can never match the Blindness curve of TM_2. These two Blindness curves will never juxtapose (the slope being 0.9 instead of 1), and certainly not at the Point of Equilibrium. This is why the Point of Equilibrium is actually a dynamic Point of Equilibrium. The best example that can be given for such a dynamic POE is the political situation that prevailed during the Bay of Pigs incident in Cuba in the 1960s, when the

United States and the Soviet Union could not move without taking the risk of starting a nuclear war, which would have destroyed both countries. Another dynamic Point of Equilibrium occurred in October 1961, when American and Soviet tanks faced each other along each side of the border at so-called Checkpoint Charlie in Berlin, where the wall had been built overnight a few months before. The confrontation was essentially a test to see who was going to blink first.

Exiting the dynamic Point of Equilibrium (POE) is tantamount to exposing oneself: the other team member is likely to take advantage of the situation. At the dynamic POE, POVs are minimized; outside, there is every reason for apprehensions to influence (and render biased) one's grasp on reality.

We will see further along in this book how moving away from the dynamic POE reduces the capacity to complete a project successfully.

5.15   *Resistance and compliance*

For sure, any attempt by a team member to move away from the dynamic Point of Equilibrium (POE) can be perceived as a threat by other team members. Hence, people naturally resist change, and this resistance is amplified by the fact that the dynamic POE represents a state of homeostasis, that is, a state of satiety given the circumstances. Change may awaken Hostile or Defensive positions among the workforce. Yet, a project is by definition a change process (as implied in the transformation phase); team members cannot escape the reality of change and thus the dynamic POE is exactly that: dynamic. Adjustments are needed every day. I have not done any extensive research on resistance, but I propose the following formula to express compliance (i.e., acceptance of change, or, put differently, the reverse of resistance):

$$\text{Compliance} = 2.3 - 2.3\,\text{POV} - \text{POV}^{\left(1-\frac{1}{k}\right)}$$

Alternatively

$$\text{Compliance} = (1+k) - (1+k)\,\text{POV} - \text{POV}^{\left(1-\frac{1}{k}\right)} \tag{5.10}$$

How I arrived at this formula is by testing my database and making wild assumptions that need to be verified in further studies.[121] Let's accept it for now and show how it appears on a graph (Figure 5.16).

Note that according to this function, compliance is actually suboptimal in the sense that it is off the dynamic Point of Equilibrium (POE).

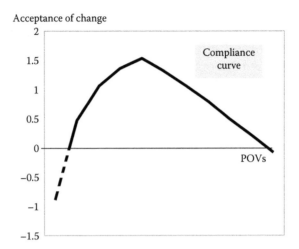

Figure 5.16 Hypothesized Compliance curve.

Compliance starts with resistance (it is negative at first, with the negative area being completely hypothetical because the system's boundaries are 0 and 2.3) but then keeps increasing as People get to know each other and become excited with the project. But a ceiling is eventually reached; some Fits (FP$_c$) become Unfits (FP$_{nc}$). The ceiling occurs slightly after the dynamic POE and below it (at [$x = 1.1$; $y = 1.26$] instead[122] of [1; 1.3], respectively). Then starts a quasi-linear waning in the motivation to accept change, which drags the project down and tends to bring the dynamic Point of Equilibrium (POE) below its optimal point (which is at [$x = 1$ and $y = 1.3$]). However, during that descent, recall that the Satisfaction curve keeps increasing: this means that people find a certain pleasure in opposing change, which, from my experience, is not far off the reality. Note also that I have tentatively linked Compliance (or the reverse of resistance) to POVs: the more the POVs, the less the incentives to be compliant. POVs bring about a resistance to change. This is normal: people who feel threatened (whose vulnerabilities are exposed) will want to protect themselves and thus resist change, the more so if apprehension is high.

Inspired by various authors,[123] I prepared Table 5.22 to exemplify different means of resistance in relation to the four Ps.

Perhaps my equation for resistance (and its opposite—compliance) is a fair yet not perfect representation of the reality of work psychodynamics in the closed dynamic systems that projects are, but it suffices to express a work dynamic as I see it.

Table 5.22 Resistance: Types and examples

Type of resistance	Four Ps	Example
Opacity (the reverse of transparency)	Planning	Hiding or withholding information
Delaying	Processes	Agreeing but stalling, procrastinating
Diverting		Going back to old methods, changing course of action
Obstructing		Sabotaging, blocking, undermining change
Arguing and denigrating	People	Finding a scapegoat
Spreading rumors		Communicating negatively
Complying reluctantly	Power	Doing the minimum, guarded, and without enthusiasm
Misguiding		Giving erroneous instructions
Terminating (even temporarily)		Taking unjustified sick leave

5.16 Culture and work psychodynamics

I have referred on a few occasions to the expression 'work psychodynamics'. The constructs of Trust and Collaboration, to take only these, do not mean much if we don't place them within the context of a project, as far as a feasibility analysis goes. I have attempted to prove that People play a key role in projects, in relation to the three other Ps that are Plans, Processes, and Power. To discount the role of People is to ignore the foundation of a project. When we put the six core competencies in interaction with each other as per my model, we actually outline what I call 'work psychodynamics' (PWP) or in a general sense 'work culture'. The question is not whether people trust each other, but rather how team members use trust to achieve the ultimate goal: completing the project on time, within costs, while meeting preset norms of quality.

This question of "How?" is crucial. A culture is, from my perspective, nothing more than how People (stakeholders) in a project use the six core competencies according to Plans. PMBOK 5 (p. 56) refers to "group phenomena." Culture is indeed increasingly complex. More and more projects call for team members that are dispersed, that belong to different nationalities, and that have highly specialized fields of expertise. This certainly affects work psychodynamics. Collaboration cannot exist without communication, which has its challenges in multicultural projects.[124] Distance makes relationships more opaque than they already are. Collaboration cannot be merely reduced to performing tasks; it entails human interactions with its swirl of emotions.[125] At the same time, diversity is good; it promotes innovation because new ideas are offered. In fact, diversity can actually be a proactive force rather than a liability.[126] It has increasingly become a key feature of large project teams.[127]

Cultural differences, including competing leadership styles and personalities, affect work psychodynamics, represent a potential POV in multicultural projects,[128] and may reduce their chances of success.[129] Indeed, as pointed out by some scholars,[130] they influence perception, and with this, apprehension (risks and POVs).[131]

In this context, *red flags* that signal poor work psychodynamics are worth paying attention to. These flags appear before events actually take place; the feasibility analyst should look for them because they imply behavioral POVs. They include dismissal,[132] a reflex that I have witnessed a few times whereby project promoters chose to brush off the reality of the facts and the possibility of a pessimistic scenario occurring; fallacy,[133] whereby erroneous assumptions or work hypotheses are formed; normalization of deviance, which can also be detected in errors present in the proposal (I often identify calculation errors that trigger my concerns—how is the project going to be carried efficiently if from its inception such errors occur?); and finally, overoptimism,[134] a phenomenon I encounter often when examining project proposals (see Chapter 7). These red flags point to a poor work culture (psychodynamics); they will most likely, sooner rather than later, transpire into stifling conflicts, poor communication, weak management, quirks, stress, the build-up of complexity, and a willingness to choose ignorance over knowledge (i.e., to tend toward the extremes of the Blindness curve).

What I am saying here is that even though the feasibility analyst cannot measure such psychological constructs as Trust and Collaboration before People become People', that is, before team members are formed and interact during the transformation phase, red flags can be sought with a trained and vigilant eye.

A sound work culture (psychodynamics) is simply assuming a Dominant strategy (DS) to its fullest.

5.17   Conclusion to Chapter 5

This chapter concentrated on the third P of project management: People. I have made a marked difference between so-called success and failure factors, something few books or articles that I have read have attempted to do.

For example, a failure factor is an overwhelming presence of Unfits (uncontrolled Forces of Production) in a high-risk context, especially when there is poor planning involved. Yet, having controlled FP in a low-risk context does not guarantee success. I have also outlined the role of bad apples, something again that few writings in project management address. I have established what Trust is composed of (Affinities,

Benevolence, Abilities, and Integrity) and what Collaboration entails (Flexibility, Sharing of information, Joint problem resolution, and Consideration, that is, working for the interest of all), something, again, I believe, and to the best of my knowledge, few books have nailed down to that level.

I have proposed a number of mathematical formulas in an effort to illustrate how humans behave in the context of projects. POVs have proved to be an integral part of human behavior.

At the back, the office where the QMA managers were overseeing the project.

5.18 What we have learned about POVs: Chapter 5

POVs …

1. Are inherent in the case of competing interests among stakeholders
2. Burgeon as $FP_{nc} > FP_c$
3. Go hand in hand with Trust in a psychodynamics context
4. Can be somewhat guessed by looking at curriculums and answers to particular psychometric tests
5. Are closely associated with the barometer of Trust
6. Have an impact on a project outcome along with risks
7. Are part of Apprehension, when internalized
8. Drive people's behaviors
9. Are often known to people, consciously or not (gut feeling)
10. Weaken a Dominant strategy
11. Relate to Trust given positive intentions
12. Influence commitment
13. Can reach a value of 2.3, at which point the project system crashes
14. That have a value of zero lead to zero Apprehension
15. Are fostered by lack of due diligence/vigilance
16. Emerge when people sit on their laurels
17. Are theoretically minimized at the dynamic POE
18. Are related to compliance/resistance to change (the more POVs, the more resistance)
19. Are sensitive to cultural contexts
20. Are an integral part of human behavior

5.19 Key managerial considerations: Chapter 5

A manager may take note of the following remarks:

1. Take the four kinds of stakeholders into account: customers, suppliers, regulators, and bad apples.
2. Do all you can to minimize the negative influence of FP_{nc} and bad apples.
3. Observe, observe, and observe again.
4. Master the project from a parametric, conceptual, and processing point of view.
5. Seek to be in control, transparent, and fair. Expect trust, collaboration, and commitment from all people at all levels of the project.
6. Hostility destroys a work atmosphere.
7. A dynamic Point of Equilibrium is more important to achieve than each and everyone's satisfaction.

5.20   Case study Chapter 5 —Africa versus Haiti

Many projects provide a clear-cut view of the six core competencies model (especially on the importance of commitment). This is no exception in Africa.

5.20.1   Africa

A large number of organizations from all parts of the world try to help various regions across the African continent, some of which are afflicted by illiteracy, a lack of infrastructure, malnutrition, and violence, among other ailments. Organizations can be formal or informal, governmental, private, or paragovernmental, based on economic profits or else on voluntary work. Many not-for-profit (NFP) organizations target the resolution of structural problems within the societies in which they operate. More often than not, a large part of the effort (the development of specific projects) rests on the ability to mobilize the local population, which should appear easy since its own survival or well-being often depends on its involvement. NFP organizations also deal with daily challenges, especially among the poorest—challenges such as access to water and hygiene—and venture into training locals for specific skills.

The World Bank and the International Monetary Fund (IMF) have set development principles that rest on a sense of partnership among the key stakeholders, such as fund providers, local and national governments, the population, and various organizations. Partnership is envisioned in the areas of financing (fund providers finance an NFP organization and let it identify and implement the project), technical assistance (fund providers supply knowledge without interfering with the project managers), and the upgrading of managerial skills.

'Collaboration' is the key word. Aid does not come merely in the form of money, but also in terms of in-the-field training. Partnership is also promoted during consultation.

The goal is to hold accountable those stakeholders who participate in any given project, to maximize the chances of success—at least this is the hope. A Canadian agency named *l'Agence canadienne de développment international* (ACDI) in French and the Canadian Agency for International Development (CAID) in English, has identified four "success" criteria: (1) a precise definition of roles and responsibilities; (2) creating new partnerships where and when necessary; (3) providing key managers with powers and tools that fit their objectives; and (4) the participation of local beneficiaries, especially women.

Many difficulties hinder the realization of African projects such as building a school or digging a well. The sluggish coordination of activities often represents a challenge. Lack of funding for specific programs slows down the process; this is exemplified by poor time management, shortages of materials (e.g., computers), and the gush of unexpected costs. From a human resources point of view, three worries are regularly highlighted: (1) lack of personnel; (2) poor staff competencies; and (3) the fact that beneficiaries have a difficult time in identifying with the project, even though it theoretically has been proposed, funded, and implemented to assist them in their development. Behaviors that accompany such drawbacks include a general lack of enthusiasm and poor quality relationships among stakeholders, all of which often lead to unsatisfactory results.

Often, the local population does not understand the nature and purpose of the project, or chooses to interpret it in its own self-serving way. It has a hard time accepting constraints and obligations. As a result, WBS are rarely implemented as planned. To make matters worse, management is not always transparent; in many countries, it tends to shun accountability.

Circumspect strategic planning, mobilizing (and motivating), and resource management (financial, human, material) are an essential part of any project in the context of African development involving NFP organizations. Often, these functions represent a burden over and above the daily requirements associated with the project, with the result that project leaders experience discouragement and fatigue, and that projects become more and more difficult to complete, and to finish on time, within the planned costs and according to the preset norms of quality.

5.20.2 Haiti

Haiti shares an island with its neighbor, the Spanish-speaking Dominican Republic, which was the first stop that Christopher Columbus is said to have made in the Americas in 1492. Its population is largely black; they are descendants of the slaves whom the French colonists had captured in Africa back in the seventeenth century. It is one of the poorest countries in the world, and certainly in the Americas.

Michel Dubuc is an architect and founder of Aedifica, a private consultancy based in Montréal (Canada) with 175 professionals working on various architectural and engineering projects, including in Haiti. I interviewed Mr. Dubuc and asked him to find parallels between the previously described African situation and Haiti.

Michel Dubuc:

> We arrived in Haiti in 2010 as part of the well-known international reconstruction effort after the devastating earthquake.
>
> There is no doubt in my mind that there are a number of similarities between the two regions. In both cases, Africa and Haiti, those engaged in implementing projects are forced by circumstances and events to get out of their zone of comfort.
>
> We discovered that Canadian or U.S. construction regulations, for example, could not necessarily and readily apply in the kinds of environments we are dealing with there. Installing sprinklers, a mandatory device in Canada and often a requirement imposed by international donors, was out of the question for us in Haiti: no one was trained to operate and fix them in case of failure—we had to palliate with other measures, such as devising additional exits. Our ways of doing things had to be adapted to the harsh reality of the living conditions we faced.
>
> Because the country has been economically depressed for so long, infrastructures and human skills are absent. We become agents of change when trying to implement projects according to our quality and performance standards.
>
> We had to find novel ways of doing things, for example, in the engineering of a low-tech air circulation system within a new hospital as an alternate to air conditioning given the high cost of fuel and the lack of electrical infrastructure. We had to plan for two corridors in a particular hospital—one reserved for the staff and one for the visitors—to make sure visitors and family, who traditionally bring food to the patients, would not interfere with staff and cause

circulation bottlenecks. Our experience taught us that success for a foreign consultant is predicated on his capacity to adapt to the local reality and culture: we had to be cautious when renovating an existing hospital because the sounds and vibrations of the heavy construction equipment alarmed the patients, who thought another earthquake was looming.

We also had to listen to our new local partners and respect our differences, which were vast on all technical accounts. It was surprising to see that understanding construction drawings was a challenge, so that we had to rethink our communication patterns to make everything crystal clear. With regard to job safety, for example, it was not rare to see workers without protective gear or doing work beside high voltage wires without any appropriate safety measures having been taken. Regretfully, in a country where basic needs are not being met, safety is often a low priority.

We had to negotiate endlessly and deal with abrupt changes all the time. In fact, if you ask me for a definition of adversity, I can provide one quite easily. The way we view it, adversity is about abrupt changes that see no ends. Adversity is about supply chain management: finding basic cement blocks of acceptable quality and locating repair parts for construction equipment proved to be a Herculean effort. It's about delays that follow as an inevitable consequence, with associated costs and tensions in human interactions at all levels.

It's also about the lack of expertise and resourceful personnel—for example, it was a real challenge to find a project manager who would want to stay in the chaotic circumstances that plagued the country. Adversity is about the lack of guidelines, of legal structures, and of benchmarks—our parameters just didn't fit in the new and devastated environment, which we faced when we arrived in Haiti.

Adversity also emerges from the contrasts between the traditional artisan do-as-you-go techniques and modern, proven, technically sound methods that we encouraged and were trained to abide to. Adversity is about a work culture: Haitians are not at ease in working in confined environments such as cavities located under the floors where wires and conduits have to be installed. Our Haitian partners responded well, but the learning curve was long and could be discouraging. This may at times have led people to losing self-confidence and believing less and less in the value of the project. Adversity is also about the lack of cultural references indeed; it is only by explaining over and over again our plans and methods and by adapting to the local culture that we could somehow arrive at realistic architectural solutions, given the context. Being sensitive to culture and to history also means being cognizant that foreigners can be perceived as selfish, given the colonial past of the island.

It takes a huge dose of determination to stay focused on a country where projects fail because of poor planning and lack of risk management. We had to rise above adversity.

The author: "Thank you for sharing your experience with us, Mr. Dubuc."

5.20.3   Questions related to Case 5: Africa versus Haiti

1. Describe the general culture of international and private aid as seen in this case.
2. Discuss how control, transparency, trust, fairness, collaboration, and commitment appear and interact in this case.
3. Highlight the POVs related to People.
4. In your opinion, are there uncontrolled Forces of Production in this case?
5. Give your definition of adversity versus chaos.

5.21   Appendix 5.1: Constructs found in the literature

Affinities: Diallo and Thuillier (2005); Dong and Glaister (2006); Zeng et al. (2009)

Being proactive: de Brentani and Kleinschmidt (2004)

Commitment: Black et al. (2000); Brière and Proulx (2013); Briscoe et al. (2001); Dainty et al. (2005); de Brentani and Kleinschmidt (2004); Diallo and Thuillier (2005); Dulaimi et al. (2003); Duncan and Moriarty (1998); Frankel et al. (1996); Halman and Braks (1999); Hampson and Kwok (1997); Hauck et al. (2004); Jung and Wang (2006); Kadefors (2005); Kerzner (1995); Kumaraswamy and Anvuur (2008); Nicolini (2002); Rahman and Kumaraswamy (2005); Stanley et al. (2006); Tailby et al. (2004); Turner (2004)

Common goals: de Brentani and Kleinschmidt (2004); Jodl (2012)

Conformity: Cheung et al. (2003); Cowan (1991); Jarvenpaa and Leidner (1998); Kadefors (2005); Kumaraswamy and Anvuur (2008); Ng et al. (2012); Thompson and Sanders (1998); Wood and McDermott (2001)

Control and leadership: Biro et al. (2006); Chevrier (2003); Turner (2004); on collaboration: Carmel and Agarwal (2001); Dong and Glaister (2006); Jodl (2012); Turner (2004); Zeng et al. (2009)

Communication: Anvuur et al. (2011); Carmel and Agarwal (2001); Chen and Popovich (2003); Clarke (1999); de Brentani and Kleinschmidt (2004); Diallo and Thuillier (2005); Duncan and Moriarty (1998); Grönroos (1994); Kumaraswamy and Anvuur (2008); Mentzer et al. (2001)

Culture: Hammer et al. (2003); Kealey et al. (2005)

Cultural distance: Chevrier (2003); de Brentani and Kleinschmidt (2004); Globerman and Nielsen (2007)

Fairness and win-win: Cleland (1999); Jodl (2012); de Brentani and Kleinschmidt (2004); Hauck et al. (2004); Jodl (2012); Kadefors (2005); Kumaraswamy and Anvuur (2008); Lusthaus et al. (2002); Rahman and Kumaraswamy (2005); Wood and McDermott (2001)

Feedback and follow-up: Lusthaus et al. (2002); Pinto and Slevin (1988); on safety: Toor and Ogunlana (2010)

Flexibility: Chevrier (2003); Cooke-Davies (2002); Kealey et al. (2005); Kerzner (1995); Lusthaus et al. (2002)

Integrity: Cooke-Davies (2002); Jarvenpaa and Leidner (1998)

Loyalty: Kumaraswamy and Anvuur (2008); Wang et al. (2005)

Listening to others: Cheung et al. (2003); Cowan (1991); Hauck et al. (2004); Jarvenpaa and Leidner (1998); Kadefors (2005); Kumaraswamy and Anvuur (2008); Nicolini (2002)

Objectivity: Black et al. (2000); Cheung et al. (2003); Halman and Braks (1999); Kumaraswamy and Anvuur (2008)

"Openmindedness": Black et al. (2000); Cheung et al. (2003); de Brentani and Kleinschmidt (2004); Halman and Braks (1999); Kadefors (2005); Kumaraswamy and Anvuur (2008); Nicolini (2002); Rahman and Kumaraswamy (2005); Thompson and Sanders (1998); Wood and McDermott (2001)

Planning: Clarke (1999); Cooke-Davies (2002); Ozorhon et al. (2007); Zeng et al. (2009)

Pride: Wang et al. (2005)

Project definition: Anvuur et al. (2011); Clarke (1999); Diallo and Thuillier (2005); Esteves and Pastor (2001); Mentzer et al. (2001); Morris and Hough (1987); Ozorhon et al. (2007); Zeng et al. (2009)

Resources: de Brentani and Kleinschmidt (2004); Dong and Glaister (2006); Jung and Wang (2006); Khang and Moe (2008); Lusthaus et al. (2002); Ozorhon et al. (2007); Toor and Ogunlana (2010); Zeng et al. (2009)

Sharing and reciprocity: Cheung et al. (2003); de Brentani and Kleinschmidt (2004); Dulaimi et al. (2003)

Performance: Zeng et al. (2009)

Stakeholders: Biro et al. (2006); Ng et al. (2012); Pinto and Slevin (1987)

Teams and team spirit: Jarvenpaa and Leidner (1998); Wang et al. (2005); Carmel and Agarwal (2001)

Technical competencies: Dong and Glaister (2006); Globerman and Nielsen (2007)

Training: Morris and Hough (1987); Vickland and Nieuwenhuijs (2005)

Transparency: Dong and Glaister (2006)

Trust: Diallo and Thuillier (2005); Jarvenpaa and Leidner (1998)

WBS: Carmel and Agarwal (2001); Clarke (1999); Zeng et al. (2009)

5.22   *Appendix 5.2: Questionnaire on a project's work*

Control	Have all steps been taken to ensure that the project meets the deadline?
	Is the organization overly bureaucratic? (Negative question)
	Are there proper measures to ensure discipline?
	Was each project element prepared with care?
	Can the project leader rest on formal and informal authority?
	Has the project leader put measures in place to ensure adherence to the project?
	Does the project leader exert self-control?
	Has the project leader made sure all of those involved understand the project?
	Has the project leader planned the tasks to be performed?
	Is the project leader respectful of norms and laws?
	Has each stage of the project been carefully planned?
	Is the project leader known to be persevering?
	Have proper measures been taken to control costs?
	Have mechanisms been put in place to ensure the full completion of the project?
Transparency	Has the project leader attributed roles and responsibilities adequately?
	Is the organization transparent?
	Is the chain of command clear?
	Is the project leader able to explain the project with clarity?
	Is the project leader overly autocratic? (Negative question)
	Have the relations between stakeholders been well defined?
	Does the organization abide by sound ethical principles?
	Is there a code of conduct, both in written and verbal communication?
	Have performance criteria been established?
	Have the tasks been designed to meet the objectives?
	Does the project leader express him/herself well?
Trust	Do the stakeholders share the same values?
	Do the stakeholders take into consideration the opinion and sentiments of each other?
	Do stakeholders understand the needs of each other?
	Is there a team spirit?
	Are the stakeholders reliable?
	Are the stakeholders accepting of other cultures?
	Are the stakeholders behaving with integrity?

Fairness	Do the stakeholders act in an impartial manner?
	Does the project leader act with diplomacy?
	Are salaries fair?
	Are stakeholders treated fairly?
	Is pay adjusted to the level of work and competencies?
	Are there mechanisms in place to reward top performers?
	Are safety and security well thought of at the job site?
	Do team members have access to the resources necessary to accomplish their work?
	Are there enough resources to complete the project?
	Is there a win-win spirit within the different teams?
	Do team members work to maximize everyone's benefit?
	Will the stakeholders gain from the project?
Collaboration	Is the project leader able to delegate?
	Is the project leader accessible?
	Is the project leader able to listen to others?
	Does the project leader aim for consensus?
	Is the project leader able to provide useful feedback to team members?
	Does the project leader show solidarity with the team members?
	Are the team members exchanging useful information?
	Are team members open to change?
	Are team members open to other people's ideas?
Commitment	Are stakeholders loyal to the project?
	Do the stakeholders go out of their way to ensure success?
	Is the team leader focused on results?
	Are adequate measures put in place to ensure delivery of the deliverables?
	Do stakeholders participate in the project at the expected level?
	Are there chances that some tasks will not be completed?
	Are all scheduled tasks likely to be completed?
	Are stakeholders meticulous in their daily work?
	Are the stakeholders likely to do what they are supposed to do?
	Is the team leader innovative?
	Is the team leader proactive?
	Can the project leader offer solutions when facing problems?
	Is the project leader optimistic with respect to the project?
	Does the project leader accept change with efficacy?
	Are there measures in place to upgrade the skills of the team members?

Distance   Does the physical distance between teams likely impede on their performance?

Are communication and IT systems between the various locations compatible?

Is the physical distance between team members likely to cause delays?

Will team integration be made difficult due to the physical distance between the teams?

Will time differences create difficulties?

5.23   Appendix 5.3: Tables of constructs and k values

U(DS) is maximized when PWP's $k = 1.3$

Value of k'	$3.1–4.6^{135}$	$3.0–1.8$	$1.79–1.1$	≤ 1.0
Observed behavior	Some tense erratic behaviors	Some erratic behaviors among group members	Fairly stable groups	Groups suffer from high turnover level

Value of the core psychological constructs	Less than 70%	From 70% to $\approx$80%	Over $\approx$80%
Trust, Fairness, Coll.	Transactional relationship	Relational relationship	Interpersonal relationship
Characteristics	Contract based Distant Nonemotional Occasional	Day-to-day Open Polite Trust based	Deep Emotional Friendship Personal history

Endnotes

1. Belout and Gauvreau (2004); Sartorius et al. (2011); Martin-Alcazar et al. (2011).
2. See Mulcahy (2013, p. 169) and our case on Sea Crest Fisheries (Chapter 4).
3. See Cleland and Kerzner (1986).
4. PMBOK 3 (2004, p. 25) and PMBOK 5 (2013, p. 30).
5. Déry, É. (Interviewed March 2015).
6. Anderson and Gerbing (1988, p. 414).
7. Oakley, G. Interviewed March, 2016.
8. This questionnaire cannot be used without the written consent of the author.
9. See, for example, Russell and Stone (2002, p. 154).

10. One probable causal link I have found over years of research is that a sense of unfairness (injustice) leads to anger.
11. Ackoff (1957, p. 7); Brewer and Hunter (1989, p. 42, 149); Brannen (1992); Neuman (1994, p. 43, 99); Cossette and Lapointe (1997, p. 49); Miles and Huberman (2003, p. 273); Buchanan and Bryman (2007, p. 494).
12. Anderson (1983, p. 19) and later Van Bruggen et al. (2002, p. 470).
13. Wieseke et al. (2008, p. 324).
14. For a substantial review ranging from 1963 to 2002, see Gond and Mercier (2005).
15. See Jepsen and Eskerod (2009).
16. PMBOK 5 recognizes the importance of collaboration as follows (p. 6); "managing a project typically includes, but is not limited to: (...) setting up, maintaining out communications among stakeholders that are active, effective, and collaborative in nature (...)"
17. See Wessinger (2011) for a substantial review ranging from 1995 to 2013.
18. PMBOK (2013, p. 394).
19. As pointed out by Meredith and Mantel (2009, p. 1).
20. See PMBOK 5 (2013, p. 51).
21. See the section dedicated to work psychodynamics.
22. International fund providers often refer to the concept of 'good governance'.
23. Cleland and Ireland (2006, p. 201).
24. Fetchenhauer and Dunning (2009, p. 264).
25. Van't Wout and Sanfey (2008); Wishniewski et al. (2009).
26. Fehr et al. (2005); Douhou and Magnus (2012).
27. See, for example, Bejarano et al. (2015).
28. See Pearson et al. (2008).
29. Raichle (2011).
30. Manikutty (2000, p. 280).
31. See Weidenbaum (1996).
32. See Ang et al. (2000).
33. Wang et al. (2011).
34. Brooks and Rose (2008).
35. See Kaye (1996).
36. See Palmer et al. (2005).
37. Campbell and Kirmani (2000).
38. The present and further sections in this chapter on People require knowledge of statistics.
39. For example, my research tends to show that Attachment is a combination of Trust, Dependence, and Distance; motivation would be formed by a combination of need and the core competencies, and so forth.
40. The questionnaire I use to assess the PWP can be made available to the readers given that certain conditions apply, such as payment of royalties and of a round-trip to the Bahamas (!), and respect of confidentiality.
41. Le Boterf (2005).
42. See Grant (1991); Prahalad and Hamel (1990); Amit and Schoemaker (1993); Peteraf (1993).
43. Buchanan (1991).
44. On average, project managers are recruited within approximately 25 days from the first interview across the main most advanced Western countries (and Australia). See PM Network, January 2016, p. 9.

45. Even if time has no polarity, I can still show the positive influence of Trust onto Collaboration as I do in Figure 5.7; Trust certainly influences Collaboration but the more profound force is time: Trust requires time to build.
46. See Gidel and Zonghero (2006).
47. From a neurobiological point of view, this makes sense as we shall see. The pituitary gland that releases hormones (such as stress, attachment, or action hormones—cortisol, oxytocin, adrenaline) directly or indirectly cannot feedback to the hypothalamus (in the HPA axis). These hormones must be released in the circulatory (blood) system, which then reach the hypothalamus.
48. Stough et al. (2000).
49. Bourne, 2015, p. XV.
50. See also: Garel et al. (2005) who link performance to commitment.
51. Deveau, J.-P. Interviewed March 1, 2016.
52. See Gneezy (2005).
53. Garbarino and Slonim (2003, p. 229).
54. The results of the SEM will somehow vary from one group to the other, but overall, each SEM result tends to emulate quite closely our core six competencies model, which I would qualify as the generic model.
55. See Rousseau et al. (1998); Bell et al. (2002); Riedl and Javor (2012).
56. Mayer et al. (1995, p. 712).
57. Lewicki et al. (1998, p. 439).
58. Deveau, J.-P. Interviewed March 1, 2016.
59. Oakley, G. Interviewed March 2, 2016.
60. Nienaber et al. (2015).
61. Such as Chua et al. (2012).
62. Ainamo et al. (2010).
63. I consider Affinities and Benevolence as emotional variables.
64. I consider Abilities and Integrity as cognitive variables.
65. See Crosby et al. (1990); Ganesan (1994); Ravald and Grönroos (1996); Doney and Cannon (1997); Smith (1998); Nicholson et al. (2001); Svensson (2001); Bell et al. (2002); Gurviez and Korchia (2002); Wood et al. (2008); Mallalieu and Nakamoto (2008).
66. For Competence/Ability/Credibility, see Wood et al. (2008, p. 29); Mallalieu and Nakamoto (2008, p. 184); for Abilities and Benevolence, see Doney and Cannon (1997, p. 36); Ganesan (1994, p. 3); for Integrity, Abilities, and Benevolence, see Gurviez and Korchia (2002, p. 53), Bell et al. (2002, p. 66); for Affinity, see Wang et al. (2011); for the four variables, see Svensson (2001).
67. As previously mentioned, there are two kinds of aggression: offensive, which is predatory in nature, and defensive, which is unplanned.
68. See Van Goozen et al. (2000).
69. Jones and George (1998, p. 536).
70. Residuals were found to behave normally as well.
71. Bstieler (2006).
72. Stahl et al. (2010).
73. Anderson and Narus (1990) with an estimator at 0.73. One noticeable difference: these authors argue that Cooperation precedes Trust on a time line, which is contrary to what the majority of studies, including mine, have shown. This discrepancy can be solved by realizing that there is a loop

arrow going from collaboration that eventually goes to trust as in our six competencies core model. See also Morgan and Hunt (1994), and Palmatier et al. (2006).

74. In many studies, the word 'cooperation' is used; in the context of project management, I use the term 'collaboration'.

75. Palmatier et al. (2006).

76. All samples have been checked for normality of populations and of residuals; outliers were eliminated where necessary, but, for the record, there were rare. The standard alpha used is $\alpha = 0.05$ or 95% degree of confidence.

77. Nunnally (1970).

78. Equivalently, 'perceived threat', a term found in the science of ecology (e.g., Johnsson et al., 2004, p. 390). The concept is well established in psychology. I give it a particular meaning in this book by emphasizing the element of surprise.

79. Ein-Dor et al. (2011).

80. Keil et al. (2000, p. 318).

81. Such as Albescu et al. (2009).

82. One project leader says: "We put in lots of efforts to inform our team members. I personally tour our plants on a regular basis and make a 'state of the union address' once a year. I can't always deliver good news, but it is better than letting rumors poison the atmosphere because suspicion leads to decline." Deveau, J.-P. Interviewed March 1, 2016.

83. Which I refer to as the law of perceived predation in the context of financial predation, the likes of the 2006–2008 U.S. predatory mortgage crisis.

84. 'Intention' refers to intention to stay with the team.

85. I will refer to this as 'Pretrust'.

86. Vlachos et al. (2010).

87. For the origin of the concept of 'attachment', see Bowlby (1973).

88. See Grönroos (1994); Mikulincer (1998); Garbarino and Johnson (1999).

89. Ochieng and Price (2009).

90. Tardif, P.M. Interviewed April 2016.

91. See Appelbaum et al. (1998).

92. Olsen (2012).

93. Bechara and Gupta (1999, p. 182).

94. For example, Houser et al. (2010).

95. Palmatier et al. (2006, p. 140).

96. Says one project leader: "We've had to create all kinds of teams over the years in order to invent and market our value-added products. Good communication between all and goodwill generate outstanding results. It's in the heart of every one of us." Deveau, J.-P. Interviewed March 1, 2016.

97. Without the regression error term, that is unnecessary to prove my point.

98. My use the sign => to signify a linear regression.

99. See PMBOK 5, p. 21 on "distant locations" as well as p. 38: "Virtual teams are often necessary for projects where resources are locate onsite or offsite or both, depending on the project activities."

100. See Lee-Kelley and Sankey (2008).

101. Yang et al. (2011).

102. Adenfelt (2010); Verburg et al. (2013).

103. Brucks et al. (2007).

104. Recall that Apprehension (perceived threat) is Defensive/Hostile positions.

105. These classes are recognized in the literature. See, for example, Freeman (1984) and Dontenwill and Reynaud (2005).
106. Durbin–Watson values are acceptable for all regressions.
107. Howell et al. (2005).
108. As the reader can see, the result is slightly different from that of the core model. Each group will generally have its own variations.
109. Youker (1992).
110. See Birnholtz et al. (2012).
111. Brock et al. (2011).
112. Chen and Popovich (2003).
113. Grönroos (1994).
114. Grayson (2007).
115. Williamson (1975).
116. Since α does not pertain to any psychological constructs I have discussed, I chose to transform it into $[k - 1]$. I hold the hypothesis, therefore, that $[\alpha = k - 1]$. This remains to be further studied, of course, but it allows me to proceed with my model.
117. This function is hypothetical; it is an attempt to portray how behaviors relating to Apprehension look. It was developed through a series of transformations operated on my core functions.
118. My graph is stylized.
119. Together, R and R_n are referred to as $|R$ and Work (T) and Knowledge (T_h) as $|T$.
120. I posit that each team member has a growth curve with a slope of 1 starting from the point of origin (at 0) for the sake of simplification.
121. This function is hypothetical; it is an attempt to portray how behaviors relating to Compliance look. It was developed through a series of transformations operated on my core functions.
122. Note: at $x = 1.1$, the team member has given an operational margin of 10% versus the optimal value of $x = 1$ set at the dynamic POE.
123. See in particular Lines et al. (2015).
124. McFarland et al. (2006).
125. See Kumar et al. (2004).
126. Stahl et al. (2010).
127. Seymen (2006).
128. See Milosevic (2002).
129. Fitzsimmons et al. (2011).
130. Low et al. (2015).
131. See Liu et al. (2015).
132. Gagnon-Bartsch et al. (2013).
133. Flyvbjerg (2013).
134. Lovallo and Kahneman (2003).
135. I had no results from my diverse researches for values above 4.6. Chaos is assumed above that level.
136. The term 'win-win' comes from the team members. It is much easier to relate to than the concept of Fairness in the context of project management. Everyone wishes to give and receive something positive in a working relationship. Many project leaders consider that being fair is part of their duties: "Our philosophy at Algues acadiennes is to treat all team members the same way: we look for abilities, efforts, experience, and results.

We're in this together, as a team." Deveau, J.-P., Interviewed March 1, 2016. PMBOK 5 states (p. 2): "When interacting with any stakeholders, practitioners should be committed to honest, responsible, fair practices and respectful dealings."

137. I discounted nationalities for which there wasn't a sufficient number of respondents.

138. Not corrected for outliers.

139. Not corrected for outliers.

140. I used Baron and Kenny's (1986) method; my analysis shows that Fairness is a partial mediating variable between Trust and Collaboration. The Beta value for Trust in the regression Collaboration = f (Trust + Fairness) is lower (at 0.297) than the Beta value for Trust in the simple regression Collaboration = f (Trust) (at 0.668) taking into consideration that all regressions between Trust, Fairness, and Collaboration are significant.

141. As one must remain equal to himself at all times, which is one of the main premises of my model.

chapter six

People (In action)

So it is, said the Wise Eagle, that those who cheat succeed, as they get to the Emptiness faster than anybody else does.

Darloz

6.1   Introduction to Chapter 6

In order to verify that a sound work culture (psychodynamic) for a given upcoming project is present, the feasibility analyst tries to detect who is likely to be a top performer, that is, who is or can be a fully efficient Force of Production (FP), as opposed to a costly uncontrolled Force of Production, FP_{nc}. Who is the Star?

There are ways to do this that I address in this chapter: we ought to examine personality from the angle of attachment theory[1] and see how the brain, under stress (submitted to some form of threat), elects Hostile or Defensive[2] behaviors—these behaviors can be so strong and enduring that they actually become part of the *modus operandi* of individuals. One refers in this case to so-called Instrumentally hostile (IP) and Defensive positions (DP). We must also look closely at egregious individuals, who are the most dangerous FP—these are people who are fully functional, but who operate based on a hidden agenda that can tarnish any aspect of the project. First, let's look at commitment.

6.2   *Commitment*

My six core competencies model suggests that people commit to their tasks. As some people say: "You won't go to jail if you collaborate!" For sure, collaboration is crucial, but there has to be a direction: that of realizing the project within the triple constraints. Otherwise, collaboration may just be a big party with no palpable results. I set Collaboration as a precedent to Commitment; cooperative team members show each other that they are committed. This tends to reinforce collaboration when the team spirit is positive and strong.

I did not measure Commitment in my various studies, but I believe it is akin to the idea of a cumulative effort as described by the *Project Management Body of Knowledge* (PMBOK), with efforts focused in particular on subduing POVs. Efforts expand over time and they tend to reach a plateau as the project nears its deadline. Essentially, I assume that the Commitment curve emulates the R_n–T_h curve, with cumulative efforts | constraints being on the y-axis and resources on the x-axis (Figure 6.1). A number of authors spanning across decades and countries have recognized that the issue of resource procurement and handling is important in project dynamics.[3]

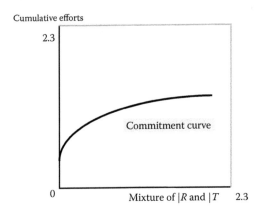

Cumulative efforts

2.3

Commitment curve

0   Mixture of |R and |T   2.3

Figure 6.1 The hypothesized Commitment curve.

As we have seen in the previous chapter, we already know the function of this curve because we know the function of the R_n–T_h curve. The Commitment curve, as I define it here, seems to make sense; if efforts are put in, this implies some form of commitment. But efforts could be spent uselessly. However, if efforts lead to an increasing use of resources, then, most likely, People are committed to the Plan and to finishing the project; otherwise, resources (total resources, that is $[R + R_n]$, or $|R$) would sit idle and team members would eventually put less effort in and even forget about their initial goals.[4] People trade off present and future knowledge (T_h), and present and future nonessential resources (R_n) as time goes on;

for example, team members (Forces of Production) learn for a while, then relax while the new information sinks in, then they use newly acquired knowledge that complements past knowledge. This process goes on at a shrinking rate because a **saturation point**[5] is eventually achieved, either because of fatigue or because the team members (FP) have learned everything there was to know in order to perform their tasks; say, to operate a particular, recently installed piece of first-aid machinery[6] such as a defibrillator. These team members (for instance, being trained in first-aid emergencies) will have to take some time off from learning and practice a bit, before learning new things about the subject and practicing again.

Squirrels in a city park can serve as an example (not in first-aid care!). They eat nuts now ($R_{present}$), yet they also hide some nuts away in anticipation of future consumption.[7] While eating at the present time (T_{now}), they cannot dig the ground and hide nuts for future consumption (T_{future}). While they dig, they cannot eat during the present time: present- and future-oriented behaviors are mutually exclusive. In fact, these behaviors compete with each other. This competition is based on a perceived threat (apprehension) and POVs: eventually, the squirrel will be hungry and will need to feed, perhaps despite the presence of predators (the squirrel will then trade off between the risks—the predators—and its vulnerabilities—hunger). Similarly, the team members (Forces of Production) need to acquire new skills because, eventually, they can be replaced by younger, more alert, and technologically advanced employees, or else by robots. External threats such as competition from the outside and internal POVs work together to force direct and indirect resources ($|R$ and $|T$) to be optimally utilized.[8] This dynamic transpires in our six core competencies model: Apprehension, which invariably exists in correlation with POVs, impacts the psychological core of Trust–Fairness–Collaboration, which then, over time, supports commitment, which amounts to efforts being made in using all of the resources efficiently in order to complete the project.[9]

I hypothesize that saturation points for present and future uses of resources are achieved at the dynamic Point of Equilibrium. Indeed, because resources (direct and indirect) have a life span, that is, as saturation points only last a certain amount of time, the Point of Equilibrium (of homeostasis) can only be dynamic.

This perspective allows me to qualify **Commitment** as a series of efforts made using a certain number of resources, whether direct ($R + R_n = |R$) or indirect ($T + T_h = |T$), that are spread along a time line (in reality, that are submitted to the utility drawback) that outlines a competition between present and future use, and that leads to at least one saturation point.

The following is the equation for Commitment, as I assume it to be, with $|R$ and $|T$:

$$\text{Commitment} = z + \beta_0 \mid R^\alpha \mid T^{(1-\alpha)}$$

or

$$\text{Commitment} = 0.3 + \beta_0 \,|\, R^{0.3} \,|\, T^{0.7}$$

or

$$\text{Commitment} = (k-1) + \beta_0 \,|\, R^{(k-1)} \,|\, T^{(2-k)} \tag{6.1}$$

Recall that I plotted the R–T and the R_n–T_h curves on the same graph to find the point of autonomy and to discuss the dynamic Point of Equilibrium. I assume that I can do the same here: the Commitment curve goes up and the Fairness curve goes down.[10] As time goes on, People express more grievances because they discover unheard of POVs that play against their own welfare and hence tend to want better working conditions; at the same time, they must perform and put in more effort (as they operate in a closed dynamic system that is bound by the utility drawback or triple constraints). The point where they feel that they are treated reasonably well while not overextending themselves forms a psychological state conducive to productivity and repellent to counterproductivity. This point is, of course, the dynamic POE (multidimensional Figure 6.2).

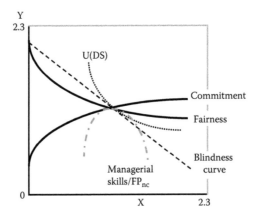

Figure 6.2 Dynamic POE, Fairness, Commitment.

It is worth noting that in recent studies (at least, those I have accessed) in the field of project management, the constructs of Control, Trust, Collaboration, and Commitment, all appear as behavioral competencies in IT projects; yet, Transparency and Fairness are ignored.[11] Numerous studies on competencies in project management fall short of identifying Transparency and/or Fairness (or the sense of a win-win situation) as being important factors, when in fact it is hard to imagine anyone wanting to commit to a series of tasks and perform at a high level while feeling something is hidden from them or that they are being or will be treated unfairly.

The feasibility analyst may pose three questions when looking at a project proposal:

1. Are adequate managerial skills in place?
2. Do all of the stakeholders trust the project and each other?
3. Are all of the necessary total resources available ($|R$ and $|T$)?

I assume that these are *sine qua non* conditions to the feasibility of a project within the triple constraints (Figure 6.3).

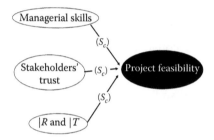

Figure 6.3 Proposed *sine qua non* conditions to project feasibility.

Overall, I have hypothesized that Commitment and Fairness share some common grounds: there is an interplay between them that allows me to plot their curves on the same graph.

6.3   *Identifying the stars (top performers)*

Functional groups are divided differently depending on the author. The PMBOK classifies what I call the "controlled Forces of Production' (FP_c) according to the stage of the transformation process: (1) start (vision), (2) planning, (3) execution (mobilization), (4) monitoring and control (deployment), and (5) closing (completion/evaluation).

Finding the right employees (the "Fits"), and especially the top performers (the "Stars," those who display the ideal DS and CS curves), for the completion of a project is key to meeting its deadline, budget, and norms of quality. A project may incorporate great employees, but it may also be subject to bad apples or Unfits. Additionally, a project is naturally prone to some forms of defects (POVs). As pointed out by some authors:

> (...) based on the interviews with project management professionals, it is clear that project-based work does produce a number of unintended pathogens that can significantly affect both the way in

which project management professionals do their
work as well as they manner in which they interact
with critical stakeholders.[12]

Not only are projects sensitive to constant changes, but the current
world itself sees dramatic turmoil (environmental, legal, and political)
that may represent clear and present danger (risk), at a rate probably never
seen before. This is even truer in the context of international projects,
where stakes are spread across cultures, distances, and time differences.

The National Aeronautics and Space Administration (NASA) has
developed a "federated approach" that states that effectiveness is (structurally) formed of people's skills and behaviors; this includes abilities
(which I have captured in the construct of Trust), alliances (which obviously are related to the construct of Collaboration); assignments (which
are included in my construct of Work or T); and attitude (which I include
in my construct of Consideration). All of this ultimately serves to differentiate between functional and dysfunctional People.

Increasingly, projects resort to cross-functional work performed in
multiple locations based on a diversity of teams.[13] The temporariness
and heterogeneity of the workforce renders the working environment
more volatile and more vulnerable than before. Inefficient employees
(FP_{nc}) may not be promptly detected. Opportunities for conflicts may
arise faster because unknown factors surface; this happens within a
pressured framework that commands specific deliverables, given time
constraints, and a predetermined budget. Many risks hinder project
developments. In addition, projects must integrate various areas of
knowledge and expertise (T_h) in unique ways. This is true for projects
of all disciplines (including software projects) from decades of project
ventures.[14]

As such, a project environment is subject to high **volatility**. One can
say that the key characteristic of contemporary projects is, indeed, volatility: a characteristic that did not exist to such an extent in the past and
that is likely to rise in intensity in the future. More and more, innovations
invade markets, projects are developed internationally, the workforce is
multicultural and multidisciplinary, and events succeed each other at
high speed given such communication tools as the Internet.

The volatile reality of contemporary project environments suggests
a potential for "shocks of personalities" to occur.[15] These clashes lead to
the rise of insecurities. Indeed, employees sometimes feel threatened by
their coworkers, their employers, or their environment. Insecure employees beget more volatility, and as the sense of threat worsens (the threat
toward one's job or safety and security, fear of negative outcomes as a
project develops, etc.), employees lose motivation, make more errors,[16] and
resort to counterproductive behaviors.

A review of actual projects across various countries reveals that both personality (profile of shareholders) and what is called an "attitude toward risks', are assumed to play a role in project outcomes.[17]

Indeed, it has been found that "the underlying cause of most disputes was the reaction to the (...) personalities of the key players (...)"[18] However, it may not be so much the personality types than the way individuals respond to threats that matter in a project environment. It thus appears important to examine what threats are all about and how individuals process them.

Some authors recognize the fact that "(...) there are at least two quite separate types of threat. On the one hand, there is a definite, localizable, actual threat (...); and, on the other, there is indefinite, diffuse, potential threat, (...)."[19] However, the authors also note that "the threat must be only mild if approach is to be a viable option."[20] Transposing this in the reality of a project environment, it may be that an employee who perceives his employer or his environment as a threat will tend to avoid it or else react in a counterproductive manner.

Inasmuch as there is high volatility in contemporary projects, certain pillars of stability provide a base for the development of the project. Managerial tools and techniques such as those promoted by the Project Management Institute (PMI)[21] are used internationally. Given that the expression of some emotions (e.g., hostility) is similar among different cultures,[22] some managerial techniques that assist with the decoding of verbal and nonverbal cues may deal with them with a certain level of success. These factors of stability help to provide a productive environment, but do not necessarily overcome difficulties arising from employees showing flimsy project capability. Additionally, they take place during the transformation stage, not before. Our objective is to detect and minimize the POVs related to People's inabilities before the project begins, at the feasibility analysis phase.

The term **project capability** can be associated with the capacity to adopt "the appropriate knowledge, experience and skills necessary to perform prebid, bid, project and postproject activities."[23] It is to a project what a Dominant Strategy (DS) is to the manager. Project capability is imperiled by exceedingly volatile environments that present previously unseen challenges, hence the importance of the proper screening of employees prior to commencing the project or even during the realization of long-term projects. I assume that there are employees with high and others with medium and low project capabilities and consider that it is important to identify them ahead of time in order to secure proper project evolution and ultimately completion within time and budget constraints.

The project management literature addresses the personality of leaders—especially with respect to leadership style—in an attempt to somewhat explain project capability. On the other hand, the Diagnostic and Statistical Manual of Mental Disorders (DSM) IV[24] (the most notorious authority in terms of personality assessment) states that personalities contain structural and functional domains, along four levels,[25] as follows:

1. The behavioral level, marked by expressive acts (e.g., impulsivity) and interpersonal conduct (e.g., exploitative).
2. The phenomenological level, which includes cognitive style (e.g., pessimistic), object representation (e.g., concealed), and self-image (e.g., discontented).
3. The intrapsychic level, with its morphologic organizations (e.g., divergent) and regulatory mechanisms (e.g., dissociation).
4. The biophysical level, referring to mood and temperament (e.g., hostility).

I choose to concentrate on a limited version of this appraisal of personality, with the key components put in parentheses—I am searching for top performers, that is, for the Forces of Production (FP) that can most probably swoop down on POVs and lead the project toward its forecasted completion.

In the context of management, it has been shown that control (being assertive, taking charge) and competitiveness (a dislike of losing and willingness to win) are two personality traits that predict whether managers will achieve higher rankings in the organization's hierarchy; this research covered a seven-year stretch.[26] Some authors found that intelligence was to be regarded as "the most important personality attribute (...), followed by physical condition, maturity, sensitivity, emotional stability (...)."[27] I argue that intelligence is not, however, a personality attribute.

For the most part, the literature focuses on the technical skills of project managers,[28] not on the interpersonal skills of stakeholders, including employees. Technical skills address one of the structural components of Trust (Abilities) but are far from encompassing other psychological variables that form Trust—Benevolence, Affinity and Integrity—or Collaboration—the capacity to be flexible, to exchange information, to solve problems jointly and to be tactful and supportive when interacting with others (so-called consideration).

Various tools are promoted for capturing managers' personalities, and countless studies have reported results based on the "Myers–Briggs personality test" (MBTI). Many authors[29] have resorted in their research to the "Big Five" personality factor theory—agreeableness (friendly/considerate), conscientiousness (efficient/organized), extraversion, openness to experience, and neuroticism (negative emotions)[30]—and to conflict resolution styles such as accommodating, avoiding, collaborating, and competing[31] as well as compromising and enforcing. Various studies report correlations and predictions based upon the Big Five. Yet, it has been noted that "although this model is generally accepted, it has not always been agreed upon that five major factors explain personality, neither has the definition of each factor received consensus."[32]

Research has supported the utility of personality assessment tests, but only to a limited degree, alongside other forms of assessment.[33] Such tests are recognized as an aid to select employees[34]; yet, they do not hold strong predictive power.

6.3.1 A personality theory

Coping mechanisms have been acknowledged by many scholars[35] by way of an established four-personality model[36] (avoiding, anxious, hostile, and stable) under the Attachment theory. Escaping/avoiding personalities tend to run away from the source of threats. Anxious personalities establish shaky relationships with their colleagues. Hostile personalities resort to aggression. Finally, secure (stable) personalities have enduring traits that are most conducive to performance. Several studies (some spanning several decades) have demonstrated that secure individuals tend to be endearing. They stay married longer—in other words, attachment is a good predictor of the length of a relationship.[37] It has been found that couples that display communication, intimacy, secure emotional commitment, and trust are more accommodating and caring than others.

Most obviously, a project leader would want to be surrounded by stable personalities and free of pugnacious ones, hence the importance of identifying them. The coping strategies underlying the attachment model have been at the heart of other research. For example, some authors have emphasized the following characteristics: confrontational coping (hostility); escaping, avoiding, and distancing (avoidance); social support–seeking (possibly anxiousness); and finally self-controlling, being responsible, effortful problem-solving, and positive reappraising (stability).[38] PMBOK 5 (p. 53) refers to "core personality characteristics."

The four personalities described in the Attachment theory (anxious, avoidant, hostile, stable) have been heavily documented. Anxious people, for example, make more errors and repeat their errors over time, thus showing poor learning from past experience.[39] In particular, research shows that contrasting personalities react differently to trust building and trust violation: more secure individuals tend to favor intimacy in the first case and to talk with a significant other in the second scenario. Avoidant and anxious people long for security in the first case and resort to denial (avoidance) or rumination (anxiousness) in case of trust betrayal.[40] Anxious people, in their need to find security, tend to overtrust, that is, to blindly fall into other people's arms without regard for their own vulnerability.

My own research shows that stable individuals form ideal candidates while unstable individuals see a 20% reduction in their capacity to trust and collaborate with others, versus stable individuals. Stable individuals are likely to deal with conflicts in a fruitful way, to display efficient tolerance to risk, to encourage a means-end approach, and to exercise smarter forms of control. In this book, I do not view being "stable" as a personality trait; rather, I believe it is a condition resulting from well-organized personality traits.

People rarely adopt one single personality trait at once, but experience bouts of these various traits (anxiety, avoidance, etc.) so that an overall balance is established between Hostile and Defensive positions, much in line with a prediction deriving from Lotka–Volterra equations.[41] In other words, people tend to be stable. Indeed, stable people make sound decisions and establish durable relationships.

Thus, different strategies aimed at building trust are necessary, with each speaking directly to the particular team member (the one exhibiting offensive[42] or defensive behaviors, or else stability[43]).

The well-known Big Five nomenclature does not allow for differentiating between so-called stable and ill-disposed personalities and hence cannot be used as a valid tool to achieve this, even though it refers somewhat to personality styles as given by the Attachment theory: agreeableness (stability), conscientiousness, extraversion, openness to experience (the reverse of avoidance), and neuroticism (anxiousness). Furthermore, to the best of my knowledge, there is no neurobiological evidence supporting the Big Five theory or other similar theories. However, I can fit some elements of the Attachment theory into the way the brain actually works.

6.4 Hostile and defensive behaviors

I simplify the aforementioned theories by proposing a dual coping mechanism, that of instrumental hostility and defensive hostility. The latter would include anxious (the "maybe"), avoidant (the "I gotta go"), and resistant (the "no") individuals. I bet that many project managers who read this book recognize these patterns.

A note here to warn the reader to differentiate between Dominant and Contingency or Short strategies, and Instrumentally hostile and Defensive behaviors nested deep in the brain (the hypothalamus). As explained in Chapter 1,[44] a Dominant strategy and a Contingency strategy refer to a managerial stance which we compared with strategies wielded by chess players: both positions have value. Hostile and Defensive behaviors refer to behaviors that occur as the result of experiencing Apprehension (a perceived threat, or a vulnerability, e.g., being hungry would lead a tiger to become a ferocious creature). The Dominant strategy, DS, is far from being malignant: it is a strategy adopted to complete the project. Individuals who master Dominant and Contingency strategies (as opposed to Short strategies) have also mastered the art of controlling hostile and defensive behaviors to serve their objectives. On the other hand, defiant behaviors disrupt projects.[45] Note that defensive hostile **behaviors** are temporary in nature.

Hostile and Defensive actions and reactions are **coping mechanisms** because they are awakened in situations of apprehension (perceived threat). Once coping mechanisms have an enduring presence, I character-ize them as (instrumentally) **Hostile position (IP)** and **Defensive position (DP)**. When the enduring presence appears regularly, at least within the realm of a short- or medium-term project, I qualify the position as **profile**. When it is part of the personality of the individual (which is divulged most often during long-term projects), I then use the terms (instrumen-tally) hostile and defensive **personalities**, which, as mentioned, come in three personality types: anxious (A-X)—the "maybe" people; avoiding (A-V)—the "I gotta go" people; and resistant-defensive (A-R)—the "no" people. To review, we have

1. Short-term coping mechanisms (reflex-based): (instrumentally) Hostile and Defensive behavioral responses.
2. Midterm coping mechanisms (including both emotional and cog-nitive efforts): (instrumentally) Hostile positions, IP, and Defensive positions, DP.
3. Longer-term coping mechanisms (with their loads of emotional, cognitive, and conative characteristics): (instrumentally) Hostile and Defensive profiles.
4. Enduring mechanisms (with their set emotional, cognitive, and conative patterns that define individuals): (instrumentally) Hostile and Defensive personalities.

As mentioned, so-called stable personalities are not personality *per se* in my model; rather, stability is a trait or state that results from people managing all four basic coping mechanisms (instrumental aggression, anxiety, A-X, avoidance, A-V, and resistant-defensiveness, A-R) in an effec-tive way, given the circumstances. It is actually risky to pretend that we can identify personalities without doing long-term, longitudinal research, which we have not done. My results thus focus on positions and profiles rather than on personalities, hence the terms "Instrumentally hostile posi-tion", IP and "Defensive position", DP.

Also, the reader, who may not be familiarized with the functioning or the anatomy of the brain, may consult Appendix 6.1 and simply view the brain as a mechanism. In Appendix 6.1, I have endeavored to por-tray the brain much like a machine with its separate parts and underlying constructs (e.g., Trust) in order to ease the understanding. The brain is a convoluted system that is controlled by some set pathways and behaviors, by a clock (in the hypothalamus[46] and in genes), and by homeostasis standards (stored in the hypothalamus[47]).

The key point of this discussion is that a project should be free of spiteful behaviors; they represent significant POVs because hostility elicits hostility (and retaliation), causes instability, and damages trust.

There is neurobiological evidence supporting the four types of coping behaviors that are generated when facing external threats or internal vulnerabilities (in both animals and humans), which, when properly balanced, are assumed to form so-called stable personalities. Consequently, and as mentioned, stable personalities are seen not as personalities *per se*, but as the result of an individual adequately balancing his reactions or coping mechanisms when experiencing apprehensions.

The hypothalamus is, indeed, the brain structure that governs the four coping responses (also called "adaptive responses";[48] again, see Appendix 6.1 to gather more details on the brain and the hypothalamus). It contains encoded (or "encapsuled"[49]) patterns that date back to our human phylogenetic ancestors.[50] Obviously, they impede on the development of trust, because by definition, they manifest themselves in the presence of perceived or real threats (that cause apprehensions). In a purely raw animal sense, the four patterns can be classified as follows[51]:

1. Instrumental aggression[52] is commonly called "hostile" (IP). In nature, one calls it predatory behavior: actions are planned and calculated to catch prey by surprise,[53] serving the predator's interest and causing that prey harm or death. One talks of an abrasive personality, or, in psychiatric terms and when pushed to the extreme, of (functional) psychopaths. From a project management perspective, this refers to hostile, feisty individuals.
2. Freezing, fleeing[54], and defensive rage.[55] These actually translate into the three following defensive positions: anxious (A-X)—the "maybe" people, avoiding (A-V)—the "I gotta go" people, and resistant-defensive (A-R)—the "no" people.

The four behavioral positions model is based on ethological and neurobiological evidence. In short, four responses to threats are recognized, one being purely offensive—instrumental aggression (or IP in the terminology used in this book) and the other three being defensive (freeze, flight, or fight—also referred to as "defensive aggression"). This is represented in Table 6.1, where A-X, A-V, and A-R combine to form a DP in contrast with an IP.

Anxiety (A-X), flight (A-V), and resistant-defensive fight (A-R) are responses to threats that are found in all animals:[56] "Fight, flight, and freezing are at the lowest level and result from a contacting or proximate

Table 6.1 The four coping mechanisms model[185]

	High (T/CRT) (steroids)	Low T/CRT (steroids)	
Low impulsivity (slow)	IP: instrumentally hostile	Freeze/Anxious (A-X: anxious)	High (5-HT/ OXT) (hormones)
High impulsivity (fast)	Resistant-defensive fight (attack to resist: A-R)	Flight (escape) (A-V: avoiding)	Low (5-HT/ OXT) (hormones)

Table 6.2 Defensive coping mechanisms found in actual projects

Mechanism	Example: Year of project
Defensive positions	1993, 1901, 19th–20th centuries, 1996, General
Disagreements	1970
Dissatisfaction	General
Frustration	1993, 2000
Hostility	1997, 1996, 2000, General
Leads to conflicts	1993, 2000
Resentment	1997
Sense of unfairness	2002, 19th–20th centuries

danger, circumstances in which there is little time for analysis of the situation or freedom of action."[57]

Defensive coping mechanisms, transferred into a human context, have been found to be counterproductive in many projects, across decades and across countries (Table 6.2).[58]

Both instrumental and defensive aggressions have been extensively studied, especially in cats, who, like humans, are also mammals. Recent research provides the evidence[59] shown in Figure 6.4.

In the model shown in Figure 6.4, an emotion leads to a response. The response is composed of two mutually exclusive reactions (hence the inhibition-activation mechanism): either offensive (IP—predatory) or else defensive (prey). The periaqueductal gray region (PAG) of the brain serves as an absolute mediator; this means that should the defensive reactions be opted out, the road to the alternative response—offensive reactions—is fully open. To use a mnemonic trick, all there is to remember is that stable people have a talent for balancing their hungry tiger side and their lonely sheep side, with the sheep freezing, fleeing, or resisting depending on which option is readily available in the face of danger. As an exercise, the reader can mentally run through his average day and check when

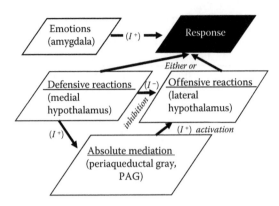

Figure 6.4 Instrumental and Defensive aggression in cats.

they adopted a hungry tiger behavior ("I planned my used car purchase well in advance so that the salesman didn't have a chance to argue with me today") or else a lonely sheep behavior ("Today, I got ripped off by an unscrupulous used car salesman who sold me a lemon")—most readers will be baffled at the result!

In humans, it remains unclear whether hostile and defensive behaviors are completely exclusive as is the case with cats. In other words, people would be able to stay in a defensive posture while also planning for instrumental attacks in the future. This would be the case of, for example, an employee sparring with his/her colleagues whom they perceive to be a threat, thus being constantly on the defensive, while planning to get them in trouble by way of various means (e.g., hiding some documents). However, the reality is that the individual is just switching from one coping strategy to the next over time rather than opting for them both simultaneously. In fact, when I plot some of my data and contrast the constructs of a Hostile position against a Defensive position, I obtain an R^2 of 0.092, which is insignificant, signaling that the two behaviors are mutually exclusive. When I plot natural log functions of the Hostile and Defensive positions, I do obtain a near ascending linear curve as opposed to an elbow curve that would tell the story of mutually exclusive phenomena (see Table 6.1).

Indeed, I can postulate that Instrumentally hostile positions (IP) and Defensive positions (DP) are of two very different natures: they are mutually exclusive (surely in cats; most probably as well as in humans[60]). I opine that instrumental hostility is simply a delayed response to a defensive condition (we will see in Chapter 7 how I put this mathematically). Indeed, instrumental hostility implies low impulsivity. Some authors offer similar explanations as to how the defensive system works.[61] Directed escape is controlled by the medial hypothalamus and undirected escape by the periaqueductal gray, while discriminatory avoidance of danger is the function of the anterior cingulate cortex[62] and avoidance by the amygdala. The defense system is viewed as a hierarchy of coping mechanisms that can, in project management terms, lead to resistance. First, an employee says "maybe," then they say "I gotta go," and finally they settle for a blunt "no."

Many brain areas are indeed solicited by hostile and defensive dynamics. Again, contrary to other personality theories, the four-encoded behavior model based on Attachment theory and various other scholarly works[63] rests on sound neurobiological foundations.

Indeed, besides from the hypothalamus, the brain structures that participate in defensive responses as well as in instrumental and defensive aggression have been identified. As an example, the amygdala (in the brain) plays a fundamental role in emotions, including in "anticipated frustration."[64,65] This is in line with findings by various authors,[66] who report that the amygdala plays a key role in the detection and avoidance of danger (perceived threat)[67] and in socialization.

Along these lines, it has been recognized that defensive mechanisms are prompted according to two conditions: the proximity of the threat and the possibility to escape.[68] According to this model, the flight response entails that the flight option is available and is more pertinent as the prey is farther away from the predator; failing that, the prey may resort to a defensive threat (resistance in the first stage) and then a defensive attack (resistance in a second stage). On the other hand, and still according to this model, the freezing response is assumed to be an available means only if flight is not an option. Again, the more intense the threat (the closer the predator is), the more the prey will resort, as an ultimate option, to a defensive threat/attack. Tests performed with cats submitted to threatening stimuli show this pattern of reactions: Cats momentarily freeze if the threat is not deemed high. If so, the cats will tend to escape if possible. If not, as the predator gets closer, they'll make defensive threats and as the predator gets within reach, they'll attack (defensively) the best way that they can, with the option of escaping as soon as it's feasible. If, on the other hand, the threat seems high and the predator is distant, cats will not hesitate and run for their lives.

Here, I am telling the story of hostility and resistance so often found in a work project environment, especially when the team is not well glued together.

Defensive behaviors in the context of project management do not represent a substantial POV to the evolution of the project as they are expressed by reactive persons who can generally be controlled through adequate managerial tools such as task design and reward. The most concerning POV for a project comes from people who intentionally plan to disrupt it. This is referred to as "Instrumentally hostile individuals" (IP) or "hungry tigers", the extremes of which are so-called functional psychopaths. Individuals adopting hostile aggression (IP) plan to stop the project's development, either by way of, say, the creation of conflicts with stakeholders, information withholding, sabotage, or voluntary delays. This leads to apprehension on the part of other team members: as previously discussed, in the context of project management, apprehension refers to the perception one has that of the *vis-à-vis* has negative intentions and/or hidden destructive agendas directed against oneself. It is deeply rooted in the brain and is at the core of social behaviors. It affects such feelings as trust and may inhibit such behaviors as collaboration.[69] Hence, viperous behaviors have the potential to poison a working atmosphere.

A factorial analysis has been offered that contrasts defensive, instrumental aggression, and stability from a personality point of view.[70] According to this model, well-adjusted individuals belong to a group that is separate from a group consisting of individuals who manifest compulsive, paranoid, passive-aggressive, and schizoid behaviors and from another group of individuals generally described as cyclothymic, hysterical, or sociopathic.

IP-associated types of individuals differ from passive (reactive) individuals, they have been described in these terms: "Given an appropriate incentive, they simply exploit whatever resources are available to gain advantage (…)"[71] Along these lines, it has also been said of them that they are "Pursuing one's concerns at the expense of the other party."[72] Hostile personalities are indeed a POV.

The reader can imagine what kind of dynamic is likely to take place within a project should the workforce be plagued with hostile individuals. A concrete example has been provided: "As a result, they developed a closer relationship with the architects and viewed the contractor's motivations suspiciously, believing they possessed a hidden agenda."[73]

In short, defensive personalities are mostly reactive. A proper environment will guarantee that they perform according to plan. Hostile individuals, however, voluntarily try to ruin the project on many fronts, most commonly in viscous ways.

Given that 70% of a project manager's time is spent communicating with people, and given that we have partly defined collaboration by the capacity to exchange information, hostile individuals—with their hidden agendas—are likely to have a strong negative impact on a project's evolution and to awaken their coworkers' defensive mechanisms. Hostile individuals strive on manipulating interpersonal conflicts (sometimes defined as "the perception of interpersonal incompatibility which includes annoyance and animosity among individuals"[74] as well as "tension, irritation and hostility among team members").[75]

While conflicts associated with tasks are sometimes found to be beneficial—as recognized by many authors[76]—they are not always productive. In fact, dysfunctional conflicts obliterate trust.

I proceeded to retrieve a database dating back to 2008 with $n = 300$ in which all four constructs of relevance (Instrumentally hostile position, IP, Defensive position, DP, Collaboration, and Dependence) were measured.[77] I performed a number of cluster analyses testing for the most-eloquent scenario. The best output is shown in Table 6.3.

Table 6.3 A cluster analysis on $n = 300$

	Class	
	1	2
Hostile (IP)/Defensive positions (DP): (k')	1.55	1.30
Collaboration/Dependence	5.67	1.22
Number of observations	5	295

Only a small percentage of individuals (5/300 = 2%) fall into the potentially hostile category (high [Hostile position/Defensive position ratio] with a relatively high [Collaboration/Dependence] ratio), which again is expected, given the voluntary setting (respondents could leave the project if they wanted to). The results show that our measuring system that employs two ratios (Instrumentally hostile position/Defensive position and Collaboration/Dependence) seems to make sense and could be utilized to assess hostility, subject to further research.

The constant k is found with the Hostile position/Defensive position ratio for group 2; k is established at 1.3 for the large majority of participants. It can be inferred that this sample population of $n = 300$ is representative of the average population.

Interestingly, the corresponding Collaboration/Dependence ratio is similar, at 1.22, to k. Overall, a Hostile position/Defensive position ratio (a proxy for Self-confidence) given a particular Collaboration/Dependence ratio seems to relate to individuals characterized as stable under the Attachment theory. This remains to be probed further.

6.5   *Hungry tigers and lonely sheep under stress*

When conditions that have favored hostile or defensive (freeze, fight, and flight) actions and reactions persist over time, what amounted to simple reflexes become more enduring traits, at least for as long as those conditions remain. Slowly but surely, individuals adopt their actions and reactions as a means of coping and eventually as a means of being. An individual adopting a DP, for example, is to a certain extent making the choice not to be instrumentally hostile. For those who have seen the movie *Silence of the Lambs*, this is easy to comprehend: Jody Foster is, for a while, the lonely sheep. While she plays the lonely sheep, she is not out actively chasing the villain. She becomes the hungry tiger toward the end of the movie, when she catches the (dysfunctional) psychopath.

Both perceived and actual threats beget stress in the eye of the receiver. As is well documented, stress may be dysfunctional. High stress levels are known to increase the occurrences of depression and cancer for both animals and humans.[78] In laboratory experiments, three-quarters of mice submitted to electric shocks with no possibility of escape eventually suffered from tumors. On the other hand, a quarter of the mice that could escape did not develop such health impairments.[79]

In the context of projects, stressful events oblige team members to stand hypervigilant, a posture that leads to fatigue, and thus, to lower productivity. Team members (just like animals) adopt two contrasting positions under stressful conditions: one of instrumental hostility and the other defensive (possibly including resistance). Indeed, findings have been reported whereby small changes in posture breed predator (IP) or defensive mental positions, alongside neuroendocrinal changes.[80] This is important to know for the project manager: Who hasn't faced resistance from team members when trying to implement a change in a project?

When I plot the ratio [Hostile position/Defensive position] as a proxy for risks against [100-Trust[81]], I obtain somewhat of a rectangular hyperbola. Recall that most of my data has been obtained in environments where stress was moderate; I predict that under more stressful conditions, the ratio Hostile position/Defensive position would grow faster than the function [100-Trust] so that the curve would eventually adopt the shape of a rectangular hyperbola.[82] We would have Figure 6.5.

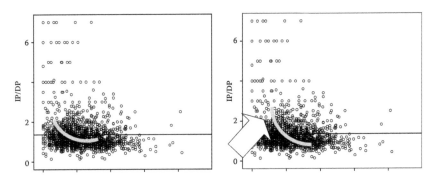

Figure 6.5 A possible U(DS) under stress

This requires further study, but again, finding teams that are willing to be tested while under high levels of stress is quite difficult.[83] However, the tests I performed and that are discussed in Chapter 7, point to the fact that under high stress conditions, the element of vulnerability (herewith somewhat expressed by [100-Trust]) is temporarily put aside in favor of an IP response, which tends to shift the U(DS) into the shape of a rectangular hyperbola.

6.5.1 A study

I conducted a study to verify whether I could differentiate between the two positions based on k' values. I included the construct of Dependence to examine its role in our six core competencies model (Dependence not being a core competency, however).

Figure 6.6 exemplifies how I went about adapting the six core competencies model to fit the two positions.

As can be seen from Figure 6.6, the model differs substantially between the two positions—Hostile (instrumentally hostile) and Defensive. Note

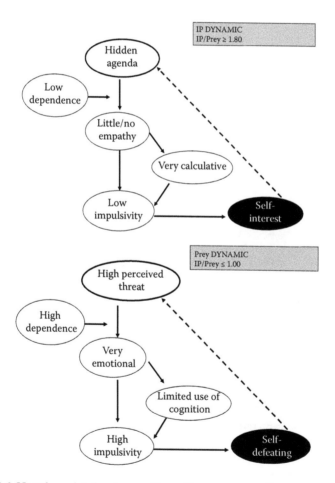

Figure 6.6 Hostile and defensive positions. Key measures: Comparative fit index (CFI) = 0.973 and Tucker–Lewis index (TLI) = 0.918: (*if > 0.9: excellent model fit*);

that I replaced Commitment with Reward because a reward can be measured in the brain: dopamine levels, for example, are an expression of some sense of satisfaction resulting from a reward. Yet, because I tested before the project was completed, I assumed that participants would have left the project if they could not reap any benefit from it. Therefore, that reward was in the end replaced by intention to stay in the project.[84]

6.5.2   Results

I ran a structural equation modeling (SEM) test on a database of 1324 participants issued from 33 different group projects that I collected over the years, using Amos software (part of my analysis consisted of separating them into two groups—controlled, CG, and uncontrolled, UTG.

Results, for the entire group, irrespective of the k' value, are shown in Figure 6.7.

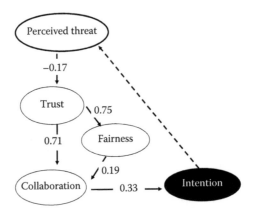

Figure 6.7 SEM on the entire $n = 1,324$ sample.

The model presents a good fit.

6.5.2.1   Single regression analysis: Trust => Collaboration
Irrespective of the SEM, I verified that the link between Trust and Collaboration was as per expectation, for the entire group. The group displays a normal distribution: in other words, it behaves normally. It shows a strong correlation between Trust and Collaboration.

The R^2 for [Trust => Collaboration] is 0.734 for the CG and 0.917 for the UTG;[85,86] both show well-distributed data forming an elliptical shape around the mean (what I call the "football" shape).

6.5.2.2 Factorial analysis

I also wanted to check whether the law of Apprehension—which is at the heart of our model—applied as expected.[87] The factorial analysis results are listed as shown in Table 6.4.

Table 6.4 Factorial analysis on the team of 1324 participants

	Population ($n=1324$)			
	Component			
	1	Tends toward	2	Tends toward
Trust	0.927	1	−0.052	0
Fairness	0.895	1	−0.028	0
Collaboration	0.923	1	0.032	0
Dependence	0.221	1	0.109	0
Hostile (IP)	0.276	≈ 1	0.795	1
Defensive (DP)	−0.198	0	0.841	1
IP versus Defensive	>		<	

Note: Varimax rotation, 3 iterations.

In short, the listed results are in line with all of the other groups I have studied over the years. This means that this sample can probably be generalized so that I could pretend it is a fair representation of the population at large (within the limits associated with my research), at least on these important characteristics.

6.5.2.3 Evidence of the two positions

Various results with respect to two ranges of k' values were obtained. I proceeded this way because my research indicated that a k' value equal to or higher than 1.80 is indicative of a hostile profile and that a value equal to or lower than 1.0 signifies a defensive profile (Figure 6.8).

Table 6.5 compares the sets of values (indicators linking the constructs).

Some differences are noticeable in the relationship between Apprehension and Trust and between Fairness (win-win) and Collaboration with respect to Hostile and Defensive positions. The Hostile individual pays more attention to changes in Trust, perhaps because he/she checks their Defensive position stubbornly, having ulterior motives (bad intentions). The Defensive individual adopts Blind trust more readily, which eventually leads to their demise. As expected, the Hostile individual is less motivated by an equitable collaborative relationship with the Defensive individual.

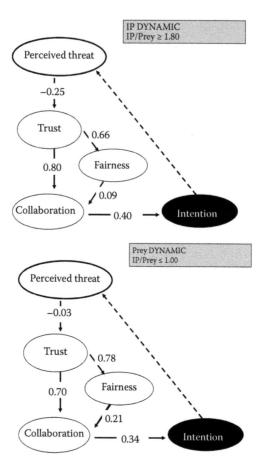

Figure 6.8 Analysis of 1324 participants using Amos. (a) IP position (*k'* ≥1.80) Key measures: CFI = 0.938 and TLI = 0.815 (*if >0.9: Excellent model fit*); (b) Defensive position (*k'* ≤1) Key measures: CFI = 0.969 and TLI = 0.906: (*if >0.9: Excellent model fit*);

6.5.3 *Another study*

I also tested a group of participants who were organizing a food fair, and

Table 6.5 Summary: Sets of indicators for *n* = 1324

	Total	Hostile profile (IP)	Defensive profile (DP)
Apprehension => Trust	−0.17	−0.25	−0.03
Trust => Fairness	0.75	0.66	0.78
Fairness => Collaboration	0.19	0.09	0.21
Trust => Collaboration	0.71	0.80	0.70
Collaboration => Intention	0.33	0.40	0.34

who had to work together before, during, and after (to wrap up the event) its implementation. I tested them before they got to know each other and afterward by way of my questionnaire.[88] In addition, I measured the level of interdependence to see how it would fit into my model. Note that the group was small, and that it is impossible to draw sound statistical conclusions from it; note also that the group expanded, in fact, it nearly doubled between the start date (sample population = 14) of the project and its completion date (sample population = 30). Yet, I feel authorized to pen down some observations, within the inherent limits of such small (and changing) populations (Tables 6.6 and 6.7).

I observe that participants engaged in the project from an interpersonal point of view (see Appendix 5.3). They were quite excited about the

Table 6.6 Main constructs *before* (n = 14)

	Population (n = 14) *before*							
	Scale of 0–100					Likert scale of 1–7		
	Trust	Fair.	Coll.	Intent.	Dep.	IP	DP	IP/DP
Value	85	76	82	84	75	4.2	3.3	1.3
Minimum	55	50	55	43	52	2.0	1.0	0.8
Maximum	99	89	96	100	100	5.8	5.6	2.5

Table 6.7 Main constructs *after* (n = 30)

	Population (n = 30) *after*							
	Scale of 0–100					Likert scale of 1–7		
	Trust	Fair.	Coll.	Intent.	Dep.	IP	DP	IP/DP
Value	80[a]	74	66[a]	76[a]	64[a]	3.7[a]	3.0	1.2
Minimum	46	36	35	15	29	1.8	1.0	0.7
Maximum	100	100	84	100	100	5.8	5.6	2.6

[a] Significant difference between the *before* and *after*, at $\alpha < 0.05$.

project and tended to trust and collaborate with each other rather readily. After the food fair, levels of most constructs declined to a relational or even an interactional level (see Appendix 5.3). Note that the Hostile position was stronger at first. In the end, the ratio of Hostile position/Defensive position remained about the same, near or at $k = 1.3$. When the level of interdependence abates, so do the values of Trust, Fairness, and Collaboration.

These results are quite typical of how people react when they engage in a project and when they eventually become part of the transformation stage of the project whereby People' then interact within the context of active and influential Plans', Processes', and Power' structures. The reduction in levels of intensity for the constructs of Dependence, Trust, Collaboration, and Intention to stay in touch with others after the event is significant between the *before* and *after* periods (at $p < 0.05$[89]).

The law of Apprehension, checked by way of a factorial analysis, proved true again in the *after* period, once team members have had a chance to work with each other, as shown in Table 6.8.

Comparing component 1 and component 2, one can suppose that as apprehension (Defensive position/Hostile position) augments, the mobi-

Table 6.8 Factorial analysis *after* ($n = 30$)

	Population ($n = 30$) *after*			
	Component			
	1	Tends toward	2	Tends toward
Trust	0.958	1	−0.073	0
Fairness	0.932	1	0.002	0
Collaboration	0.980	1	−0.111	0
Dependence	0.762	1	0.149	0
Hostile (IP)	0.276	≈ 1	0.865	1
Defensive (DP)	−0.294	0	0.879	1
IP versus Defensive	>		<	

lization of the vital resources by team members fades out.[90]

I cannot relate these findings directly to Hostile and Defensive positions, but I can "guesstimate" that there is an element of such positioning in the way that team members in the food fair project engaged and disengaged with it. When put in an environment that does not require hostile or defensive reflexes, People still deeply rely on the core principle underlying these two contrasting coping mechanisms by way of adopting, most probably in an unconscious manner, either a Hostile or a Defensive position, or, more realistically, a nonsimultaneous mixture of both (leading to an ideal k ratio of 1.3).

6.6   *Identifying hungry tiger (IP) profiles*

Ideally, projects must be exempted from hostile individuals (so-called hungry tigers); they represent a considerable POV. It is thus important for the feasibility expert to be equipped with a means of detecting them.

I conducted a study with the objective of measuring Hostile (PIP) and Defensive *profiles* (A-X, A-R, and A-V), with the former (instrumentally hostile individuals or hungry tigers) measured by four known traits: calculative, cold, egoistic, and cunning. I also sought to measure such constructs as Apprehension, Trust, Fairness, Cooperation, and Intent to stay in the project.

6.6.1 The study

I approached 343 artists from 9 different one-time shows and asked them to evaluate their director using my questionnaire. I merged the results because the responses were almost identical across the nine directors, although I recognize the limits of such a procedure. The questionnaire had been used in a number of studies and had been meticulously tested for its psychometric value. For all constructs, except for the Hostile position/Defensive position, results are given in percentage, computed from a 7-point Likert scale. The constructs of the Hostile position and Defensive position are measured through five items each and results are computed on a 7-point Likert scale, but not transformed into percentages. Results for the Hostile position are also posted on a 7-point scale. The code "stable" refers to a self-assessment of stability.

The research took place during the 2010s in Canada. Fifty-two percent of the participants in the sample were female, 42% male, and 6% did not respond to the question about gender. Forty-two percent were aged between 18 and 30, 15% were aged between 31 and 44, and 43% were between 45 and 65 years old. The 343 participants were composed of nine groups, most of them musicians and other artists as well as support staff (e.g., stage set) who engaged in the project of producing a one-time show involving a choir and/or an orchestra. The team had rehearsals once a week for nearly a year and twice a week in the few weeks prior to the show. I distributed the questionnaire during a break between two sessions of a rehearsal three weeks prior to the concert with the approval of the project director. The team members were to evaluate their director and, embedded in the questionnaire, I included questions about the respondents' profiles.

The following constructs were measured with respect to how the team members viewed the project director: the Hostile position, the Hostile profile (related to how much Apprehension the project director generated), Trust, Fairness (a sense of a win-win relationship), Cooperation, Conflict, and the Intention to stay in the project. The following constructs were measured with respect to the participants (how they themselves felt toward the project director): the Defensive position, A-X, A-R, A-V, the Stable as well as the Hostile position (potential for instrumental aggression), and Dependence.

My reasoning was that I would see the impact of a Hostile position on the perception of the director taking into account two measures of Apprehension: Hostile position/Defensive position and profile (PIP or "Profile of Instrumental hostility").

The questionnaires were collected on the spot and analyses were performed using a "Statistical Package for the Social Sciences" (SPSS) 15 as well as Partial Least Square (PLS 4).

6.6.2   Results

I provide an analysis of my key findings segued from identifying the basic emerging model to examining the role of a Hostile position (the way individuals respond to perceived threats by using uncivilized behaviors), the creation of a project capability index (PCI), and the identification of Stars.

6.6.2.1   Main statistics

The main statistics obtained for the entire group was as shown in Table 6.9.[91]

Table 6.9 Main statistics for the entire group ($n = 343$)

	IP	DP	IP/DP	PIP	Trust	Fair.	Coll.	Confl.	Intent.	Dep.
	On 7-point Likert scale				In %					
Mean	3.86	3.30	1.30	2.60	75	68	70	29	70	59
Std.	1.00	1.08	0.56	1.66	14	15	14	13	22	21
Min.	1.00	1.00	0.26	1.00	23	21	12	3	14	10
Max.	6.60	6.60	5.00	7.00	100	100	100	81	100	100

Note: IP = Instrumentally hostile; DP = Defensive; PIP: Instrumentally hostile profile; Confl. = conflict level; Intent. = intention to stay in the project; Dep.: level of dependence. IP, DP, IP/DP, and PIP are on a 7-point Likert scale; Trust, Fairness, Collaboration, Conflict, Intention, and Dependence are on a 100% scale.

These results indicate that for the entire team, their relation to the project director is mostly in a transactional/relational mode (with Trust, in particular, being at 75%; see Appendix 5.3), which means that the project director was mostly regarded as a "business" colleague rather than as a friend. I also observe that the Hostile position/Defensive position ratio (the reverse of Apprehension) of 1.3 corresponds to the standard k-value found in other similar research.

6.6.2.2   SEM model

I ran dozens of structural equation modeling (SEM) scenarios with the best model appearing as shown in Figure 6.9.[92]

The so-called estimators (β values) and the degrees of significance (p-values) indicate that this model is acceptable. The model reads as follows: self-confidence (the ratio of Hostile position/Defensive position, the reverse of Apprehension) has a positive impact on Trust. As Trust builds up, collaborative behaviors increase.[93] This is encouraged by a sense of a

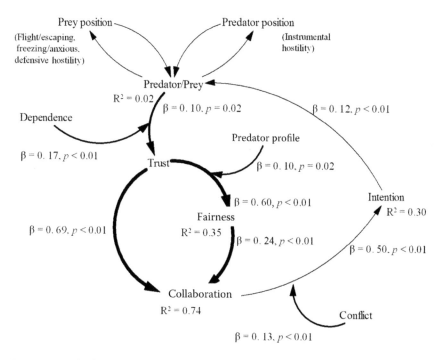

Figure 6.9 The best model using SEM ($n = 343$).

win-win situation (Fairness) between the parties. As Collaboration takes place, the Intention to stay in the project gains ground; this, in turn, has a positive impact on Self-confidence. Dependence on the director (and hence on the project) serves as a moderating variable.

Conflict also serves as a moderating variable, this time between Collaboration and the Intention to stay. This may be somewhat surprising because many models assume conflicts have a direct negative impact on a group dynamic. However, as noted previously, conflicts within projects can have two sides: functional or dysfunctional. Therefore, it is logical to see that Conflict serves as a moderating variable because it will affect the Intention to stay positively or negatively depending on its nature and intensity. The reader may think of conflicts as if they were hurricanes. North of the equator, hurricanes rotate counterclockwise, but south of it, they spin clockwise, yet they both are hurricanes, which may cause heart-breaking devastation.

These results corroborate findings extracted from various projects that spread across cultures and decades. In the case of the Pendjari National Park Project in Benin in 1993, for example, it has been noted that a sense of trust leads to positive results while conflicts muster delays.[94] Countless scholars highlight the fact that trust promotes team integration

(Collaboration).[95] Some recognize the fact that dysfunctional conflicts bear additional costs.[96] Hence, at least two of the three elements of the triangle of constraints (time and budget) are afflicted by the dynamic exemplified in Figure 6.9.

Also of interest is the fact that the Hostile profile (PIP) equally serves as a moderating variable, this time between Trust and Fairness (the sense of a win-win situation). This means that a Hostile profile (in the sense that a profile is composed of a certain number of traits) does not cause conflict by itself, but rather impacts on the emotional (Trust) and cognitive (evaluation of Fairness or the sense of a win-win situation) views of the team dynamic. This also makes sense: for some people, certain personalities will be conducive to better interpersonal relationships, while for others, the opposite view prevails. Hence, profile by itself cannot explain hardships that may arise during the course of a project.

Rather, what seems to count more directly are the levels of self-confidence (the Hostile position/Defensive position ratio) and of the Hostile profile considered jointly. This is what I examine next.

6.6.2.3   *The role of hungry tigers*

I computed and compared all profile types (IP, A-X, A-V, and A-R; adding Stable as a check measure[97]). Anxious, escaping/avoiding and resistant-defensive profiles all showed a normal distribution curve when plotting the percentage of individuals versus intensity. PIP (instrumentally aggressive) and stable profiles displayed opposite trends, as shown in Table 6.10.

Table 6.10 Comparing PIP (instrumental aggressive) and stable populations ($n = 343$)

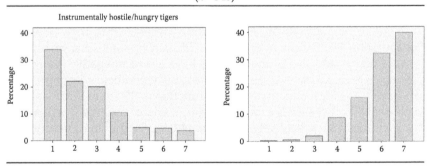

Unequivocally, when most people claim to be stable, the same people contend not to manifest instrumental aggression.

I ran some factorial analyses[98] and found that Hostile position was an excellent discriminator between the groups, as shown in Table 6.11.

I observe that when the Hostile value is higher than the Defensive value, scores for Trust, Fairness (win-win), and Collaboration tend

Table 6.11 Group segregation
separation using IP (*n* = 343)

Construct	1	2
Trust	0.869	0.054
Fairness	0.765	0.262
Collaboration	0.897	0.094
Intention	0.743	0.108
Profile (PIP)	−0.185	0.050
Hostile (IP)	−0.121	0.841
Defensive (DP)	0.264	0.760

toward 1. However, when the Defensive value is higher than the Hostile value, (i.e., when one feels very vulnerable), the scores for these same constructs tend toward 0. Interestingly, in the first group, the Hostile position is negative and in the second, it is positive.[99] This could well be an indication, albeit subject to more verification, that Hostile position is a key marker of the project capability construct.

A discriminant analysis further tended to confirm the intricate part that the construct of a Hostile position plays in our six core competencies model (Table 6.12).

Table 6.12 Discriminant analysis for the entire group
(*n* = 343)

Construct	A	B	C
Trust	58	72	84
Fairness	50	64	79
Collaboration	54	67	80
Intention	33	65	90
Hostile (IP)	3.28	3.79	4.15
Defensive (DP)	3.00	3.45	3.23
Hostile (IP)/Defensive (DP)	1.27	1.21	1.40
Profile (PIP)	2.78	2.79	2.31
Number of observations	51	154	138

I find in the analysis that the Hostile position/Defensive position ratio is quite close for class A and class B; the same applies for PIP. For class C, however, the Hostile position/Defensive position ratio diverges by about 15% versus classes A and B, and the value of the Hostile position differs by about 18%. Furthermore, each class represents the three known interactional modes with class A being transactional,[100] class B being relational, and class C interpersonal (see Appendix 5.3).

6.6.3 An index to separate the Stars from the average team members

Based on these findings, I developed an algorithm (the project capability index)[101] that takes into account the Hostile position and the variance between the functional and actual Hostile and Defensive values, functional and real Hostile position/Defensive position ratios, and the three core psychological constructs: Trust, Fairness (sense of a win-win situation), and Collaboration. Since we had measures of Hostile and Defensive positions and Hostile profiles (PIPs), I suspected that I had enough information to separate exceptional ("top performers" or Stars) from average individuals ("others").

Results show that the project capability index (PCI) sharply divides the top performers (the Stars and incidentally the most stable people) from the others, as shown in Table 6.13.

Table 6.13 Overall summary of the two distinct groups generated by the PCI ($n = 343$)

	PCI	PIP	Prof	IP	DP	IP/DP	Trt	Fair	Col	Cfl	Int	Dep
	On 7-point Likert scale						In %					
Stars (Top Performers)												
M	24	1.30	2.75	4.69	3.30	1.56	88	87	88	13	93	70
Sd	12	0.56	0.87	0.75	0.95	0.57	6	5	6	5	12	21
Mn	12	1.00	1.00	3.00	1.20	0.81	79	79	74	3	52	22
Mx	49	3.00	4.50	6.00	5.00	3.38	98	100	100	22	100	100
Others												
M	76	2.69	3.45	3.80	3.30	1.28	74	67	69	30	69	58
Sd	27	1.67	1.02	0.99	1.09	0.56	14	15	13	12	22	21
Mn	50	1.00	0.50	1.00	1.00	0.26	23	21	12	5	14	10
Mx	147	7.00	6.00	6.60	6.60	5.00	100	100	96	81	100	100

Note: M = Mean; Sd = Standard deviation; Mn = Minimum; Mx = Maximum; Prof = Hostile Profile; Trt = Trust, Fair = Fairness; Col = Collaboration; Cfl = Conflict; Int = Intention to stay in team; Dep = Dependence.

As can be seen in Table 6.13, the PCI discriminates well between the Stars (the top performers) and the others. Values of Trust, Fairness, Collaboration, and even the value of Intention, correspond to the most stable individuals, using an interpersonal mode (above 85%). This means that these individuals recognize the efforts for trust building, the sense of a win-win situation, and the collaborative endeavors sustained by the project director and "buy" into his/her performance; their Hostile position/Defensive position ratio is functional, at $k' = 1.56$ (see Appendix 5.3). Their PCI value is approximately one-third that of the "other" group, those who are not Stars (not top performers). Standard deviations for the constructs do not have as wide a spread for the Stars (top performers) as for the others, which indicates that the former are more focused.

A factorial analysis further confirms the utility of the PCI, as shown in Table 6.14.

Table 6.14 Factorial analysis in consideration of the PCI ($n = 343$)

	All participants		Stars (Top perf.)		Others	
	1	2	1	2	1	2
PCI Index	0.232	−0.832	0.035	−0.900	−0.800	0.004
Trust	0.832	−0.066	0.911	−0.039	0.735	0.502
Fairness	0.457	0.658	0.553	0.561	0.083	0.699
Collabr.	0.810	0.286	0.794	0.427	0.577	−0.066
Intention	0.268	0.668	0.338	0.738	−0.077	0.855
Mean	0.592	0.387	0.649	0.422	0.330	0.497

I observe that the "'Stars 1'" (top performers) have a higher construct mean (0.649). Also, when the PCI index is positive (0.035), values of Trust, Fairness, and Collaboration tend toward 1. Hence, the PCI index respects the law of Apprehension, which predicts that a functional Hostile position/Defensive position ratio leads to higher values for these important psychological constructs.

I performed another discriminant analysis, this time using the PCI and comparing the three groups (all participants: Stars and others), with the results in Table 6.15.

Table 6.15 Discriminant analysis based on the PCI ($n = 343$)

	Class									
	A	B	C	D	E	F	G	H	I	J
Trust	91	89	87	87	94	89	93	79	89	88
Fair.	86	85	86	80	86	88	82	82	89	83
Coll.	86	82	88	91	92	89	85	74	89	87
Int.	74	95	83	86	95	95	71	52	100	67
IP	4.88	4.61	5.07	5.70	5.60	4.46	3.20	4.00	4.54	3.93
DP	3.96	3.43	2.93	3.30	4.00	3.06	3.20	4.00	3.14	2.60
IP/DP	1.26	1.50	2.11	1.75	1.41	1.50	1.00	1.00	1.55	1.51
PIP	3.00	3.33	1.00	6.00	5.50	3.00	5.00	1.00	1.07	3.67
CPI	52.40	71.10	19.63	143.4	85.79	50.16	117.5	28.19	17.66	72.22

Since, according to my algorithm, the PCI must be under the value of 25, one can see that the top performers/most stable people are found in classes C (3 participants) and I (14 participants) for a total of 17 individuals out of 343, or 5% of the total participants. These two classes also record the best measures for all constructs as well as for the Hostile position/

Defensive position ratio (albeit quite high for class C) and the score for a Hostile position (IP), which is expected since the PCI algorithm is based on these measurements.

Lastly, I decided to reconfirm that the two groups (Stars and others) were indeed significantly different, by performing a one-way analysis of variance (ANOVA). Groups (top performers/stable vs. others) are indeed found to be significantly different at $p < 0.05$ on all constructs except Defensive Position DP at *sig.* = 0.999).

The PCI seems to effectively delineate those who are stable (called "top performers" or "Stars") from those who are not as stable. This PCI allows us to establish subcategories among the "others" group, so that I can identify, say, the next 10 or 20 superior performers below the actual top performers.

6.6.4   Discussion

My results indicate that there are two distinct groups, called "top performers" and "others" (Table 6.16).

Table 6.16 All participants: Key indicators ($n = 343$)

	CPI	PIP	Prof	IP	DP	IP/DP	Trt	Fair	Col	Cfl	Int	Dep
	On 7-point Likert scale						In %					
All Participants												
M	49	2.60	3.40	3.86	3.30	1.37	75	68	70	29	70	59
Sd	33	1.66	1.03	1.00	1.08	0.56	14	15	14	13	22	21
Mn	12	1.00	0.50	1.00	1.00	0.26	23	21	12	3	14	10
Mx	147	7.00	6.00	6.60	6.60	5.00	100	100	100	81	100	100

Note: M = Mean; Sd = standard deviation; Mn = Minimum; Mx = Maximum; Prof = Profile; Trt = Trust; Fair = Fairness; Col = Collaboration; Cfl = Conflict; Int = Intention to stay in team; Dep = Dependence.

Top performers (Stars) display high values for the constructs of Trust, Fairness (win-win), and Collaboration. They tend to want to stay in the project longer than the "others" group. They feel a sense of attachment epitomized by high values of both Trust and Dependence.[102] Their value for the Hostile position/Defensive position ratio is 1.3, which equals the standard *k*-value, and this value is half that of the "others" group (at 2.69). They also display much less standard deviation than the "others" group for all constructs, which means they are more focused. They show an excellent balance between the core constructs that are Trust, Fairness (win-win), and Collaboration. They are not geared toward conflicts.

Overall, the so-called top performers are the ideal employees in projects of any kind. They truly are the Stars.

Table 6.17 shows that for the same Hostile position/Defensive position ratio of 1.3 (the *k*-value), the two groups diverge quite substantially. The Hostile position and the Hostile profile appear to play a role that could explain this difference.

Table 6.17 A comparison between top performers and "others" given $k=1.3$ ($n=343$)

	PIP	Prof	IP	DP	IP/DP	Trt	Fair	Col	Cfl	Int	Dep
	On 7-point Likert scale					In %					
Top	1.33	2.25	4.83	3.77	1.29	89	86	88	12	94	74
Oth.	2.88	3.18	4.20	3.25	1.29	78	76	75	15	75	63
Df%	−54	−29	15	16	0	15	14	17	−22	26	17

Note: Top = Top performers; Oth.= Others; Df % = Difference in percentage.

It can be noted that when the Hostile position/Defensive position ratio equals or tends to equal the value of the Hostile profile (1.29 in the last case, close to the *k*-value of 1.3), top performers stand out. In other words, stable employees are those who manage to find a fair and harmonious balance between their self-confidence level (Hostile position/Defensive position ratio) and their tendency to respond to threats (Apprehension) by way of instrumental hostility. Those employees are most likely to face project volatility with a mature and efficient response.

It can also be observed that for the same *k*-value of 1.3, the Stars (top performers) display construct values that are on average approximately 20% higher than that of the "in between" group. The Stars (top performers) tend to see the world in a more positive way and to have, from the start, a lower hostile profile. In other words, when the hostility level doubles, the perception of the world changes by approximately −20%. This tends to confirm a generally accepted view that unstable individuals likely view the world in more negative terms than stable individuals.

Most particularly, stable individuals have been found to share a number of characteristics that are highlighted in my results and discussion. High project capability individuals rank well on Trust, Fairness (win-win), and Collaboration. They are focused; they are well balanced emotionally, intellectually, and from a behavioral point of view (Trust, Fairness, and Collaboration); and they maintain their relationships for longer, and display little resentment. In addition, they have a functional level of self-confidence and are not geared toward conflict; they have a strong sense of attachment with the project and enjoy a positive vision of the world. This likely indicates that they will perform better (hence, my labeling of "top performers"). All these characteristics are in line with theories about secure personalities. In harmony with my argument, some authors make the following comment:

> A large literature indicates that secure people engage in positive, intimacy-promoting, and tension-reducing interpersonal behavior, and have a positive memory bias for interpersonal exchanges. In contrast, avoidant people are relatively disengaged during social interactions, easily become bored or tense, and are likely to forget their partner's feelings and remarks.[103]

The authors mention that stable people "try not to become either positively or negatively emotional." Antsy people have a hard time forgiving; they juggle between positive and negative reactions and debate between their need for closeness and their need for self-defense. They resort to negative self-assessment.

Stable individuals are flexible and respond to the needs of others. They privilege people first, rather than tasks. They are predictable and use a heuristic logic, taking into account a large view of the world rather than a padlocked, somber view of it. They do not display episodic mood changes or unexpected variations in their decisions. They are usually better at developing their own talents and tend to excel academically and/or in extracurricular activities. They are able to concentrate, are not bothered by negative thoughts or souvenirs, and don't feel victimized by events or by others.[104] Rather, they show control in their cognitive, emotional, and conative processes. Some authors report quite succinctly on the difference between stable individuals and the "others" by way of the following results obtained from their own research.[105] For what I judge the construct of Trust to be, the following traits are proposed (with a valence evaluation in parenthesis): trustworthy (0.95), caring (0.91), responsible (0.91), attractive (0.79), and confident (0.63). For Fairness: emotionally stable (0.91). For Collaboration: sociable (0.90) and intelligent (0.70). Finally, for Hostility: dominant (-0.30), unhappy (-0.70), aggressive (-0.75), mean (-0.78), threatening (-0.78), and weird (-0.85).

In summary, stable individuals have a good grip on their hostile tendencies, both in terms of action (k) and their general demeanor or general level of hostility (IP). These people are likely to outperform others. They represent the ideal candidates that a manager would want in a volatile environment, the likes of which are common in contemporary international projects. They are also the ones that should most likely be identified from the get-go and rewarded so that others can follow their tracks. As a manager, wouldn't you agree?

I strongly suspect that my measures associated with the Hostile position/Defensive position ratio and Hostile profile are not only ingrained in the brain (most particularly, in the hypothalamus and the PAG for the coping mechanisms), but that they are part of the evolution of human nature.

As DSM-IV notes when referring to human beings at an early age: "From the evolutionary model presented earlier, trust and mistrust represent facets of the pleasure and pain constructs, generalized to "adaptational" venues within the physical environment (...)."[106]

People who have a better perception of the quality of trust to attribute to others are likely to be less defensive, that is, they are likely to be more open and to find instrumental hostility unjustified. This behavioral disposition, in turn, generates a climate that is conducive to a friendly atmosphere,[107] which then forges a better working environment.[108] People who gain trust and respect interdependence are likely to face risks (and thus the threats that are part of volatile environments) and apprehension most effectively. As it has been pointed out in the context of supply chain management: "Trust must exist in an alliance since there is interdependency between the parties to mutually achieve goals (...) trust needs to exist for allies not only to share critical strategic information."[109]

Some scholars have demonstrated that two characteristics of stable individuals—meticulous problem-solving and positive reappraisal—lead to satisfactory outcomes in a significant way when it comes to interpersonal relationships.[110] In a more recent study, it has been shown in the context of buyer–supplier industrial relations, that levels of trust are highly correlated to positive performance.[111]

Overall, it appears fair to say that stable employees are likely to perform better, or at least to engage in more productive relationships with their peers. The PCI seems to be a strong predictor of such a positive outcome.[112] Of course, the challenge for the feasibility analyst is to identify the Stars (top performers) ahead of time; this is only achievable by examining past records or by developing tests based on scenarios that would highlight hostile personalities and hostile profiles, as well as through interviews.

6.7   Functional and dysfunctional FP_{nc}

The previous section examined top performers, that is, Forces of Production that excel—the Stars. I now discuss the Unfits (uncontrolled Forces of Production, FP_{nc}): the Forces of Production that may block the development of a project. The Unfits (FP_{nc}) may perform a variety of actions that prevent others from completing their tasks in three ways: they affect their mobility by delaying the action plan or by transmitting erroneous information, they hold possession of assets such as valuable information or even material or equipment, and they slow down or attempt to stop activities scheduled for the day. There are two types of FP_{nc}:

1. Those that are functional
2. Those that are dysfunctional

Functional ones have a hostile profile and have found a way to adapt to their environment in order to maximize their selfish gains.

Dysfunctional FP_{nc} present anxious, avoidant, or resistant-defensive profiles that are completely distinct from general behavioral norms and that cannot bring about a positive contribution to a project. Functional FP_{nc} are hard to detect while dysfunctional FP_{nc} stand out. Overall, both types of individuals are Unfits.

Different views ...

Often, functional$_{nc}$ are willing to wait in order to get a sense of reward from their action. They plan for long-term, self-centered gains. They often go undetected, thus representing a significant POV, because they can at times excel at executive functions and in displaying goal-oriented behavior (which is why they are functional). Functional FP_{nc} display weaker responses than others to disturbing emotional information[113] that could affect a project. Because they at times do not seek material gain, longing instead for other forms of gain such as mere power,[114] they are not necessarily sensitive to financial rewards and may not be willing to adjust their behaviors to fit the needs of the project. They have their own agendas— they belong to the Instrumentally hostile category. Functional but uncontrolled in some sense, these Forces of Production respect social norms, but will brush them off when convenient to do so.[115] They are often charming. They may dream of taking charge of the project, but they don't tend to reveal their intentions; they may play the "diva" when this meets their objective. They are known to excel at conceptual and linguistic tasks. This alone separates them from the dysfunctional Forces of Production,[116] as the latter tends to fuss rather than to charm. If I were to test the functional FP_{nc}, I would probably find that they rate low on the Hare psychopathy checklist—revised (PCL-R) scale[117] on dysfunctional psychopathology. Yet, functional FP_{nc} share the four fundamental traits of psychopathology as mentioned before: they are (1) calculative (manipulative) in their collaborative efforts, (2) cold (they show little or no empathy) in their approach to control and transparency, (3) selfish when committing to projects, and (4) sneaky (deceitful) in their trust-based relationships. Some authors refer to Machiavellian tendencies.[118] Perhaps this terminology resonates with the reader; in any case, most project managers or team members have been confronted with such types of individuals and their shenanigans.

In stark contrast, dysfunctional FP_{nc} cannot hide; they exhibit their antisocial behavior right away.[119,120] They are reactive in nature—they belong to the defensive category as defined previously. They don't plan; they merely react emotionally to the situation, at times in the extreme.

Recall my model of the six core competencies (Figure 5.4). I found that the construct of Control/Transparency (or equivalently, the Hostile position/Defensive position construct) had a direct one-way link to Trust. There was no direct feedback arrow to the construct of Control/Transparency (Hostile position/Defensive position or the reverse of Apprehension) from the construct of Trust. This forces the entire psychological mechanism to go through the process that sees Trust leading to Collaboration with time, and with Fairness being a mediating variable. Collaboration encourages (C+) Commitment, which then is followed by a feedback loop stretching from Commitment to the construct of Self-confidence (the opposite of Apprehension). This means that functional and dysfunctional Forces of Production (FP) eventually get exposed at some point, most likely during the most pragmatic stages of a project, which call for commitment and whereby tasks are to be completed imperatively. Indeed, there is no other way around it because, again, Trust does not feedback directly to the construct of Self-confidence (or its reverse: Apprehension). I found this by way of numerous macroanalyses spread over many years of research. Interestingly, the anatomy and physiology of the brain corroborate this flow (see Appendix 6.1). As seen before, the hypothalamus (the center for apprehension) does feed the pituitary gland (from which emanates, directly or indirectly, the key hormones relating to the four coping mechanisms), but the pituitary gland does not feed back to the hypothalamus. Hormones have to go through the entire blood system before reaching the hypothalamus again. What I am saying here is that the essence of the six core competencies model may be supported by neurobiological evidence. For me, this is good news: I estimate that the brain works much like a production process[121] with its specific parameters (e.g., those ingrained in the hypothalamus such as body temperature, energy level, heart beat, hours of sleep, and sexual desire) and flows of activities (like that of hormones). Naturally, this flow has to transpire in the way people (read: FP) behave, and this is exactly what I have discovered.

As an example, it has been found that antisocial individuals (FP_{nc}) typically show significantly reduced prefrontal activity;[122] this is in line with what is to be expected from our model. On the other hand and to paraphrase some authors,[123] functional FP_{nc} perform differently; they can meet "specific computational demands" and excel at "representation, valuation, action selection, outcome evaluation and learning." There are other marked idiosyncrasies between functional Hostile (functional FP_{nc}) and defensive, dysfunctional FP_{nc}. Appendix 6.1 compares hostile (instrumentally hostile individuals) and defensive psychological constructs and their accompanying brain structures. Hostile individuals show little changes in their vegetative system before attacking: a slow cardiac rhythm allows them to tolerate stress more effectively. The ability to keep their "cool" in the presence of high stress conditions may even encourage the employers to hire such individuals as project managers. Indeed, panic leads to more errors in decision-making:[124] dysfunctional FP_{nc} threaten projects and, additionally, tend not to learn from negative experiences.[125] Overall, functional FP_{nc} seem much more attractive than dysfunctional FP_{nc}; the problem is that they are a potent source of headaches because of their hidden agendas. They are, indeed, invisible POVs.

6.7.1   A study and its results

In one of my studies, I found the distribution for a group of 191 participants (shown in Figure 6.10).

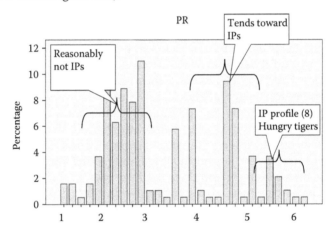

Figure 6.10 Distribution of hungry tiger profiles (PIPs) ($n = 191$).

Where the x-axis is the self-perceived hostile profile. Eight participants out of 191 definitely fall within the "pure hostile profile" (but none appear at the extreme value of more than 6.75/7.00; this is to be expected, if it were the case, they would have been kicked out of the team and maybe would already be spending time in jail!). This represents roughly 4% of the sample population, that is, 4% of people admitting (unknowingly) to have a pure hostile profile. This is in line with our observation made in Chapter 1 that every project contains at least 4% POVs. Interestingly, 5%–10% is the average known level of violent individuals in most societies since the sixteenth century.[126]

The majority of people prefer to be seen as "low-hostility individuals", or else are "low-hostility individuals" because their response is below the middle point of 4 on the Likert scale of 7 points. Yet, the majority of people are neither overly feeble nor overly bellicose. In either case, they would be completely dysfunctional within the project. In fact, most answers range between 2.5 and 4.5, with the majority being close to 3.5 on average, signifying a docile temper that again is to be expected within project teams. I conducted various statistical studies on this sample group ($n = 191$); I present the results in the next sections.

6.7.1.1   Main construct values

Comparing the values of the main constructs between the eight hostile individuals and the rest of the team, I found a significant difference[127] (Table 6.18).

Table 6.18 Main constructs measurements ($n = 191$)

Construct	IP	DP	IP/DP	Trust	Fair.	Coll.
	On 7-point Likert scale				In %	
8 IP	4.2	4.3	1.0	84	86	82
183 others	3.5	3.5	1.2	84	79	79
Difference	+0.7	+0.8[a]	+0.02	0	+7	+3

[a] Significant change between 8 IP and 183 non-IP at $p = 0.05$.

Using a p value of 0.05, I note that the only significant difference pertains to the Defensive position construct. However, the ratio Hostile position/Defensive position or its reverse (Apprehension) is not significantly different: hostile individuals (functional FP_{nc}) mix well with other people and pass by undetected.

6.7.1.2   Factorial analysis

I obtained Table 6.19 by performing a factorial analysis[128] where two groups appear.

Table 6.19 Factorial analysis of the groups of $n = 191$

	Component for $n = 191$			
	Group 1	Tends toward	Group 2	Tends toward
Trust	0.861	1	−0.027	0
Fairness	0.832	1	0.065	0
Collaboration	0.871	1	0.074	0
Hostile (IP)	0.228	0	0.798	1
Defensive (DP)	−0.285	0	0.741	1
IP versus Defensive	>	–	<	–

The law of Apprehension is respected yet again: when Apprehension is low (Defensive position < Hostile position), that is, when it tends toward 0, the constructs of Trust, Fairness, and Collaboration each tend toward 1. This can be interpreted as follows: when team members feel vulnerable (Defensive position ≥ Hostile position), they tend not to trust, they feel they are treated unfairly, and they are hesitant to collaborate.

6.7.1.3   Structural equation modeling

I performed an SEM on the sample consisting of 191 participants. The model parameters were found to be relatively poor: the CFI = 0.796, which is smaller than the critical value of 0.9, but the "root mean square error of approximation" (RMSEA) = 0.305, which is greater than the critical value of 0.10. The resulting model is exhibited in Figure 6.11.

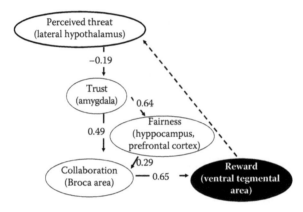

Figure 6.11 SEM ($n = 191$).

The model is in line with other studies that I have conducted, as explained previously.

6.7.1.4   Cluster analysis

I referred to the profile types that I presented in Section 6.4[129] and drew Table 6.20.

Table 6.20 Cluster analysis on $n = 191$

	Class				
	1	2	3	4	5
Trust (in %)	–	91	69	84	79
Coll. (in %)	–	89	66	79	73
IP/DP	–	1.3[186]	0.98	1.2	1.0
# Observations	–	53	17	66	55
Likely profile type	Hostile	Stable	Avoiding	Anxious	Anxious
% Population	–	28	9	35	28
Quadrant[187]	Dominant-repulsive	Dominant-seductive	Dominated-repulsive	Dominated-seductive	Dominated-seductive

There is no hostile class that stands out when doing a cluster analysis. This may mean that the eight hostile individuals are included in the stable or anxious populations, which accounts for 91% of the sample. Again, I know that they are there, but I am not necessarily able to detect them; this characteristic alone is typical of POVs.

I decided to introduce a new construct formed of the reverse of the interplay of the **core psychological constructs** that are Trust, Fairness, and Collaboration. I temporarily call this reversal of the core psychological construct "Propensity to hostility".[130] I obtained a triangular distribution, which is typical of a moderating variable, as seen in Figure 6.12, where the x-axis is the Hostile profile (PIP) and the y-axis the Propensity toward conflict, by percentage.

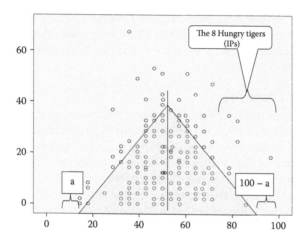

Figure 6.12 The triangular distribution resulting from hungry tigers (IP profile) vs. propensity to hostility.

This triangular distribution tells us a story. Very weak hostile individuals avoid conflict: they don't feel they can win the battle. Most people experience antagonistic situations at various levels. We would assume that these levels are functional in a project environment. The propensity toward conflict rarely exceeds the threshold of 40%–60%; at most, 60% of interactions between members contain some level of disagreement. This seems to be in line with what most of us experience in a project setting. People with hostile profiles long for confrontation and escape the triangular distribution.

I computed the edges of the triangle and obtained these equations:

$$\text{Propensity toward conflict} = \left(\begin{array}{c} (-i+1.14x)\ if\ a \le x \le 50 \\ ((100-i)-1.14x)\ if\ 50 \le x \le (100-a) \\ 0\ \text{otherwise} \end{array} \right) \quad (6.2)$$

The slope of 1.14 turns out to be the average of the Hostile position/ Defensive position ratio seen earlier. We could hypothesize that the equation is therefore (e.g., left side of the triangle):

$$\text{Propensity toward conflict} = \begin{pmatrix} (-i + k\,x)\ if\ a \leq x \leq 50 \\ ((100-i)-k\,x)\ if\ 50 \leq x \leq (100-a) \\ 0\ \text{otherwise} \end{pmatrix} \qquad (6.3)$$

I found a relationship between the Hostile position/Defensive position construct and the Hostile profile; I had developed an algorithm based on that same observation in the section on Stars (top performers).

6.7.1.5   Discussion

Based on the sample available, my research shows that about 4% of average teams consist of FP_{nc} of the functional type (Instrumentally hostile individuals). Conflict (Propensity to hostility), when it remains functional, can be good or bad depending on the circumstances and who is involved. A feasibility analyst can expect 4% of any team to be composed of potential hostile individuals, that is, of hostile personalities; these people will likely interfere with the project at some point when this fits their hidden agendas and as such, they represent a POVs right from the start.

6.8   Conflicts

As can be now recognized, the transformation phase, which consists of the four Ps interacting with each other, may have its load of challenges, especially when it comes to People'. The core of the dynamic of the Forces of Production, whether controlled or uncontrolled, rests on the tight interaction between three psychological constructs: Trust, Fairness, and Collaboration. Once this is somehow established, then the team members really commit themselves; they cannot commit to work diligently until that core has been set somehow, even to a minute degree. However, this triumviri is very sensitive to a

Not much happiness here …

psychological phenomenon called "Apprehension", or, if I reverse it, "Self-confidence". Hence, the psychological core could be seen as a process, as shown in Figure 6.13.

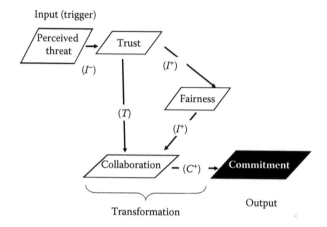

Figure 6.13 The psychological core as a process.

As a process, Figure 6.13 can be written as: Input (Apprehension, or its reverse: Self-confidence) (I^- or its reverse: I^+) → Transformation: Core (C^+) → Output: Commitment. An understanding of the psychological meaning of Commitment is provided by a project leader I interviewed, as follows:

> Team members and managers alike embrace a prosperous future. We encourage ownership of the work; we recognize that everyone at all levels of the projects and at every step of it plays an important role. We are part of a team; we are proud. The combination of these positive feelings drives us to outperform ourselves every day: we do our best work day in and day out and look forward to a larger goal. This sense of accomplishment is well represented during milestone events such as the launch at sea of a new vessel, when workers bring in their families and rejoice.[131]

The psychological foundation works hand in hand with Processes, which I examined in Chapter 4. Ignoring this state of affairs is assuming that Processes are not subject to human intervention, which is not a possibility.

The issue is that the entire psychological process is very sensitive. As apprehension gains momentum, as trust and collaboration become hard to develop, and as a sense of fairness trades places with a sense of inequity, frictions arise. Conflicts are ubiquitous in any project.[132] With the increasing complexity of projects—making Planning more tedious and Processes more sensitive to excruciating details, and as People work in multiple remote locations, with various cultural backgrounds, devoting their time to highly specialized tasks—comes the possibility of developing misunderstandings, delays, and frustration. Along these lines, some scholars have pinpointed four interactional challenges, mostly related to international contexts, that impact teams negatively[133]: (1) contrasting attitudes toward hierarchy; (2) difficulties with language, accents, and fluency; (3) divergence in leadership style; and (4) indirect communication.

Some authors suppose that the level of conflict varies with the stages of the project's life cycle.[134] Confrontations would be at their peak during the planning stage, then somewhat to a lesser level at the vision and deployment stages and to an even lower level at the completion/evaluation stage.

I question this view because it does not espouse the Commitment curve that I have found at the beginning of this chapter. Since Conflicts stand in sharp contrast to Trust, Fairness, and Collaboration, and since Commitment results from a savvy combination of these three constructs, I would expect Conflicts to somehow mimic the pattern of the Commitment curve, especially as qualms accumulate (as compliance recedes). Furthermore, I feel that all tensions are related, directly or indirectly, to resources—recall that I set Commitment as a function of efforts and resources. Not all resources are accessed or expensed in the initial stage, far from it. They are rather used mostly during the deployment stage. Evidently, if commitment is low, little achievement is to be expected.[135]

The presence of antagonistic demeanors at the beginning of a given project is certainly a good indicator of the fact that deadlock situations will pop out thereafter. Quite naturally, some scholars note that conflicts are a major concern for project managers.[136] Some authors[137] have interviewed a number of managers and found that the following are short sighted actions that can nurture conflicts (in parentheses are the percentage number of respondents). For Plans: unrealistic due dates (67), hazy goals (63), lack of planning (56), and inappropriate changes of goals/resource allocation (42). For Processes: inadequate resources (69). For People: a lack of commitment (59), communication gaps (54), and conflicts between teams (35).

Not surprisingly, Power is not listed—managers don't blame themselves. Not puzzling either, Apprehension does not appear: People simply do not admit that they feel threatened by others; it is much simpler to blame communication mishaps or a lack of motivation. On the lack of commitment, it is interesting to note that even in the animal kingdom, it is cause for dispute, as pointed out by some scholars: "chimpanzees attack allies that do not support them in third party conflicts, and queen naked mole rats will attack workers that they judge to be lazy."[138]

The following comment has been made, which describes well the dynamics underlying difficult projects: "often there is frustration, unease and suspicion because of the complexity of the situation, which is one of total dependency, lack of control and uncertainty."[139] Somehow, apprehension is linked to uncertainty, which is taunted by the levels of risks (external threats).

Many observables that pertain to conflicts can only be detected once the project is in marching order, not during the feasibility analysis period. These include complaints and grievances, errors, and turnover. We are left with guessing the level of turmoil ahead of time.

Conflict is not a dead construct: it is indeed dynamic and has a life of its own. I contend that conflicts are resource based and that they are composed of four main *sine qua non* characteristics: First, they burst out because they have an opportunity to do so (an opportunity, which is linked to the availability and importance of resources). Then, generally, bad luck, lax managerial measures, or some network arrangements that disfavor a particular team or a team member over another, which fuels confrontation, jealousy, and resentment, are also involved. However, this is not enough. There has to be a means of expressing the discontent; it can be a meeting, an e-mail message, or some rumors. Additionally, there has to be a motive; People do not engage in confrontations without an ulterior motive, which is to gain a position that allows them better access to or control of said resources. Finally, strong, negative feelings arise because People do not see a solution; they stick to the obstacle. Hence, a conflict is a means of finding a solution to a particular problem, with the risk being that it may turn out to be more dysfunctional than functional. Figure 6.14 illustrates the components of conflicts.

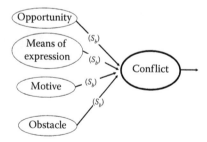

Figure 6.14 Structural components of conflicts.

Falling short of determining ahead of time what dilemmas will likely rotten the project, the feasibility analyst can nevertheless check whether some of their *sine qua non* components are present. Identifying motives, obstacles, opportunities, and tools all at once is one way of predicting their inevitable emergence.

I have expressed the fact that conflicts revolve around resources. This is certainly true in organizations as well as in the animal kingdom.[140] We have also seen in Chapter 4 that resources come in two forms: essential (R) and nonessential (R_n). In fact, I argue that work (T) and knowledge (T_h) are a form of resources, because work and knowledge eventually lead to the acquisition of R and R_n. A well-paid job, for example, allows the worker to buy fancy, unnecessary assets that he/she may enjoy during their leisure time. Hence, by extension, quarrels arise as a result of a scarcity of R, R_n, T, and/or T_h. Often, boorish behaviors degenerate in a spiral with tit-for-tat behavioral games; aggression may even take place.[141] Scarcity implies apprehension: it threatens the homeostasis of the individual. The more the apprehension gains in intensity, the more valuable $|R$ and/or $|T$ become. To draw a parallel in the animal kingdom, the "valuation of cover increase(s) with perceived threat."[142] Brown trout subjected to prior predatory attacks (hence, which have accumulated knowledge about a threat) show more aggressiveness in protecting their territory than other trout precisely because they have acquired knowledge about a real threat. We can therefore suggest that the conflict is a function of resources, in the four forms that they can take $|R$ and $|T$.

Because I have established that Commitment is expressed by efforts that are made versus the uses of resources, because I have just determined that resources can take four forms (directly R and R_n, and indirectly T and T_h), and because I have recognized the fact that Conflict can be functional at times and dysfunctional at other times, I can therefore set that Conflicts serve as a moderating variable between Collaboration and Commitment. This is actually what I have found in some of my research (see Figure 6.15).

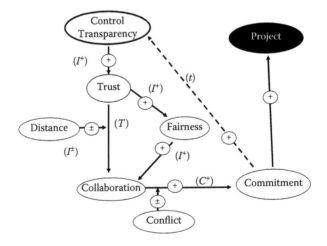

Figure 6.15 Conflict as a moderating variable in the six core competencies model.

As the reader can assess, I am completing the initial model to include Dependence, Distance, Profiles and Conflicts, as moderating variables. All of these variables can have either a positive or a negative impact on the project, depending on circumstances.

Grudging face-offs are generated by the inability of team members to deal with scarce resources, whether they come in direct ($|R$) or indirect forms ($|T$). It exacerbates apprehension, as People feel endangered by the fact that they are not likely to sustain their level of comfort (or homeostasis) for much longer if resources are depleted. This in turn has a negative impact on Trust and Fairness, and then on Collaboration; in short, on the psychological core. This, then, reduces the ability to deal with scarce resources and is compounded by interdependence, the presence of Unfits (uncontrolled Forces of Production, FP_{nc}), and Distance. A negative cycle is triggered that can cause the project to capsize.

6.8.1   Vulnerability and conflicts

I assume that there is a correlation between perceived vulnerability (or perceived points of vulnerability) and conflicts. POVs are related, in essence, to $|R$ and $|T$: the more an individual possesses each of these direct and indirect resources, the more invulnerable he/she becomes. Points of vulnerability present two interesting features:

1. They occur with a relative stable frequency from the mere fact that production, during the transformation phase, is partly repetitive.
2. The occurrence of an active point of vulnerability is independent from the time span of past occurrences.

These two features obey the conditions inherent to the law of Poisson; this means that over time, as POVs occur, confrontations tend to normalize. Put in terms of human experience, people get more and more accustomed to living with their differences, expecting that there will be occurrences of points of vulnerability as time goes on and accepting their stochastic occurrences. From this perspective, POVs tend to foster the normalization of conflicts; put differently, a system (a project) where fighting is prevalent points to the existence of POVs. This is the order of things: as people anticipate their vulnerability (i.e., as they feel their job is threatened by failures in the system they cannot or would not address), they become more defensive, less willing to trust and to be fair (as their own interests prevail over that of others at this point), and less inclined to collaborate.

As just mentioned, the normalization of conflicts, which is a function of the occurrences of POVs, is best described by the law of Poisson, as shown in Figure 6.16.

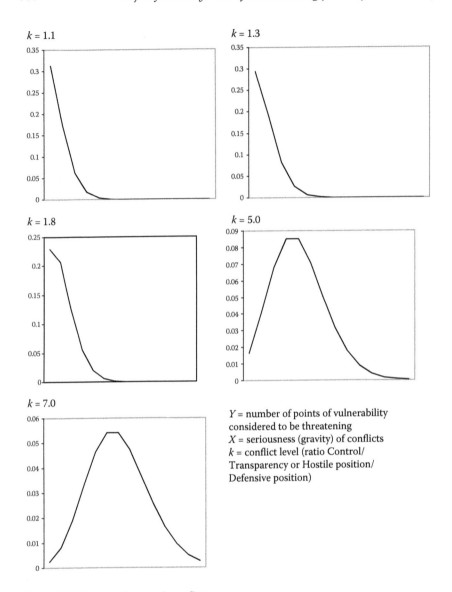

Figure 6.16 Poisson law and conflicts.

At a value of 7, which is the maximum level of Conflict on our scale (Hostile position > Defensive position), Conflict is deemed normal as the curve mimics a near normal distribution pattern (bottom left corner in Table 6.22). This explains one particularity of human behavior: the tendency to work amid or even to nurture confrontations, especially when POVs keep occurring. Put differently, there is a process of **habituation** (a phenomenon that is well documented in neurophysiology): a large number of POVs are initially deemed to be threatening, but the work environment is somewhat mellow.[143] As time pressure mounts (there is a deadline for the project), the situation becomes more and more critical, yet points of vulnerability that were once considered to be discomforting are now seen as being of lower importance because people got habituated; they developed coping strategies whereby they view the problems as trivial (left tail), or whereby they regard them as serious (average), or whereby they consider them to be very serious (right tail). As time progresses, however, this tendency for normalization (habituation) keeps being reinforced. On average, in the end, the team pays less attention to potential difficulties induced by POVs,[144] and fewer individuals join to counter them while confrontational episodes keep growing. Of course, if the occurrences of POVs could be predicted, the dynamic of human behavior would differ: the fact that team members in a given project know that POVs will inevitably keep popping up, but that the occurrences are not predictable because there are no time span correlations between past and present occurrences, creates an irritating environment. This provides a fertile ground for misunderstandings while, at the same time, team members develop coping strategies and habits, resulting in a culture of conflicts. From a project management perspective, this is the worst possible scenario, because things will only get worst.

An example of this dynamic is provided by the Stanford Prison Experiment.[145] In this famous case, which changed the course of research in the academic community, a number of candidates (roughly 30) were chosen to play prison guards and inmates in a setting that emulated a prison. It took only a couple of days before the guards started to abuse their artificial power, trying to go around the rules of conduct that were placed upon them, and before some of the fake inmates began to truly believe that they were being victimized. The situation festered rapidly and soon became out of control. Much to the scientists' surprise and dismay, it showed how quickly humans adopted predator and prey positions and stuck to them, at times vehemently, if not violently altogether.

This observation is important for project managers. People will tend to get accustomed to skirmishes as POVs keep occurring, with the result that they are expected to emerge and to become a privileged tool in trying to address POVs, a managerial approach that, of course, is counterproductive.

Conflict intensity may be particularly more acute in the early stages of the implementation of the project (but not in the vision or planning stage); as the deadline fast approaches, they are often tied to technical problems and procedures, prioritizing, and scheduling, but not to costs.

This leads me to the fifth law of project feasibility.

6.8.1.1   *Fifth law of project feasibility: Law on conflicts*

The more conflicts among stakeholders are intense, frequent, and cover critical issues, the less likely the project is feasible.

6.8.2   *Solving conflicts*

The PMBOK addresses conflicts to some extent and makes various recommendations for project managers: developing a vision, establishing effective communication, "getting things done" (i.e., being committed), motivating, negotiating, and solving problems.

Table 6.21 is a rendition of some various authors' advice completed by my own comments.[146]

Table 6.21 Conflicts and recommended measures

Stage	Source of conflict	Recommendation	4 Ps
Planning	Priorities	Define plans	Plan
		Encourage joint decision-making	
		Set contingency plan	
		Focus on goals	
	Procedures	Develop detailed procedures	
		Respect schedule	
Mobilization and Deployment	Priorities	Set technical measurements from the start	Plan
		Define tasks and priorities	
	Procedures	Favor feedback	Power
		Encourage collaboration	
		Check Work in Progress (WIP)	
Completion and Evaluation	Priorities	Prepare reallocation plan	Plan
	Procedures	Inform stakeholders	People
		Promptly solve problems	Processes/ Power

Some authors also computed the recurrence of conflicts in a list that I adapted to the four Ps as follows[147] (in parenthesis are the number of conflicts at the start, during, and at the end of the project). In relation to Plans, the sources of problematic situations are priorities (35, 24, and 16), schedule (29, 36, and 30), and cost estimates (13, 15, and 11). For Processes, they are administrative procedures (27, 15, and 9) and technical changes

(26, 31, and 11). Finally, for People, we have employees (25, 25, 17) and personality (19, 15, 17). At the start, there are a total of 174 sources of conflict; during the project, 160; and at the end, 111. Overall, the setting of priorities and procedures seems to be the redundant sources of problems; they both have an intimate link to resources.

As can be seen, the majority of conflicts take place during the starting (vision) stage of the project according to this research. This contrasts with my viewpoint probably because they are not defined in the same way: most likely, so-called conflicts at the vision stage are, in fact, benign disagreements; otherwise, it is unlikely that the project would have been given the go-ahead.

The theme that keeps recurring is "priority". Indeed, the scarcer the direct and indirect resources are, the more dissensions are likely to surface. Put differently, the more critical (prioritized) $|R$ and $|T$ are, the more stress is generated and the more conflicts find a fertile ground from which to bloom. A point of vulnerability exists here with respect to the criticality of the access to resources. Thus, to enlarge my definition, such tensions are really a function of resources given the triple constraint, with time being exemplified in Table 6.21 by way of the notion of priority. Conflicts relate primarily to the three sides of the Bermuda Triangle: as the utility drawback looms larger and larger, the project itself is threatened, with people's jobs and reputations being at stake.

As mentioned, various authors provide advice as to how to deal with these hellish situations. Some authors make extensive lists that include avoiding aggressive comments and tone, vilifying, and making judgmental inferences; building true (and not fake) trust; identifying oneself; saying "hello" in the morning; showing empathy; and so forth. I provide hereafter a few examples extracted from various sources.[148,149] The strategies for physical conflict are being swift and ascertaining oneself; containing, dividing, and conquering; intimidating the opponent; leaving no vulnerable position open; sizing up the opponent; staying on guard; and verifying the facts. The strategies for emotional quagmire are given as follows: avoiding making judgments; letting down one's guard; limiting futile argumentation; opening up and exposing vulnerabilities (I am not sure I concur with this recommendation); individualizing the response; pacing the conflict; seeking equality; and lastly, showing confidence by being open to vulnerability (Hmmm, not sure about that either!). Other recommendations have been made: in the case of divergence → resort to commitment; of guarded, censored, withheld information → encourage open communication; of impermeable feelings → work through disagreements; of people taking advantage of one's vulnerability → try mutual trust (I am not sure this is the right strategy here; I can hardly see how a lonely sheep would want to entertain a hungry tiger on the virtues of friendship); of a cold, strained, unpleasant, formal atmosphere → nurture a positive atmosphere (easily said, but difficult to implement).

As can be seen from these lists, vulnerability (POVs), as well as Trust, Fairness, and Apprehension, comes into play when providing advice on conflict management.

6.8.3   A small study

I performed a small study to see how a resource could affect Conflict. I could not, for obvious reasons, deprive team members from a valuable resource and examine what would likely happen. Instead, I operated in a different way. I introduced a resource that I could control, that of a negotiating skill identified as T_h and verified its impact on a small team *before* and *after* it was introduced, for two groups, one that received the training in negotiation (TG) and one that did not (UTG). This research was conducted in 2011 and 2012.

6.8.3.1   Core values

The two groups, UTG and TG, which received training or equivalently T_h, were compared to see if there were substantial differences between them.

The UTG and the TG *before* and *after* values for the key constructs of Hostile-Defensive positions are shown in Table 6.22.

Table 6.22 Main constructs UTG and TG ($n = 95, 54$)

Construct	Hostile	Defensive	IP/DP	Trust	Fairness	Coll.
	7-point Likert scale			In %		
UTG ($n = 95$)						
Before	3.7	2.9	1.4	74	71	69
After	4.0	3.0	1.4	74	69	79
Difference	0.3[a]	0.1	0	0	2	10[a]
TG ($n = 54$)						
Before	3.7	3.0	1.3	71	69	69
After	3.5	2.5	1.5	74	73	73
Difference	−0.2	0.5[a]	0.2[a]	3	4	4

[a] Significant change at $p < 0.05$.

I observe that some of the Hostile position/Defensive position values changed significantly[150] for the Hostile position construct and the Collaboration construct. There is a tendency for the 95 members of the Untrained Group (UTG) to develop Intrumentally hostile (IP) behaviors in the absence of training (T_h). Collaboration levels were expected to rise since the members had a duty to collaborate. For both groups, the values indicate transactional relationships.

6.8.3.2   Structural equation modeling
I conducted an SEM[151] for the *before/after* data, as illustrated in Figure 6.17.

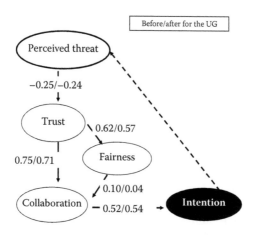

Figure 6.17 An SEM modeling of the UG (*before/after*).

I found that Apprehension (expressed as Defensive position/Hostile position) has a significant (negative) impact on Trust right from the beginning of the relationship between the members of the Untrained Group (UTG).

The 95 UTG members were invited to be part of a negotiation setting without prior preparation (without the indirect resource T_h). While they tried to collaborate, their hostility "traits" spiked upward significantly. Perhaps this means that short of appropriate resources, people tend to resort to hostility, that is, to adopt a more hostile approach as a means of winning the negotiation. If this holds true, this is an indication that resources such as training are a good starting point to minimize hostile tendencies; in other words, a project feasibility analyst will look at the presence of training programs in the project plan as a means of assessing whether certain POVs can be kept under control.

Note that the 54 members of the Trained Group (TG) received a substantial amount of training in negotiation (T_h), and the group had a similar sociodemographic profile to that of the Untrained Group (UTG). The Defensive value changed significantly; the participants with training chose to position themselves in a less victimized way (defensive) and ended up necessitating less hostile tactics.

The SEM[152] for the TG (*before/after*) is shown in Figure 6.18.

The ratio of the Defensive position/Hostile position (Apprehension) construct versus Trust in the *before* and *after* periods was not significant ($p = 0.416/0.635 > 0.05$).

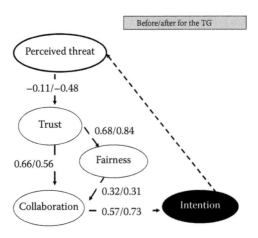

Figure 6.18 An SEM modeling of the TG.

6.8.3.3 *Discussion*

One probable explanation for my results delves into the concept of an "asymmetry of information" (see Chapter 7): the trained individuals (TG) did not feel as threatened as their untrained counterparts (UTG) probably because they had better information. The cure to uncertainty is transparency; the cure to chaos is control.

Recall that I have connected T_h with R_n (nonessential resources) and T (work) with R (essential resources) and that I have equated Commitment with efforts (direct ($|R$) and indirect ($|T$) resources). My results point to the fact that a surge in T_h leads to better commitment, so that this would entail requiring a diminishing input of R_n during that stage, and, presumably, an increased input of R and a decreased input of T. In other words, POVs that are tackled by way of training also demand, at the same time, a rise in the supply of $|R$; the feasibility expert would be advised to look at training programs throughout the entire project, in not only one single stage or the other.

6.9 Chaos

Transparency—by way of training, for example—is key to a project's success. At the same time, a certain level of control needs to be exercised, because the four Ps, once in the transformation phase, tend toward chaos. This is because the project is a closed dynamic system: there are theoretically no possible exits for behavioral outliers (e.g., FP_{nc}) so that the entire system may become infected, so to speak, by antagonistic behaviors.

My model sets the boundary value of the Hostile position/Defensive position ratio at 4.6 (2.3 + 2.3) with [2.3 = 1 + k], past which it is assumed that the system becomes disorganized, even chaotic. The value 4.6 is, incidentally, near the Feigenbaum constant (4.669)[153] in a logistic map function, which expresses chaos.

Based on some of my research, I set some equations we have already seen with respect to Trust and Collaboration into a logistic map function (which expresses a disjointed state and which is formulated as $[x_{n+1} = r\, x_n (1-x_n)]$), as follows[154]:

$$\text{Collaboration}_{t+1} = k\left(0.3 + 0.9\,\text{Trust}_t\right) + \left(1 - \left(0.3 + 0.9\,\text{Trust}_t\right)\right) \quad (6.4)$$

Doing so, we obtain Table 6.23.

Table 6.23 A logistical map function for Collaboration in the future (Coll_{n+1})

$k=0.9$		$k=1.3$		$k=3$		$k=4.7$	
Trust_t	Coll_{t+1}	Trust_t	Coll_{t+1}	Trust_t	Coll_{t+1}	Trust_t	Coll_{t+1}
1.00	−0.216	1.00	−0.312	1.00	−0.720	1.00	−1.128
0.9	−0.110	0.9	−0.159	0.9	−0.366	0.9	−0.574
0.8	−0.018	0.8	−0.027	0.8	−0.061	0.8	−0.096
0.7	0.059	0.7	0.085	0.7	0.195	0.7	0.306
0.6	0.121	0.6	0.175	0.6	0.403	0.6	0.632
0.5	0.169	0.5	0.244	0.5	0.563	0.5	0.881
0.4	0.202	0.4	0.292	0.4	0.673	0.4	1.055
0.3	0.221	**0.3**	**0.319**	0.3	0.735	0.3	1.152
0.2	0.225	0.2	0.324	0.2	0.749	0.2	1.173
0.1	0.214	0.1	0.309	0.1	0.714	0.1	1.118

Recall that when I examined, in Chapter 5, the relationship between Trust and Collaboration, I found that they were substantially correlated, with an R^2 (or estimators for SEM studies) of roughly 0.75 in most groups I (and other authors) analyzed. As can be seen from Table 6.23, the k-value that best meets the requirement of a high correlation between the constructs of Trust and Collaboration for an initial value of [Pretrust=0.3] as seen in Chapter 5, is indeed 1.3, the constant that is at the basis of my model.

6.9.1 A study

I looked at a database of my own, consisting of some 834 individuals spread across 19 different projects. I examined the way each of these 19 groups operated by attending their daily work once a week over several weeks. My observations are shown in Table 6.24 (see also Appendix 5.3).

Table 6.24 Critical k' values for 19 groups ($n = 834$)

k' value	Observed behavior
4.7	Some tense erratic behaviors
$3- \approx 2$	Some erratic behaviors
$\approx 2–1$	Fairly stable groups
<1	Group 7 and Group 11: These two groups showed a high turnover level

I actually could find some of these groupings done by using the k' value by way of a cluster analysis (Table 6.25).

Table 6.25 A cluster analysis on $n = 834$

	Class (rounded-up numbers)			
	1	2	3	4
Trust (in %)	90	81	84	64
Collaboration (in %)	90	79	83	63
Hostile (IP) 7-pt Likert scale	5.5	3.3	5.2	3.3
Defensive (DP) 7-pt	1.3	2.3	4.7	3.5
Hostile/Defensive positions	4.7	1.5	1.2	1.0
Number of observations	57	315	223	239
% of population	7	38	27	29

The class with a high Hostile position/Defensive position ratio (5.5 over 1.3 for a ratio of 4.7) is, of course, indicative of hostility. Over time, groups where the value of k' was hovering over 4.7 or where the value tended to sit under 1.0 displayed dysfunctional behaviors: some members quit the project and a high turnover was observed.

The k' values and corresponding behaviors that I witnessed tended to approximate a logistic map. As k' approaches the Feigenbaum constant, the group became quite dysfunctional and sustained a lot of stress. I set the following equation, [Collaboration$_{t+1}$ = k (0.3 + 0.9 Trust$_t$) (1−(0.3 + 0.9 Trust$_t$)]:

$$\text{Hostility}_{t+1} = k \text{ Hostility}_t + \left(1 - \text{Hostility}_t\right) \qquad (6.5)$$

where "Hostility$_{t+1}$" is a behavioral manifestation of instrumentally hostile aggression (planned aggression) and "Hostility$_t$" represents defensive aggression (present reactive aggression).

It should come as no surprise that behaviors adopt a certain mathematical logic. Our biological body operates with some form of mathematical logic, the circadian rhythm controlled by the hypothalamus and red blood cell reproduction patterns being prime examples.[155]

Equation 6.5 entails that future hostility is a function of present hostility, which seems to make sense; this is indicative of retaliation. Unless proper conflict management is introduced in a closed dynamic system, such as a project, people will tend to become more and more aggressive in an effort to shield their gains and to avoid losses.

The logistical map function discloses the presence of loops. A loop, as it happens, is included in our six core competencies model (the *t* loop going from Commitment to Apprehension). Recall that we established that our model had sound neurobiological foundations (see Appendix 6.1). As it turns out, the brain is indeed equipped with self-regulatory loops. As seen before, there is efferent circuitry, leaving from the hypothalamus and going to the pituitary gland; however, there is no afferent circuit coming into the hypothalamus from the pituitary gland. Hence, hormones secreted by the pituitary glands must go through the entire blood system before reaching the hypothalamus, which will regulate their emissions and levels. Also, as another example, so-called Renshaw loops, permit immediate feedbacks once the neurotransmitter acetylcholine is produced in order to modulate muscle action and avoid overreactions.

From this perspective, my model, which was initially developed at the macroscopic level—examining groups of individuals and their psychological constructs—finds correlates in the way the brain functions.

This may be good news for the model: it stands to reason at both the macroscopic and microscopic levels. From a managerial point of view, project managers are spread A door to the future amid gray areas.
between two opposite ends: one tending toward Collaboration and one toward hostility (read "Conflicts") [Hostility$_{t+1}$ = k Hostility$_t$ (1–Hostility$_t$)]. The presence of intense hostility traits among the Forces of Production or of extreme hostile profiles or personalities is a marker of future problems. In the context of a project, there are only two options, as shown in Figure 6.19.

The level of one option gives the level of the other. In other words, the probability of success in terms of People only (irrespective of the three other Ps), rests in the maximization of functional collaboration and the minimization of dysfunctional conflict. Some scholars put it this way[156]: Hostile → neutral → friendly → teamwork. As it turns out and as mentioned, some of my research points to Conflict as being a moderating variable between Collaboration and Commitment. Some other researches

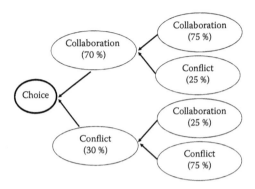

Figure 6.19 Road map to success or failure.

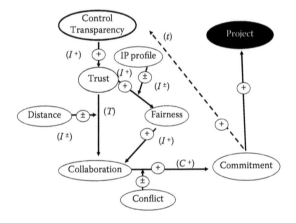

Figure 6.20 Moderating role of the hungry tiger's profile.

posit that hostile profiles (hungry tigers) act as a moderating variable between Trust and Fairness, as shown in Figure 6.20.

These last two observations require more research; however, it is obvious that some level of strife may be constructive given specific circumstances.

6.10   Conclusion to Chapter 6

This chapter oversaw the third P of project management: People, in a dynamic context. People cannot be reduced to human resource management, although proper employee compensation and recognition are certainly a source of motivation, which can potentially lead to improved performance. People live with behaviors (and attitudes), beliefs, heuristics, and personalities; they clash; they engage in trust building and collaborative efforts; they judge based on fairness and performance; they display hostility or stability depending on their circumstances; and they generally long for some sense of pride in what they are doing. We have seen that Apprehension may exercise a strong influence on the psychological core that is Trust–Fairness–Collaboration.

Overall, projects are the making of People. POVs can affect any parts of the projects; in the end, it is People who identify the POVs, treat them, and achieve success. Top performers can be identified through an in-depth feasibility analysis where such effort is justified.

6.11 What we have learned about POVs: Chapter 6

POVs…

1. May play against People's own welfare.
2. Diminish with People's commitment.
3. Are related to the constant k.
4. Are best tackled by top performers.
5. Are linked to hostile behaviors.
6. Are, at their peak, composed by the people wanting to intentionally disrupt the project (hungry tigers).
7. Are related to $|R$ and $|T$.
8. Foster the normalization of conflicts.
9. Affect Trust, which, being intimately bonded with Collaboration and a sense of Fairness, impacts Conflict levels.
10. Foster blind management when left unchecked.
11. Are, in part, unpredictable.
12. Demand employee training and the supply of $|R$.

6.12 Key managerial considerations: Chapter 6

1. Equilibrium in project management is a dynamic event; it is constantly challenged by circumstances and people.
2. Three simple questions may be posed to rapidly assess a project:
   a. Are adequate managerial skills in place?
   b. Do all the stakeholders trust the project and each other?
   c. Are all the necessary resources available ($|R$ and $|T$)?
3. Top performers (Stars) must be quickly identified and encouraged for what they bring to the team.
4. Weed out instrumental hostility; deal with hidden agendas and dysfunctional Forces of Production.
5. Address problems and conflicts right away; do not let things rot (they inevitably will).

6.13   Case 6: The MID: The best managed project in the world

Interview with Clément Demers, architect (*Ordre des architectes* [OAQ], *fellow de l'Institut royal d'architecture du Canada* [FIRAC]), urban planner (*L'Ordre des urbanistes du Québec* [OUQ], Institut Canadien des Urbanistes [ICU] or Canadian Institute of Planners [CIP]), and project manager (project management professional, PMP). He is the cofounder and was the general manager of the Montréal International District (known as QIM for *Quartier international de Montréal* in French, but will be referred to here as MID—"Montréal International District"—to avoid confusion with the "Montréal Olympic Stadium" [MOS] and the "Québec Multifunctional Amphitheatre" [QMA]). He is also a part-time professor and coordinator of the graduate studies program in project management at the School of Architecture in the Faculty of Environmental Design at the University of Montréal. The interview was completed in May 2016.

6.13.1   Project description

- Question 1 (The author): "Please give us a description of the MID project:"
- Answer 1 (Clément Demers):

"The MID project was (and still is) the largest urban redevelopment project to have ever taken place in Canada. It covers an area of 27 ha right in the center of Montréal, just between Old Montréal and the central business district. We planned 8 ha of pedestrian space such as parks, sidewalks, and streets. The project forecasted three stages with stages I and II to be completed by remodeling 7 ha of urban development. The stages included coverage of an existing highway (named *Ville-Marie*)—a trench that separates parts of the downtown business core of Montréal from Old Montréal—over 1300 feet.

Public squares and sidewalks were to increase by 40% with the creation of close to a mile of new interior pedestrian corridors embellished by 500 mature trees. The current area was to be completely refurbished and the highway fully covered. By comparison, the Big Dig Project in Boston was 25 times larger and its cost was 250 times higher.

The project eventually attracted a larger number of investment properties bordering the facilities. The modern core business center and the business center of Old Montréal used to be quite active until the 70s, before the Ville-Marie highway was built and cut it in two sections. Old Montréal features stone-paved alleys, large stone building with high ceilings and monumental wooden doors, all of which offer a cachet of old France. It is home to inimitable restaurants, numerous businesses, and a large church

("Notre-Dame"), which transforms into a media magnet when celebrities marry or when sorrowful and touching eulogies are pronounced. A 15 min walk suffices to reach the more modern Montréal business center, but one must cross over the highway.

A new business center was eventually developed around Place Ville-Marie (a set of high-rise towers). The Convention Center, built in 1983, forms a so-called bridge building over the Ville-Marie highway but does not provide sufficient access to Old Montréal."

6.13.2   *The project's key calendar of activities*

- Question 2: "Could you give us an overview of the milestones of the MID project?"
- Answer 2:

"The important dates were as shown in Table Case 6.A.
The calendar was fully respected."

6.13.3   *Need*

- Question 3: "To what need did the MID project respond to?"
- Answer 3:

"The urban space now known as the MID is an area that provides access to Montréal for countless motorists coming from the United States and the rest of Québec and Canada.

The issue of the redesigning of the various highways that meet in the downtown core was originally raised because of climatic concerns, not because of pollution or business interests. The highways, built in the 1960s, aged prematurely due to the many cycles of freezing and thawing during the harsh Montréal winters and the consequent use of deicing salts. The economic development of the metropolis was certainly a priority. Indeed, Montréal had just come out of a recession and had an urgent need for a big project that could help revive its economy and image. We saw an opportunity to requalify this portion of the city to strengthen the bond between the main business district and Old Montréal."

6.13.4   *Goals*

- Question 4: "What were the primary goals of the projects?"
- Answer 4:

"The first objective was to palliate the aging infrastructures. The second was to create a world-class district that would benefit Montréalers as well

Table Case 6.A MID important dates

1997	
January	Prefeasibility study of the MID project
May	Prefeasibility study presented to the public

1998	
January	Feasibility study
June	Feasibility study presented to the public

1999	
July	MID business plan
November	Project announcement to the public

2000	
January	Stage I: Start
February	Start of construction, over Ville-Marie highway, expansion of the Convention Center and of the business offices of the *Caisse de dépôt et de placements du Québec* (CDPQ)[188]
April	Creation of the association of residents bordering the MID

2001 (West Section)	

2002	
December	Opening of the business office of the CDPQ

2003	
December	Stage I: Completion (C$76 million)
	Stage II: Start

2004	
December	Stage II: Completion (C$14 million)

2005	
January	Stage III (McGill street): Start

2006	
December	Stage III: Completion (C$23 million)

as visitors and international tourists. In this area, the Convention Center was already attracting a large international clientele. The quality of its equipment ranked second in North America. Several international organizations, including the International Civil Aviation Organization (ICAO), took offices there.

A third objective was the development of a sustainable urban environment. Initially, this project was not limited to cover the motorway; it was mainly intended to transform and enhance the urban environment. Coverage of the highway was therefore a means rather than an end in itself.

Also, we wanted to showcase our local know-how in the field of architecture, industrial design, landscape architecture, planning, project management, and, of course, urban design. Our creators had encountered some difficulties to break through internationally in these creative fields. Many people thought that Québecers were only able to stand out in the show business—think of Celine Dion and the Cirque du Soleil, for example—when in fact they are also very adept in other artistic fields.

Another objective was to promote walking and public transportation.

In summary, the objectives of the MID were multiple: we wanted to offer an attractive, upgraded district that would attract real estate development and benefit from a leverage effect—as infrastructures improved, so would business investments in the area. We realized that it was also essential to develop indoor parking and pedestrian access, and to upgrade public spaces.

Between 1999 and 2006, the increase in traffic in the subway stations was of a tiny 1.52% in downtown Montréal. After the completion of the project, the traffic in the MID area rose by 15%, almost 10 times more than in the rest of downtown, all thanks to the new real estate development, which included offices and some 2000 residential and hotel units."

6.13.5   Challenges

- Question 5: "What were some of the challenges you faced?"
- Answer 5:

"The project was a first for Québec's Ministry of Transportation. It contained many challenges, especially with the covering of the existing highway. From an engineering point of view, the implications are very different depending on whether a highway is a trench or a tunnel. Among the issues that concerned the ministry were the operational costs of the tunnel, the air circulation, and the dust and noise levels. Most notably, passage of dangerous goods in tunnels are forbidden. In a city like Montréal (which is an island), alternative routes for the transportation of hazardous materials by trucks are rare. The work had to be done at night, which caused some noise and kept nearby residents awake. To handle all that traffic on the highway was a very tortuous procedure indeed.

The initial investment and annual operating costs did not come with a guarantee that business-related revenues would follow suit. Other elements to take into account included allowing for the risks of accidents in the tunnel; establishing a fund to cover maintenance costs; finding support from sponsors to cover future operating costs; mitigation measures during construction to ensure the work could be done efficiently; securing title deeds in order to provide access to the site at all times; and securing the capacity to maintain the infrastructures. Costs included the initial investment—not only the operating costs but also reinvestments in infrastructures. As an example, while tiles are sturdy, communication devices and lighting have a limited life span of 5–10 years, especially in Canada due to salt spraying during winter. Thus, they must be replaced regularly. Typically, the operating cost of a covered highway is five times higher than that of a trenched motorway."

6.13.6   Stakeholders

- Question 6: "Who were the key stakeholders and what were their roles?"
- Answer 6:

"This project was conceived as a public–private partnership that saw many partners working closely together. The promoter, the Caisse de dépôt et placement du Québec (CDPQ), had been involved in the project of the World Trade Centre in Montréal (not the one in New York City), built in the same area. The largest landowner was the City of Montréal, which owned all public and civic spaces in the area. As for the Government of Québec, it owned the highways. For its part, the Government of Canada provided financial support to the project. The main fund providers were therefore public. Local owners formed a group named the *Association des riverains du Quartier international de Montréal*, which helped with the project with great enthusiasm.

The project enjoyed the support of many partners and one not-for-profit organization; the company the *Quartier international de Montréal* (MID) was born thereafter. It was mandated by the city to coordinate the project. Other stakeholders included a board of directors, the management team and a group of professionals under contract with the architecture and design firm named *Daoust Lestage Inc.* (which submitted the original idea of the project). Partners such as some of Montréal's large corporations and even foreign investors poured funding in for the implementation of some of the project components. Some of the key players included those shown in Table Case 6.B. We all worked together to attain our goals."

6.13.7   Stages

- Question 7: "Describe to us the different implementation stages of the project."
- Answer 7:

"The project developed in three stages: the first stage took place between 2000 and 2003 and the second between 2003 and 2004. The first deliverables consisted in improving the infrastructures. The highway, the exchanger,

Table Case 6.B MID Partners

The founding partners and some of their financial contribution in millions (C$)	
La Caisse de dépôt et placement du Québec (CDPQ)	0
The Government of Canada	30
The Government of Québec	30
The Association of residents of the MID	8
The City of Montréal	14
Various partners	5
Banque Nationale	1
Power Corporation	1
Hydro-Québec	1
Gaz métropolitain	1
Bell Canada	1
Four associate partners	2
Nine other partners	1

and the metro were underground. This sector already included the largest concentration of optic fibers in Canada; major sewers and high-pressure water systems were present.

The highway's coverage was the second deliverable; this included the construction of bridges and the building of a parking lot. The Convention Center was expanded; protected pedestrian sidewalks were extended and a cultural showcase area was designed to promote the works of art by local artists of international caliber. Finally, a pedestrian signage system was put in place and we renovated the accesses to the subway network.

The third deliverable would see the redevelopment of the public spaces and streets covering a total surface of 8 ha, part of which represented a 40% improvement compared with the existing pedestrian area. This increase was reflected in the widening of sidewalks, the installation of new public squares (Place Jean-Paul Riopelle), the execution of an important canopy consisting of 500 large-caliber trees, as well as the installation of public fountains and urban art. In the end, this endeavor offered an attractive center that encouraged property development. Soon, access to the metro, adjacent to a sophisticated landscaping area, almost tripled."

6.13.8   Impacts

- Question 8: "What were the impacts of the MID?"
- Answer 8:

"More than C$700 million of real estate developments in five years and more than C$2 billion after ten years have enriched the district with such business infrastructures as brand new hotels.

The first investment by the City of Montréal, which was of C\$14 million, brought in C\$20 million a year in property taxes immediately at the completion of stage II, in 2004. As for the Canadian government, its initial investment of C\$30 million led to revenues, by way of direct and indirect taxes, of C\$42 million plus an additional recurring revenue of C\$6.6 million/year. The project has attracted foreign currencies as international congresses have multiplied. Tourists have discovered the city and real estate investors have invested in Montréal. The project allowed those who participated, including designers, to obtain contracts abroad.

The project accumulated a striking number of awards in a variety of fields: in all, 32 national and international awards were received covering 15 different areas—arts and design; business, economic development; engineering, environment, and project management including design and urban development; landscape architecture; real estate, innovation, industrial design; sustainable development; tourism and urban design; transportation; and urban planning. In 2005, it received the "Project of the Year" award given by the Project Management Institute (PMI) as it was judged the best managed project in the world."

6.13.9   Key Success Factors

- Question 9: "In what way was the project a success?"
- Answer 9:

"The project respected the schedule; was completed under budget and all the high-quality objectives were met; this was achieved without conflicts or complaints.

A number of factors or actions can be associated with our success.

A favorable context (Plan): The MID project was the result of a long process. It corresponded to the needs and expectations initially set forth; it had the support of stakeholders who were eager to develop an international neighborhood.

The initial capital (Plan): An initial investment of C\$35,000 paid for a small commissioned team to design the project, allowing it to make the project appealing to the City of Montréal, the financial institutions, the governments, the neighbors, and the potential sponsoring partners. A second down payment covered the feasibility study. It provided a pristine vision of the project and of its targeted quality standards.

Sustainable development (Process): Our goal of ensuring sustainable development was supported by a strong will emanating from the designers and managers involved; our focus was on quality—the quality of the concepts from start to finish, the quality of materials, the quality of realization, and finally, the quality of operation and maintenance.

Competencies (People): Professionals were hired on the basis of experience, talent, and skills, rather than on their tenders. There is a big difference between skills and talents. It is the soul and creativity, the art of thinking and thinking outside the box that typifies the talent and the special quality of a project. These criteria were favored in the choosing of the project's various teams.

Autonomy (People): The various teams enjoyed great operating autonomy. A clear vision and a limpid decision-making process made this feasible.

Commitment (People): Team members made the necessary compromises and avoided divergent orientations. People committed wholeheartedly to the project.

Leadership (Power): A project of this nature must be driven by a strong and committed leadership. The CDPQ played a predominant role. Its president at the time, Jean-Claude Scraire, manifested an ongoing adherence to it from beginning to end, and encouraged consistency of efforts, as well as the promotion and the pursuit of the project's objectives."

6.13.10   Lessons learned

- Question 11: "What have you learned throughout this project in terms of its value?"
- Answer 11:

"I learned that it was possible to grow together, as a team, even though our project was ambitious. We knew it was inspiring and feasible. That was our source of motivation; we maximized our calendar of tasks and activities and aimed for efficiency. We individually and collectively put all the energy and passion toward this project. It was a win-win situation for all the participating stakeholders, including the managers, the City of Montréal, the various institutions, and the users.

Such a project is nothing other than a great source of pride for everyone. It just feels good."

The author: "Thank you, Mr. Demers, for having shared your enthusiasm with us."

6.13.11   *Appendix A of Case 6: Pictures of the MID*[157]

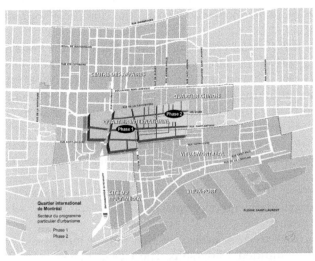

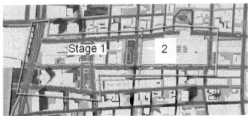

1999

6.13.12   Appendix B of Case 6: Awards won by the MID[158]

2008   *Mention d'honneur—Concours international de design urbain: Biennale de Quito 2008*

Prix Brownie 2008—meilleur projet toutes catégories: Canadian Urban Institute

Prix pour le meilleur projet d'aménagement—Place des designers québécois 2008: 9ème conférence mondiale de la FIV sur le vieillissement

Grand prix d'excellence en transport 2008—Prix Transport collectif: Association québécoise du transport collectif

Urban Leadership Award 2008: Urban Renewal Canadian Urban Institute

2007   Mention 2007: *Mouvement québécois de la qualité*

2006   *Prix Fondateur—Élexir 2006*: PMI Montréal

Prix national de design urbain 2006: Institut royal d'architecture du Canada

2005   *Mention spéciale 2005*: METROPOLIS, Berlin

Project of the Year Award 2005: Project Management Institute (PMI), Pennsylvania

Certificate of recognition 2005: Project Management Institute (PMI), Pennsylvania

Prix Ulysse 2005—Attraction touristique—50,000 visiteurs ou plus: Tourisme Montréal

2004   *Prix d'excellence en architecture—Design Urbain 2004–2005: Ordre des architectes du Québec*

Prix du Choix des enfants 2004–2005: Ordre des architectes du Québec

Prix spécial du jury 2004: Nouveautés touristiques Bienvenue Québec

Prix Thomas-Baillargé 2004: Ordre des architectes du Québec

Prix Entreprise 2004: Institut de design de Montréal

Prix Canada (urbanisme) 2004: Institut de design de Montréal

Prix de la Métropole (architecture du paysage) 2004: Institut de design de Montréal

Prix infrastructures urbaines 2004: Association du génie-conseil québécois

Prix d'excellence en immobilier 2004: Institut de développement urbain du Québec

Gold Award 2004: Urban Design National Post Design Exchange Awards

Prix Orange 2004: Sauvons Montréal

Prix aménagement 2004: Les Arts et la Ville et Télé-Québec

Mention prix orange 2004: Transport 2000

Gold Award 2004: Landscape Architecture National Post Design Exchange Awards

Prix du public 2004: Nouveautés touristiques Bienvenue Québec

2003   *Prix du design industriel 2003: Salon international des inventions, Genève*

Médaille d'or 2003: Salon international des inventions, Genève

Prix Canada (design industriel) 2003: Institut de design de Montréal

Plaque témoignage 2003: Destination centre-ville

6.13.13   Questions related to Case 6: The Montréal International District

1. Please find the POVs in this case, if any.
2. Please explain in what way a project that is well designed, well planned, and that has the support of all the key stakeholders produces tangibles that exceed its original goals and targeted intangibles that can help foster economic development.

6.14   Appendix 6.1: The brain and the hypothalamus

Some scientists provide a useful description of the hypothalamus,[159] which in humans is the size of an almond[160]: "Almost every major subdivision of the neuraxis, or central nervous system (CNS) communicates with the hypothalamus and is subject to its influence". The same authors add: "hypothalamic structures are thus able to recruit a sequence of motor acts that appear in logical order."[161]

The hypothalamus is a very special brain area: it is an archaic structure that generates strong reactions, which are, however, exempt of full cognitive control. It is autonomous and yet it is in partnership with the autonomic nervous system (ANS), which controls, among other vital functions, heartbeats and sleep patterns. It communicates with other brain structures through hormones and neurons and produces its own so-called pleasure chemical—nitric oxide, NO. It operates on a two-speed system: a slow one (neuropeptids) and a fast one (neuromediators/modulators).

The hypothalamus is involved in the body's visceral and most fundamental activities: (1) blood flow (cardiac, renal function, etc.), (2) energy metabolism (body temperature, food, water, and sleep), (3) growth[162], (4) reproductive functions (sex, pregnancy, and lactation), and (5) responses to threats (stress hormones, sympathetic, or parasympathetic tone, etc.).

As previously mentioned, the hypothalamus is in charge of the body's homeostasis; it can deploy defensive rage, also called "sham rage"[163] through its ventromedial section when needed. It has been explained that "the *medial* hypothalamus (...) provides a mechanism that modulates feeding, generates affective processes, such as rage behavior, in animal and people."[164] The medial hypothalamus also promotes escape as a coping mechanism for dealing with threats while the lateral hypothalamus is in charge of instrumental aggression.[165]

The hypothalamus, the center for the four core coping mechanisms (one offensive, three defensive) receives inputs from various parts of the brain and sends outputs to various parts of the brain in return. In some cases, there are reciprocal connections, in some others, such as with the pituitary gland, this does not occur. This last comment is crucial for the model I have presented: the hypothalamus sends efferent and vascular connections to the pituitary gland, and both, together with the adrenal glands, form the "hypothalamic–pituitary–adrenaline," (HPA) axis, in charge of fast responses to stressful stimuli. However, as discussed before, the pituitary gland does not reciprocate to the hypothalamus; the hormones produced in the pituitary gland (some under the influence of the hypothalamus) do not go right back to the hypothalamus; they must enter the vascular system (the blood), which, circulating through the entire body, will eventually reach the hypothalamus where they will pass on important information on the state of the body. My model takes into account this biophysiological dynamic: there is no direct arrow going from the construct of Trust (associated with the amygdala) to the construct of Apprehension (associated in part with the hypothalamus).

The amygdala is an important brain structure related to sensory perceptions through the thalamus and to the memory through the hippocampus, and is linked to attachment (and thus to trust). Indeed, the amygdala plays a central role in emotions such as trust and attachment, in socialization and in the expression of contextual fear and hostility avoidance.[166] The anterior, cortical, medial, and basomedial amygdala are related to defensive rage, but not to predatory attack; the central, lateral, and basolateral amygdala are associated with predatory aggression but not with defensive rage.[167] The thalamus is connected to the prefrontal cortex where cognitive functions take place. The HPA axis is coupled with the HPG axis, where G stands for gonads— this refers to sexual activity and testosterone release in particular. Predation cannot be understood from the HPA alone; it must include the HPG axis as well. In the animal kingdom, withdrawal or loss of necessary food resources activates the HPA[168]; this may lead to potential, calculated aggression (search for meat for carnivores). Imminent threats lead to defensive reactions.[169]

Table Appendix 6.1 provides some clues as to how the brain works.

6.15 Appendix 6.2: An understanding of hostility

Defensive aggression (rage) is indeed different from instrumental aggression (predation). Animals that experience rage hiss, lunge, and attack.[170] Predators show their teeth, bite, and attack. Some authors[171] emphasize the difference between the two types of aggression as follows: affective (defensive) aggression is impulsive, overt, and unplanned; while instrumental aggression is without emotion, goal directed, planned, and sneaky. It is not preceded by autonomic arousal, such as an acceleration of heart rate. In sharp contrast, "anger entails physiological arousal (e.g., increased heart rate and blood pressure) and preparation for verbal and physical aggression."[172] Studies with cats (which display easily measurable reactions such as hissing and hair erection) show that irritants do indeed cause defensive rage.[173]

Table Appendix 6.1 A preliminary view at the brain

HPA and HPG Axes

Instrumentally hostile behavior (at the moment of attack) or defensive (under attack)	Some of the chemicals involved:	Cortisol, adrenaline, testosterone (involved in dominance), arginine vasopressin (linked to territoriality)
	Some of the brain structures involved	Hypothalamus; pituitary gland, adrenal glands; cerebellum; autonomic nervous system (sympathetic and parasympathetic)
	Others	Leukocytes, microglia

Core Psychological Constructs

Apprehension	Some of the brain structures involved:	Enteric nervous system Thalamus, hypothalamus (e.g., olfactory bulb)
Trust, Attachment[189]	Some of the chemicals involved:	Oxytocin or so-called love hormone (released by the pituitary gland under the control of the hypothalamus)
	Some of the brain structures involved:	Amygdala, anterior cingulate cortex (ACC), subgenal anterior cingulate cortex (subACC)
Fairness	Some of the chemicals involved:	Acetylcholine
	Some of the brain structures involved:	Hippocampus, prefrontal cortex, caudate nucleus, accumbems
Collaboration	Some of the chemicals involved:	Pheromones (mostly in animals)
	Some of the brain structures involved:	Thalamus, Broca area, primary motor cortex, associative prefrontal cortex, inferotemporal gyrus, striatum, orbital frontal cortex (OFC)
Reward	Some of the chemicals involved:	Dopamine, serotonin
	Some of the brain structures involved:	Nucleus ambiguus; all functional systems: immune, autonomic, endocrine, neuropeptides, etc.

(Continued)

Table Appendix 6.1 A preliminary view at the brain

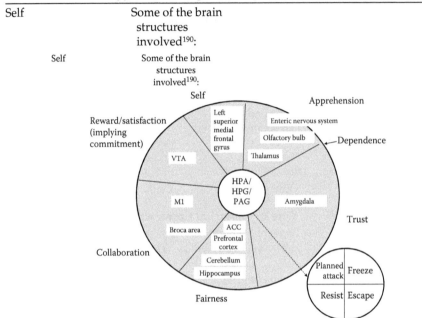

In the case of both defensive aggression (rage) and instrumental aggression (predation), levels of testosterone (T) and cortisol (CRT) play a crucial role. The T-to-CRT ratio is a marker of readiness to social aggression.[174]

Serotonin is linked to impulsivity in both animals and humans.[175] Low levels of fast-acting serotonin encourage high impulsivity, and promote lower levels of fear and anxiety[176] while provoking hyperresponsiveness by inhibiting prefrontal control. High levels of serotonin foster prosocial behavior, which begs for some prefrontal cortex activity[177] when in a non-threatening context. The determining differentiator between defensive and instrumental aggression is therefore impulsivity. A high T/CRT ratio with low serotonin levels leads to defensive aggression. High levels of serotonin in a threatening context encourage instrumental aggression,[178] which entails highly focused activities and strategies.[179] About 10% of the serotonin, which is a neuromodulator, is found in the neighborhood of neurons (but not in the synapses) of the central nervous system (CNS), which regulates appetite, mood, and sleep. The balance of it (90%) is found in the enterochromaffin cells in the alimentary canal (gut) where it regulates intestinal movement. The enteric nervous system, which is a network of neurons attached to the intestinal walls, is connected to the brain by the vagus nerve. Thus, there is an important role played by the gut in experiencing apprehension. This is something we have all experienced through "gut feelings."

Testosterone is a steroid hormone (produced through the hypothalamic–pituitary–gonad axis—HPG); it is linked to aggressive, antisocial behavior and sensation seeking.[180] As put forth by some authors: "Testosterone inhibits the function of the HPA axis and associated autonomic systems, reducing sensitivity to punishment. In contrast, cortisol is thought to empower the sensitivity to fear through suppression of the hypothalamic–pituitary–gonad (HPG) axis."[181]

Cortisol is a steroid hormone that is strongly associated with social withdrawal and a reduction in the activity of the immune system. It is initiated by the hypothalamus in response to stress (through the hypothalamic–pituitary–adrenal axis—HPA). A high T/CRT ratio suggests a high propensity for aggression (either defensive or instrumental) and a low tendency toward socialization.

A high T/CRT ratio with (in part) high serotonin levels leads to Instrumentally hostile aggression: this is characterized by highly focused activities and strategies.[182] Therefore, what differentiates defensive from instrumental aggression is time, expressed as fast (impulsive) versus slow speeds.

I hypothesize that impulsivity is not only a function of serotonin levels, but also of the neuropeptide hormone oxytocin, which is produced by the posterior pituitary gland under the control of the hypothalamus.[183] Oxytocin has been found to be involved in the nurturing exhibited by mothers toward their infants.[184]

A social individual with a very low level of dependency will likely display less impulsivity because their life is not deemed to be threatened, as they are independent. However, someone who depends on others will likely respond very rapidly and strongly to a threatening stimulus.

I hypothesize that the serotonin (5-HT) to oxytocin (OXT) ratio, less known than the steroid-based T/CRT ratio, can be regarded as a measure of impulsivity. Both hormones play a role in impulsivity: low levels of serotonin mean low levels of sociability and low levels of oxytocin are linked to a lower sense of attachment. Their ratio can serve as a marker for impulsivity. I propose the model shown in Table Appendix 6.2, which we have discussed previously.

Table Appendix 6.2 A model for the four core hypothalamic responses to threats[191]

	High (T/CRT) (steroids)	Low (T/CRT) (steroids)	
Low impulsivity (Slow)	Predation	Freeze	High (5-HT/ OXT)(hormones)
High impulsivity (Fast)	Rage (resistant-defensive)	Flight (escape[192])	Low (5-HT/OXT) (hormones)

This information will become handy when discussing my neurobiological research in Chapter 7.

Endnotes

1. Bowlby (1973).
2. Recall that contingency and short behaviors are two different positions.
3. Examples: Al-Maghraby (2008), project name: "itSMF Egypt Chapter," year of project: 2007; Brière et al. (2013), "Equality within the Moroccan public administration," 2005; Consoli (2006), "Australian prisons," 2000; Genus (1997), "Channel Tunnel" (Europe), 1994; Graham (1996), General; Hodgins et al. (1978), "Temagami," 1901; Musali (1998), "Mt. Elgon region" (USA), 1997; Vam Marrewijk (2007), "Environ Magaproject": 1996.
4. The human brain is equipped to organize information along a temporary dimension, which involves memory (e.g., mediated for episodic events by the hippocampus), the present, and the future (e.g., the dorsolateral, ventrolateral, ventromedial anterior prefrontal, and anterior cingulate cortex systems)—see Lewis (2005).
5. The hypothalamus controls the sense of food satiety, for example.
6. Recall the Italian Floortrite example.
7. Food foraging in animals is planned for immediate consumption as well as future consumption (e.g., by hiding food) depending on the level of threats exercised by nearby predators—see Hinds et al. (2010).
8. The less one has useful information (T_h), the more one perceives external stimuli as potential risks (see Wang et al., 2011).
9. The case on Sea Crest Fisheries relates this to quality: the core constructs of Trust, Fairness, and Collaboration, when functional, lead to commitment to quality efforts.
10. Recall the function of Fairness as being $[z' - \beta_0 \, Coll^\alpha \, Trust^{(1-\alpha)}]$.
11. See Dillon and Taylor (2015, p. 100); PMI, 2002; Napier et al. (2009); Skulmoski and Hartman (2010).
12. Pinto (2014, p. 385).
13. Clark and Wheelwright (1992); Hobday (2000).
14. Thakurta and Suresh (2012).
15. Barbin and de Oliveira (2006, p. 2325).
16. See Thakurta and Suresh (2012).
17. See Al-Maghraby (2008), project name: "itSMF Egypt Chapter," year of project: 2007; Barbin and de Oliveira (2006), Brazil, General; Brière et al. (2013), "Equality within the Moroccan public administration": 2005; Chen and Partington (2004), General: circa 2000; Consoli (2006), "Australian prisons": 2000; Creasy and Anantatmula (2013), General; Genus (1997), "Channel Tunnel" (Europe): 1994; Graham (1996), General; Hodgins et al. (1978), "Temagami": 1901; Musali (1998), "Mt. Elgon region" (USA): 1997; Ogunlana et al. (2002), Bangkok (Thailand), General; Pheng and Leong (2001), "Tee Tong Pte Ltd (TTPL)": 1970; Senaratne and Udawatta (2013) Sri Lanka, General; Vam Marrewijk (2007), "Environ Magaproject": 1996.
18. Consoli (2006 p. 77).
19. Gray and McNaughton (2000, p. 43).
20. Gray and McNaughton (2000, p. 44)

21. PMI (2008).
22. Eibl-Eibesfeldt (1971); Plutchik (1980).
23. Davies and Brady (2000, p. 62).
24. The recently published DSM-V could be reformatted by using my modeling system: the Instrumentally hostile position/Defensive position construct would fall into the category "schizophrenia spectrum and other psychotic disorders; paraphilic disorders"; the Trust construct (including confidence) (emotions) would belong to the "bipolar and related disorders; depressive disorders; anxiety disorders; obsessive-compulsive and related disorders; trauma- and stressor-related disorders"; the Dependence construct would relate to "substance-related and addictive disorders"; the Fairness/win-win construct (cognitive) would be integrated into the "neurodevelopmental disorders; somatic symptom and related disorders; neurocognitive disorders"; the Collaboration construct (social) would be associated with "disruptive, impulse-control, and conduct disorders"; the Commitment/Reward construct would relate to "feeding and eating disorders; sleep–wake disorders; sexual dysfunctions," and finally; the Self construct (the project identity in a project context) would fall into the "gender dysphoria; dissociative disorders; and personality disorders."
25. DSM-IV (1996, p. 138).
26. Dulewicz and Herbert (1999).
27. Ogunlana et al. (2002, p. 392)
28. Creasy and Anantatmula (2013).
29. For example, Costa and McCrae (1992).
30. Norman (1963); Goldberg (1990).
31. Thomas and Kilmann (1975); Wood and Bell (2008).
32. Tyler and Newcombe (2006, p. 37).
33. Barrick et al. (2001).
34. Tett et al. (2003); Waldman et al. (2004).
35. For example, Bowlby (1973).
36. Sometimes referred to as avoiding/escaping, secure, anxious/preoccupied, and hostile/resistant.
37. Mikulincer et al. (2002).
38. Folkman et al. (1986)
39. Lösel and Schmucker (2004); Bakamitsos (2006); Mobbs et al. (2009).
40. Mikulincer (1998); Johnson et al. (2009); MacLaren et al. (2010).
41. These equations apply to predator–prey dynamics; in essence, the population of each one is dependent on the other and varies over time to ensure that overall both populations survive.
42. I shall explain in more detail that there are actually two types of hostility: instrumental and defensive. Instrumental hostility is referred to as IP or "Instrumentally hostile position" and defensive hostility as DP (for "Defensive position").
43. See Mikulincer and Shaver (2007).
44. Contingency position is ignored in this section. The brain relies on two general positions under perceived threat: hostile and defensive aggression.
45. See Mulcahy (2013, p. 392).
46. The hypothalamus acts as a biological clock "A principal function of the hypothalamus is to impose temporal organization on hormonal and behavioral processes by virtue of the timekeeping properties of the biological clock in the rostral hypothalamus" (Squire et al., 2003, p. 901).

47. The hypothalamus is equipped with sensors that detect heat, hunger, libido, thirst, time of day, and so forth.
48. Mobbs et al. (2009, pp. 12, 236); see also Squire et al. (2003); Purves et al. (2004).
49. Woody and Szechtman (2011).
50. MacLaren et al. (2010).
51. We will see that instrumental hostility is, in fact, and from a certain perspective, a defense mechanism; the difference between it and the purely defensive mechanisms (freeze, flight, and resistant fight) is that it is spread over time and involves advanced cognitive capabilities.
52. Adams (2006).
53. Predation is closely linked to the action of neuropeptides that are released by the hypothalamus and that support "specific mammalian agendas (e.g., nurturance, predation, defence" (Lewis, 2005, p. 181), and that are connected to aversive behaviors (fear).
54. Siegel and Sapru (2011).
55. Bear et al. (1996).
56. Eibl-Eibesfeldt and Sütterlin (1990).
57. Gray and McNaughton (2000, p. 95).
58. for 1901, see the Temagami project (Hodgins et al. 1978); for 1970, see the Tee Tong Pte Tld (TIPL) project (Pheng and Leong, 2001); for 1993, see the Pendjari National park (PNP) project in Benin (Idrissou et al., 2013); for 1994, see the Channel tunnel in Europe (Genus, 1997); for 1996, see the Environ Megaproject (Van Marrewijk, 2007); for 1997, see the project M. Elgon region in the USA (Musali, 1998); for 2000, see the Australian prisons projects (Consoli, 2006); for 2002, see the Ironic Infrastructure project in Australia (Rose and Manley, 2010); for nineteenth to twentieth century, see the new Acropolis Museum (Fouseki, 2006), and; for General, see the case of Sri Lanka (Senaratne and Udawatta, 2013).
59. Gregg and Siegel (2001).
60. At least in cats, defensive and offensive (instrumental) aggression cannot occur simultaneously: the medial amygdala suppresses predatory attack behaviors (originating from the lateral hypothalamus) (Han et al. 1996).
61. Gray and McNaughton (2000, p. 7).
62. We will see these brain parts again when discussing an *f*MRI study in the chapter on power.
63. See Terburg et al. (2009); Gray and McNaughton (2000); and Siegel and Sapru (2011) in particular.
64. Amsel (1992).
65. Siegel and Sapru (2011, p. 215) write that: "the *medial* hypothalamus (…) provides a mechanism that modulates feeding, generates affective processes, such as rage behavior, in animals and people." The central amygdala has connections with the lateral hypothalamus (seat of instrumental aggression in cats—see Siegel and Sapru, 2011) and the medial hypothalamus (for defensive aggression in cats) (Campbell, 2008). Bear et al. (1996, p. 452) mention that "displeasure centers, or negative-reinforcement sites are located in more medial portions of the hypothalamus and lateral parts of the midbrain tegmental area." It is noted that stimulation at these locations may elicit fear or activate an urge to flee in a negative-reinforcement situation. The PAG is "activated by an immediate threat, by an immediate dominant conspecific,

by pain" (Gray and McNaughton, 2000, p. 31). The PAG "may then be the most important node through which information passes during fight, flight, and freezing." (Gray and McNaughton, 2000, p. 32). The central amygdala entertains key connections with the lateral hypothalamus (responsible for predation in cats—see Siegel and Sapru, 2011, and nociceptive modulation in rats—see Safari et al., 2009) and the medial hypothalamus (for defensive aggression in cats) (Campbell, 2008).

66. Amaral (2002).
67. Perceived threat can affect the brain and induce strong, automatic reactions.
68. Blanchard and Blanchard (1989, p. 75).
69. Rosebloom et al. (2007).
70. Plutchik and Platman (1977, p. 420).
71. Graham (1996, p. 68).
72. Wood and Bell (2008, p. 130).
73. Consoli (2006, p. 81).
74. Senaratne and Udawatta (2013, p. 60).
75. Senaratne and Udawatta (2013, p. 60).
76. Ohlendorf (2001); Creasy and Anantatmula (2013).
77. Dependence was measured with two questions. Of course, I recognize that this makes the evaluation of dependence rather weak.
78. Bowles (2000).
79. Pelletier (1993).
80. Carney et al. (2010).
81. With Trust measured here on a scale from 0 to 100.
82. This remains largely hypothetical but serves my purpose of trying to explain human behavior in the context of projects.
83. I have tried many times over but it has proved to be quite challenging.
84. I measured intention only, not whether participants actually stayed in the group or not.
85. SEM does not check for normality of population.
86. Durbin–Watson = 1.701; $F = 3{,}651.781$, $p = 0.000$; normality of residuals.
87. SEM analyses cannot be used to check the law of apprehension (perceived threat).
88. My questionnaire has gone through extensive psychometric testing and reliability tests, and was even checked by a lawyer to ensure that it met its objective.
89. Independent t-tests were used.
90. This is certainly true in the animal kingdom (see Botham et al., 2006).
91. PIP refers to an IP personality. IP refers to an IP behavioral response, not a personality.
92. Note: in the model of Figure 6.9, "intention" is considered to be an expression of "commitment." I did not measure commitment *per se*, but rather intention to stay in the group. Intention is a very subjective and diluted construct but it is commonly used in marketing research.
93. Jones and George (1998).
94. Idrissou et al. (2013)
95. For example, Hodgins et al. (1978; in the Tamagami case), Rose and Manley (2010; Iconic infrastructure project in Australia), Ogunlana et al. (2002; Bangkok example), and Creasy and Anantatmula (2013; general case).
96. Newcombe (2000; High-Tech Projects in Thailand).

97. As mentioned, I posit that "stability" (as in stable persons) is not a personality trait but a state that expresses the fact that the individual has found a way to manage all four basic preset coping mechanisms.
98. Varimax rotation with Kaiser normalization, 3 iterations.
99. This is referred to as the *law of perceived threat* or *law of Apprehension*, as we have seen in Chapter 5.
100. See Pels et al. (2000).
101. For obvious proprietary reasons, I am not prepared to divulge this algorithm or to provide the complete data that would allow one to find it.
102. Attachment can be linked to both trust and a sense of dependence: "Attachment theory has obvious relevance to social support processes (...) because the attachment behavioral system will be automatically activated in response to stressful or threatening events (Mikulincer et al., 2000; 2002)." The authors also point to the fact that attachment implies implicit and explicit expectations about others being emotionally responsive to their needs (Collins and Feeney, 2004, p. 364).
103. Mikulincer and Shaver (2007, p. 279).
104. See Magdol et al. (1998).
105. Todorov and Engell (2008).
106. DSM-IV (1996, p. 110).
107. Collins and Miller (1994).
108. Mayer et al. (1995).
109. Ferrer et al. (2010, p. 422).
110. Folkman et al. (1986).
111. Gundlach and Cannon (2010).
112. I did not find an effect for age. I had assumed that older people would perform better on the PCI given their experience. I did find an effect for gender, with women outperforming men; this, however, requires a more advanced study, the ultimate top performer group being too small of a group to generalize on the effect of gender.
113. See Glass and Newman (2009).
114. See Reidy et al. (2011)
115. See Hoff et al. (2009).
116. See Patrick et al. (1994).
117. Hare (1991).
118. Verbeke et al. (2011).
119. Morgan and Lilienfeld (2000).
120. Antisocial individuals display lower levels of serotonin compared with a normal population. When not in a hostile mood, IP personalities tend to have lower skin conductance and lower heart beats.
121. Sporns (2011, p. 76).
122. Yang and Raine (2009).
123. Vartanian et al. (2011, p. 26).
124. Mobbs et al. (2009).
125. Lösel and Schmucker (2004).
126. Muchembeld (2008) found that most violent acts are the fact of 17–24 year old males, a group that, incidentally, has a high level of testosterone.
127. I used an independent *t*-test with Levene test for equality or inequality of variances; with $p = 0.05$.

128. Varimax rotation, 3 iterations.
129. See Mikulincer and Shaver (2007).
130. This is discussed in more details in the next section on conflicts.
131. Oakley, G. Interviewed March 2, 2016.
132. Meredith and Mantel (2009, p. 11).
133. Brett et al. (2006).
134. Meredith and Mantel (2009, pp. 220–221).
135. See Jap et al. (1999).
136. Meredith and Mantel (2009).
137. Meredith and Mantel (2009, p. 158).
138. Seymour et al. (2007 p. 306).
139. Soila-Wadman and Köping (2009, p. 39).
140. See Gherardi (2006)
141. See Lewicki et al. (1992); Gherardi (2006).
142. Johnsson et al. (2004, p. 390).
143. Habituation is somehow rendered in Tuckman's model whereby teams are formed, then go through a turbulent episode (which involves hostility), followed by *normalization*, a performance, and a dissolution phase (Tuckman, 1965).
144. A form of blindness well exemplified by Hitler's decisions toward the end of WWII when he decided to attack the Eastern, Southern, and Western fronts all at once.
145. See, for example, Zimbardo (2007).
146. See, for example, Meredith and Mantel (2009, p. 220).
147. Meredith and Mantel (2009).
148. See, for example, Sizemon (2008).
149. See, for example, Graham and Englund (2003).
150. Significance levels were set at 95% (or $(1 - \alpha) = 0.5$). To compare means, independent t-tests were used after checking for the normality of populations (Kolmogorov–Smirnov and Shapiro–Wilk tests) and the normality of residuals (Levene test).
151. CFI = 0.998 (*before*) (>0.9)/0.886 (*after*). RMSEA in the *before* is at 0.029 (<0.05) but found to be at 0.199 in the *after* experiment.
152. CFI = 0.923 (*before*)/0.970 (*after*) (>0.9); it is thus acceptable. RMSEA = 0.210/0.144 (>0.05) indicating poor fit in both the *before* and *after* periods.
153. "The future states of a dynamic system are a function of its present sate, as modified by its own activities" (Lewis, 2005, p. 173).
154. Here, Trust$_t$ refers to Pretrust. In the theory of deterministic chaos, systems are indeed quite susceptible to initial conditions (May, 1995).
155. Mackey and Glass (1977).
156. Graham and Englund (2003).
157. Photos graciously provided by Daoust Lestage Inc. architects—urban designers, and Stéphan Poulin.
158. I kept the original names.
159. Squire et al. (2003, p. 897).
160. Gregg and Siegel (2001, p. 97).
161. Squire et al. (2003, p. 897).
162. McCullough et al. (2007).
163. Bear et al. (1996, p. 446).
164. Siegel and Sapru (2011, p. 215).

165. It may be that the right lateral nucleus of the hypothalamus is a stronger center of Instrumentally hostile behaviors than its left counterpart. The right hemisphere of the brain plays an important role in a variety of functions such as agreeableness, fear processing (including cortisol control), inhibitory capabilities in regard to affective, behavioral, and cognitive processes, "maintenance of positive social relationships, collaboration and attention to social norms", and social behavior. This would have to be probed further.
166. Alvarez et al. (2008); Amaral (2002).
167. Gregg and Siegel (2001, p. 96).
168. Woody and Szechtman (2011).
169. Mobbs et al. (2009).
170. Adams (2006).
171. Gregg and Siegel (2001, p. 93).
172. Reimann and Zimbardo (2011, p. 175).
173. Siegel and Victoroff (2009).
174. Terburg et al. (2009).
175. Ueda et al. (1999).
176. Woody and Szecthman (2011).
177. Verbeke et al. (2011).
178. Montoya et al. (2012).
179. Meloy (1997).
180. Stenstrom and Saad (2011).
181. Reidy et al. (2011, p. 520).
182. Meloy (1997).
183. Siegel and Pradu (2011, p. 215).
184. Purves et al. (2004, p. 486).
185. The T-to-CRT ratio has been considered as a key marker for social aggression (Terburg et al., 2009). Testosterone (T) and cortisol (CRT) are associated with avoidance behavior and distrust; serotonin (5-HT) and oxytocin (OXT) are linked to approach behavior and with trust (Riedl and Javor, 2012). Table 6.1 contains hypotheses I have formulated years ago on the role of hormones, based on contemporary research data.
186. Interestingly, the stable "class" exhibits a k value of 1.3.
187. Shapiro (2010) uses two axes: one for autonomy, one for affiliation. Dominance and seduction are two main axes associated with animal behaviors (see Bowles, 2000, p. 870 in particular).
188. A C\$80 billion financial institution in Québec.
189. As an example of the inevitability of attachment, newborns depend completely on others for body heat, food, fluids, and hygiene).
190. See Goldberg et al. (2006).
191. Testosterone (T) and cortisol (CRT) (along with arginine and vasopressin) are typically associated with avoidance behavior and distrust, while serotonin (5-HT) and oxytocin (OXT) (along with estrogen and dopamine) are typically associated with approach behavior and trust (Riedl and Javor, 2012, p. 73).
192. Under the control of the medial hypothalamus (Di Scala et al., 1984).

chapter seven

Power

So it is, said the Wise Eagle, that even a hurricane keeps an eye on what it does.

Darloz

7.1   Introduction to Chapter 7

In an interview reported in a project management magazine,[1] an emphasis is put on the importance of being able to finish a project before it turns into a catastrophe. Decision excellence is thought to be the result of acceptance of the interests of stakeholders, fitting process, looking out for political agendas (true motives), proactive agility, and the presence of qualified people (informed and able to drive the project), all of this with due consideration of the risks. In another interview, a famous management expert is quoted as saying that managers have to learn to manage in situations where one has no authority, is not under control from anyone and is not yet able to control.[2] Power is exercised in different forms and across different project organizations, be they of top-down classical structure or else in a matrix format, each of which have their advantages and disadvantages.

We have seen that knowledge is an important component of project management. *The Project Management Body of Knowledge* (PMBOK) refers to a book of knowledge (BOK) developed once a project is brought to its conclusion; my model includes knowledge (T_h) as an indirect resource necessary for Commitment. Acquired knowledge from previous projects leads to more efficient planning (Plans), improved control of Work Breakdown Structure (WBS) (Process), more efficient communication (People), decreases in resource consumption (Process),[3] faster task execution (Process), and enhanced problem-solving (Power). Project leaders acquire knowledge and thus make informed decisions.

Despite countless articles and books on leadership in project management, the critical importance of people and the proper training of leaders are often underestimated.[4] As we saw in the general introduction, project feasibility analysts must scrutinize the competency of project promoters before committing funds.

There is more to project management than adhering to a plan, checking WBS, and negotiating with the forces of production. A leader relies on experience and memory; they must apply managerial tools, evaluate progress and results, and comprehend and fathom out situations, and, not to forget, devise new ways of dealing with challenges, day in and day out.[5]

At the same time, project managers are prone to errors, either because of their own weaknesses or because they lack access to valuable information.[6] By experience, I can affirm that many tools or reports that are conveyed to project managers are laden with errors; some are related to psychological factors that influence the behaviors of the Forces of Production, FP,[7] such as jealousy.

Leaders face risks with stoicism and determine how to deal with them. Some authors suggest that risks act as moderating variables between control exercised by project managers and performance.[8] In low-risk environments, control would significantly enhance project performance, while in high-risk environments, control would have little impact on project performance. This probably stems from how risks are defined and from the fact that control is dissociated from the concept of "transparency." As seen in our six core competencies model, project managers need not only to get a grip on things, but also to be able to convey their vision, their requirements, and their rules; that is, they must be transparent. Failing to find a right balance between Control and Transparency can surely nurture points of vulnerability (POVs). For one, some of the management and team members (the Forces of Production) may become less motivated and increasingly careless when working under either lax or shrouded management.

This chapter is about understanding the fourth P, that of "Power."[9]

7.2   *Understanding power*

There are countless definitions of leadership and power. Leadership, for example, is at times classified according to four patterns, to name just a few: cooperative (likely transformational or so-called authentic leadership); dictatorial (traditional), which, having seen it in action in many countries ourselves, I can confirm still exists; moderator (so-called charismatic or situational); and transactional. Various theories, such as trait theory and the theory of expectancy, have capitalized on the notion of leadership. The latter implies qualities such as the capacity to conceptualize, to forecast, to listen, and to persuade, as well as the capacity to offer "healing" and knowledge, to show empathy, and finally to promote commitment and community building.[10]

My goal is not to engage is a nomological debate over these two concepts, that of leadership and power, which are different in nature. Roughly speaking, power is about the capacity to enforce action by way of formal or informal authority; in some cases, it may fuel a dynamic of abuse.[11] Project leaders with informal power resort to psychological influences rather than to contractual structures and organizational charts (formal lines of authority, which imply, for example, job descriptions, fringe benefits, privileges, and salaries); they adopt a vision for the project,[12] share it, and delegate tasks accordingly. Says one project leader:

> When a company grows, it faces numerous challenges, yet flexibility is crucial just as is the setting the rules up front. When preparing a project, we discuss with our teams in an attempt to seek their approval; we then drive and complete the project. To supervise and to delegate go hand in hand.[13]

Power has a lot to do with informal lines of authority. A project manager is the one holding the "real" power.[14] Capturing a project manager's informal power requires an understanding of work culture (psychodynamics), which is challenging in a world where the concept of "authority" varies with cultures. Power rests not only with people in authority, but also with the team members, who are empowered by the project organizations. To take the example of an operation such as a Toyota factory, it is known that an employee on an assembly line has the real power to stop a production line should he detect a problem. The **real power**, from my point of view, is the capacity to address POVs.

7.3   *Project manager profile*

The feasibility analyst can assess power by looking at the ability to deal with adversity and to complete projects with minimal utility drawbacks, and by examining contractual terms of employment, curriculums, organizational charts, past projects performances, presence of training programs, and references.

Personalities (which we have studied in Chapter 6), style, and goals are certainly other elements to take into account. According to some sources,[15] there is a gap between what contemporary leaders prioritize and what people think these leaders prioritize with respect to concerns for employee growth, development, and well-being, and meeting short-term and long-term financial goals. Some scholars[16] explain that managers are expected to exhibit capacities to communicate first and foremost, to deploy interpersonal and personal qualities, to handle changes, to provide leadership, to show ethical behavior, and generally and less importantly, to adapt.

Various authors propose the following qualities: having an ability to implement and to prioritize, and to display accessibility, accountability, credibility, and visibility (walk the talk).

Grosso modo and based on numerous readings, I can state that managers must be able to set goals (Plans); manage resources and facilitate WBS (Processes); manage discords and expectations as well as promoting an efficient and motivating work environment (People); and monitor changes and represent the teams, provide direction, and set standards (Power). The underlying motivation is in part to provide the team members (Forces of Production) with a sense of inclusion, pride, respect, reward, and security, and to instill a culture of information sharing and career development.

As I have discussed in Chapter 1, management is often judged at fault by way of adhering to opaque objectives, changing priorities, lacking a sense of direction, providing insufficient support and resources—all of which relate to Plans—and, interestingly, by being disconnected from reality.

Another way of looking at Power is through the Product, Organization, and Work (POW) lens. Project managers are said to have to ensure quality of products (products), to mediate disputes and manage budgets (organization), and to take care of planning and scheduling, coordinating activities and day-to-day operations, as well as keeping work on track (WBS).

In terms of POVs, they are to deploy strategies to secure resource procurement and allocation, to drive the project while controlling and monitoring for vulnerabilities, and to mediate contentions. Various authors have proposed ways to decrease conflicts.[17] Avoiding, compromising, confronting, facilitating, forcing, problem-solving, smoothing, and with-drawal are a few examples.

Recall that in Chapter 4, I referred to the six Ps of strategic project management: Plans-Processes–People–Power (four Ps); pessimistic, real-istic, and optimistic (PRO) scenarios; Product–Organization–WBS (POW); Present Work Psychodynamics (PWP) (six core competencies model); points of vulnerability (POVs) using the strengths, vulnerabilities, oppor-tunities, and risks (SVOR) framework; and dynamic Point of Equilibrium (POE), an illustration of which appears in Figure 7.1.

Figure 7.1 Six Ps of strategic management.

In Table 7.1, I take a brief look at what some of the literature suggests Power should entail,[18] where applicable.

For the so-called mechanical qualities, I observe that Control occupies a large portion of the mechanical characteristics of Power. Little attention is given to control of POVs and none to securing equilibrium.

For the so-called human qualities, I observe that, indeed, the human aspect of projects has a role in the implementation of Power. Fairness and Commitment are given little weight (by number of times cited). POVs are lightly covered; Points of Equilibrium (POE) not at all. All key com-petencies relating to psychological constructs are listed: Control and Transparency, Trust, Fairness, Collaboration, and Commitment.

Of course, these tables are not exhaustive, but they provide some structural and functional variables for the concept of "Power." They offer a flavor of perceptions that scientific authors and field experts have of Power. It seems obvious that a strong definition of Power must be given, one that takes into account the six strategic Ps. A project manager who would not perform in any one of the six strategic areas would unlikely fully meet the project stakeholders' expectations, especially in high-risk environments.

I thus propose the following general definition of Power in the context of project management:

Table 7.1 Tentative summary of Power qualities according to the six strategic Ps

Qualities	PWP	Four Ps	PRO	POW	POV	POE
Mechanical aspect						
Connecting to reality	Commitment	Plans	P	O	POV	—
Managing budget	Control		R	W	—	—
Planning and scheduling			O	W	—	—
Setting priorities			R		—	—
Ensuring quality of products	Control	Processes	R	P	—	—
Keeping the work on track			R	W	—	—
Managing day-to-day operations			R		—	—
Securing resource procurement and allocation			R		POV	—
Human aspect						
Adapting/handling changes	Collaboration	People	R	POW	—	—
Communicating			R	O	—	—
Sharing information			R		—	—
Being accessible	Transparency		R		—	—
Being accountable			R		—	—
Reinforcing credibility	Trust		R		—	—
Mediating conflicts	Collaboration	Power	P	O	POV	—
Fostering career development	Commitment		R	O	—	—
Coordinating activities	Control		R	W	—	—
Prioritizing	Control		R	O	—	—
Controlling and monitoring for vulnerabilities	Control/ transparency		P	POW	POV	POE
Driving the project			R	O	—	—
Showing ethical behavior			P	O	—	—
Motivating the FPs (reward, pride, respect, consideration, security)	Fairness		O	O	—	—
Managing expectations	Trust		R	O	—	—

7.3.1 Definition of power

From the perspective of project feasibility, the assessment of Power allocated to a project leader implies their ability to exercise control, to be transparent and fair in their relations with stakeholders, and most particularly their ability to foster trust, collaboration, and commitment on the part of both management and team members (FP). The project leader aligns their efforts along six strategic axes: (1) the coordination of Plans, Processes, People, and Power itself as key components of the project (the four Ps); (2) the careful examination of PRO scenarios; (3) the supervision of the products, the organization itself, and the tasks at hand (POW); (4) the identification of strengths, opportunities, risks, and POVs, along with appropriate responses in order to minimize their impacts; (5) the search for a POE that maximizes the chances of success of said project (POE); and (6) the giving of due consideration to PWP, including the management of conflicts and commitment.

I believe my definition may provide some initial insights into best practices because of its emphasis on monitoring deviations on all four Ps: Plans, Processes, People, and Power itself.

PMBOK 5 (p. 47) offers an example of the exercise of Power that is in line with my view: "A primary function of a PMO (Project Management Office) is to support project managers in a variety of ways which may include, but are not limited to: (...); coaching, mentoring, training, and oversight (...)"

In their attempt to fulfill the requirements of their jobs, as set forth in the definition I gave of Power, project leaders can rely on three forms of authority: (1) the internal and formal authority that is recognized by the organization; (2) the informal authority, which includes their own charisma; and (3) the legal mechanisms that surround the project. In the case of the Montréal Olympic Stadium (MOS), for example, laws that governed the project within the eight constituents of risks as seen in Chapter 2 included the zoning regulation (environment); corporate laws (financial/organization); the Labor code (human resources/sociocultural); the Consumer Protection Act (marketing); and the Building code, the CSA, and other standards such as ISO 9000 (technical).

7.3.2 In search of value

The project leader is not only concerned with perceived value—that is, value that consumers allocate to the project, which I defined in Chapter 1 as [(Functionality + Design)/Costs]—but evidently with added and residual value. The former is the value that results from adding individual efforts (T) at each stage of the transformation phase. In the case of the

Québec Multifunctional Amphitheatre (QMA), carpenters, electricians, masons, and plumbers all bring their input, which compounds to engender the added value (often established as book value).

A truly successful project occurs when the perceived value exceeds producer (added) value. This means that the project leader is constantly trying to minimize the added value while maximizing perceived value; any changes to the project are made with the aim of reducing costs while offering more to the end consumer.

In terms of Power, we envision that adequate authority and control lead to the maximization of success, or, put in marketing terms, of stakeholders' satisfaction.[19]

The example of the engine oil filter seen in Chapter 2 is revealing: the added value is relatively high (e.g., arranging for proper insurance coverage, installing the oil/filter case in the car, and securing a second one when the first is being replaced) while the proposed solution may seem, to a vast majority of potential consumers, more difficult to implement, more expensive, and more troublesome than a routine change of oil. The perceived value of the product is most likely quite low. Add to this that POVs (everything that can go wrong) are numerous.

To determine perceived value, it is not the accumulated costs associated with each stage of production that are taken into account, but the sum of the critical attributes that ultimately define functionality and design. Let us revert to the example of the cat litter seen in Chapter 4 (Table 7.2).

Table 7.2 Cat litter case: Value analysis

Type of product	Brand X cat litter	Brand Y cat litter
Ease of cleaning		
Manufacturer's reputation		
Odor control		
"Repulpable"[144]		
Retail price		
Weeks of use		

Perceived value cannot be assessed without fastidiously selecting the attributes deemed essential for the consumer who will use the project or buy the product resulting from the project (e.g., cat litter). This requires a marketing research; put differently, it requires a needs analysis. In addition, perceived value is always subjective; it is the value of a project compared with another one that counts. In the case of the cat litter, its capacity to control odors and the possibility to recycle[20] the packaging materials are criteria that may be instrumental.

Any project that has not defined its key attributes, and done so in comparison to other projects, is hiding POVs, consciously or not. A project leader cannot possibly fulfill their job if they do not understand its nature. By experience, I can affirm that many project proposals that I have seen fall short of the evaluation of perceived value while emphasizing added (book) value.

7.3.3   In search of control and metrics

My six core competencies model calls for a prime role for the construct of Control. According to some authors, project leaders need to adapt it to the size and complexity of projects; they must apply cost diagnostic techniques, commit to them, engage personnel resources to implement projects, recognize and understand projects, and secure owner/contractor contractual arrangements. PMBOK 3[21] recognizes a series of activities that includes planning, deployment, and action. Other sources[22] propose another cycle of action that includes acquisition of knowledge, evaluation, and improvements; yet others[23] emphasize the priority of corrective actions.

In essence, **Control** is about the capacity to implement change—change ingrained in the transformation phase itself, and changes that appear mandatory when a Plan meets reality. It has been posited that three documents are generally used to secure control-related actions:[24] (1) routine documents; (2) documents addressing contingency situations; and (3) special analyses for particular needs. Certainly, the feasibility analyst should check whether documents exist to control processes under the Dominant strategy (DS) and the Contingency strategy (CS).

A control panel with cameras.

Based on my experience, large portions of measurements that are presented to feasibility analysts contain errors in either the pure data entry and calculations or else in psychodynamics evaluations. As an example, in a recent study I performed, the market data provided by the promoters dated back four years earlier. However, the product's life cycle was exactly four years, so that the data submitted was, essentially, useless.

Because each project is different, and because each project requires a marketing assessment (that implies some psychological constructs such as satisfaction and intention to use the project's deliverable), project leaders must either develop their own scales or rely on measurement scales and metrics developed by others (that could be faulty or that may not adapt particularly well to the reality of their own projects). Forty years ago, it has been written: "Strictly speaking, one validates not a measuring instrument, but rather some use to which the instrument is put. For example, a test used to select college freshmen must be valid for that purpose, but it would not necessarily be valid for other purposes (…)"[25] I feel this statement has value: instruments are valid in the context in which they apply.

Many scales for assessing project feasibility have been developed without, however, having passed a reality test.[26] This is tantamount to evaluating POVs with a deficient instrument.

Scales that employ opposing terminology, such as "unsatisfactory" on the one hand and "fully satisfactory" on the other, are erroneous. Dissatisfaction and satisfaction, to take only this example, are two different constructs.[27,28] As a consumer, I may not be satisfied with a product (because it doesn't meet my needs) without, however, being dissatisfied ("I used it and it disappointed me"). The proper scale in this case is "not satisfied at all' on one end and "completely satisfied" on the other end: the word "satisfied" has to be repeated; the same construct must be used.[29]

Many scales suffer from lack of parallelism:[30] choices that are given along the scale do not respect the same logic. A scale that offers the following alternatives—"a lot," "not at all," "enormously"—is not properly designed. The term "enormously" falls into a different psychological field than "a lot" and "not at all." Project leaders who base their decisions on faulty scales are prone to vulnerabilities. The scales may engender biases, which later can explain why a particular project was wrongly evaluated and given a "go ahead" when it shouldn't have.[31]

As put forth by some scholars,[32] questions that can be part of a questionnaire assessing stakeholders' interest in a project, as an example, must be measurable, objective, pertinent, and significant, and finally use an economy of language. I also contend that they must be understood by the participant[33] (using concepts and wording that the participant is not familiar with will bias responses) and stick to one idea at a time.

Referring to marketing research done when evaluating the potential of a project, I contend that the use of so-called additive questions is misleading. This occurs when the same questions are rephrased in slightly different forms. This procedure does not test the answer, but rather whether or not the respondent is consistent.[34]

Many sources call for the use of **Key Performance Indicators** (KPIs)[35] as a metric. KPIs should relate to best practices in the sector[36] and to results. They should be determined before the project starts, but will be put to use throughout the project and, in particular, during the project closure stage.

I am of the opinion that KPIs should include indicators that relate to POVs and in particular to their description, their presence, and their importance. I discuss various methods to achieve this in Section 7.4.

7.3.4   *In search of simplicity*

As mentioned in the introductory chapter, projects are becoming increasingly more complex. Complexity encourages POVs; in fact, it can even hide them, much as a dense orchard would hide a tree with rotten apples. Project complexity has been associated with capital investment, geographical locations, nonlinearity of project flows, number of activities/tasks required, and size.

Some authors[37] propose a model whereby project complexity is a function of interdependencies within the project (both as mediating variables), as well as a project's size and variety, and contextual elements. The latter is the only variable to have a negative influence on complexity, which seems odd since the contexts are precisely known for their multiple risks. Nevertheless, interdependencies, size, and variety are certainly determining components of complexity. Generally accepted characteristics that make a complex system include

1. Changes that cause instability
2. Dynamic elements (e.g., a large number of tasks to be done in a very short time)
3. Interdependence between project components
4. Number of elements
5. State of the system—open or closed
6. Uncertainty[38]

As can be seen from this list, each element has so far been discussed in this book. These elements behave by way of multiple interactions, meaning that they reinforce each other. Indeed, projects are complex dynamic systems; with complexity comes a greater propensity for errors.

To account for complexity, project leaders work around **operating margins**, which, as the project gets more and more constrained by its critical path, tend to shrink, thus inducing errors. Operating margins take into account pessimistic and optimistic scenarios; in other words, the strategic PRO is an indirect measure of operating margins and of complexity. An example is provided by way of the MOS, as is seen in Table 7.3.

The observations we have just covered lead me to propose an additional law with respect to project feasibility.

7.3.4.1   *Sixth law of project feasibility: Law of complexity*

> The more complex a project is, the riskier it is and the more POVs it contains; the higher the probabilities of failure.

Overall, project leaders are well advised to divide complex elements of the project into simple, manageable units. Based on my experience, I have

Table 7.3 Analysis of complexity for the Montréal Olympic Stadium (MOS)

Elements of complexity	Proofs
Interdependence between project's components	Tower with retractable roof
Change factors	Weather, retractable roof
Number of elements in project's system	Inclined tower, retractable roof
Uncertainty	New technology, weather
Project's elements dynamism	Multiple
Closed system	Bound by time constraints, costs, and norms of quality
Overall complexity	Very high

concluded that doing otherwise inevitably leads to catch-up efforts, which influence the *g*-spread of the project.

7.4   Decision-making

So far, my discussion has attempted to show that Power, or persons with authority, are not perfect! They stumble over many challenges that make them prone to errors while having to make decisions that can dramatically affect their project.

A plethora of tools are available to the project leaders and decision-makers (those who will provide financing for the project), some of which, of course, come in software programs. I briefly discuss a few of them.

7.4.1   Risk assessment and radial maps

Evidently, there are a number of ways to assess risks; many companies employ their own methods. Scientific authors emphasize their own take on risks, whether financial or sociocultural. On the latter, for example, it has been proposed to use radial maps to counter various barriers.[39] These include (1) cultural hurdles: language and values; (2) economic hurdles: exchange rates, inflation, market conditions, taxes; (3) environmental hurdles: climate and weather, housing availability, infrastructures (transportation, hospitals, etc.), pollution; and (4) politico-legal barriers: activism, corruption, crime, espionage, labor movements, riots, violence, wars, and work ethic.

In fact, my six core competencies model and its associated radial map[40] can also be employed and has the advantage of being supported by two questionnaires, one of which is included in Appendix 5.2. Figure 7.2 illustrates how such a radial map can be put to use.

Other tools are available; however, the reader should note that they are not intended to detect POVs, but rather to manage the entire project or only parts of it.

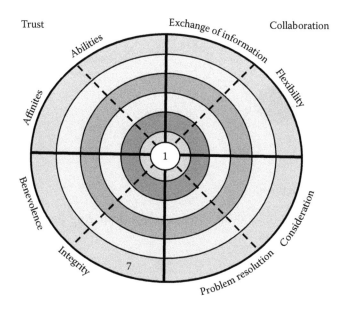

Figure 7.2 Radial map for some key constructs.

7.4.2   Analysis by comparative tables

As seen when I discussed perceived value, comparative tables consist in comparing the attributes of at least two projects. I recommend resorting to contrasting cases as well: projects that are dissimilar—most particularly, projects that have succeeded and others that have failed.

This approach is most convenient when doing a prefeasibility study because it provides a data set against which measurement parameters can be forged, keeping in mind that these measurements should be as error-free as possible.

7.4.3   Product tree analysis

This method breaks down particular products according to a hierarchy of characteristics, as shown in Figure 7.3. This method cannot be used to identify causality so that it cannot serve directly toward the identification of POVs.[41] However, because it outlines the key attributes of the products (or projects) in a hierarchical manner, it is useful when assessing priorities and underlying psychological constructs (e.g., in this case, the underlying concept is flying, which raises the question of a need for transportation, which raises the question of the reason for the need for transportation). As seen in Chapter 1, each project shares some common structural variables; these may be expressed differently depending on the project, so that the

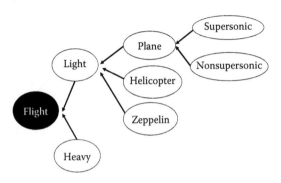

Figure 7.3 Product tree.

product-tree type of analysis can assist in prioritizing these variables as the project progresses.

7.4.4   Multicriteria analysis

Multicriteria analysis is a method by which the criteria chosen to support a managerial decision are listed and given a particular weight (see Table 7.4).[42] Generally, one criterion must be sacrificed in favor of another, so to speak. This is evident in the case of the utility drawback (or triple constraints): most commonly, accepting to change the design of a project while in the transformation phase implies costs. The manager has to weigh whether the change is worth the increase in costs. However, the implementation of the change itself may compromise the deadline, which again is a criterion to take into account. In fact, the triple constraints encompass all the different possibilities because, by way of the norms of quality, project attributes (or product attributes as in the case of the product tree shown in Section 7.4.3) are a given.[43] By definition, therefore, a project implies a multicriteria analysis in about every phase of the project (inputs, transformation, and outputs) as shown in Table 7.4.

Table 7.4 Multicriteria analysis (example)

Criterion	Option 1	Option 2	Option 3
Functionality	9	5	9
Design	4	5	10
Cost	9	10	5
Total	22	20	24

Note: Weight   assigned   to   each   criteria   (0 = none; 10 = maximum).

Quality itself (as in norms of quality) implies a compromise between functionality and design. Any product developer will attest to the fact that better design generates increased design costs, and that an improved functionality sometimes forces a sacrifice on design. Products that excel in the market (such as the iPad) have had the ingenuity to reach the best balance between functionality, design, and selling price given the customers' needs. Likewise, projects with the greatest perceived value are generally those with optimized trade-offs between functionality, design, and costs.

POVs have a greater chance of developing when there is excessive weight given to a particular criterion versus others. For example, a project that is far over its cost ceiling provides an indication that something went wrong, and if something went wrong, probably some POVs played a part in it. The feasibility analyst can check the promoter's project plan to see if the criteria of [calendar–costs–norms of quality] and the criteria of [functionality–design–cost] are reasonably weighted.

7.4.5   Analysis by decision tree

The analysis by decision tree lays down scenarios by answering "Yes" or "No" to questions raised on each point reached along the decision process.[44] It is regularly used in product development (Figure 7.4).

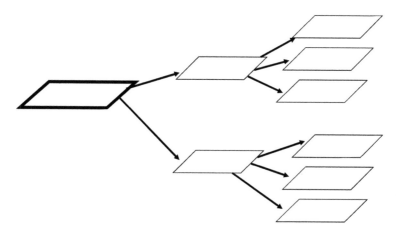

Figure 7.4 Decision tree.

Of course, this method finds a perfect application in the PRO (Pessimistic, Realistic, and Optimistic) scenarios strategic effort. It is not uncommon to associate probabilities along each path so that better decisions may be made. The method can also be elected when deciding between two mutually exclusive projects, evaluating in the process the net present value (NPV) and the internal rate of return (IRR).[45]

The way to use it to detect POVs is by assessing whether the particular path that seems most likely to lead to success or to the desired outcome is realistically feasible. The more the elected path (pessimistic or optimistic) shies away from reality, the more POVs will obstruct the chances of success. Through this method, it is not necessary to identify the nature of the POVs; rather, the analyst infers them from the course of actions and outcomes that emerge. It is best practiced by someone outside the project, that is, by someone who is unbiased and preferably experienced.

7.4.6 Analysis of cause-to-effect relationships

This analysis[46] is at the core of my proposal set out in Chapter 4 to identify POVs (positive causality: C^+, negative causality: C^-). The key to the analysis of cause-to-effect relationships is to pose the question "Why?" at every step along the way. While it may seem quite annoying (and by experience, I can testify to the fact that it often is!), eventually, tracking the causal links leads to the source of the problem.

Let me take an example. Why has the project not been finished on time? Because there was a strike by the electricians. Why was there a strike by the electricians? Because they were not happy with their working conditions. Why were they not happy with their working conditions? Because they believed they were exposed to potential danger. And so on. We arrive at the conclusion in this hypothetical case that adequate security measures were not planned appropriately on the construction site.

Not all causes are within the project leader's control—some may point to internal elements (POVs), others to outside forces (risks); however, one of my tenets is that risks and POVs go hand in hand by way of the [Risks = k/POVs] relationship. Hence, the feasibility analyst must look at the two sides of a probable cause tenaciously: external and internal. Blaming external forces only is forgetting or refusing to admit accountability, all the while ignoring them is putting unnecessary blame onto the Forces of Production.

7.4.7 Sensibility analysis

This method attempts to measure uncertain aspects of the project. A change in a factor of production such as the addition of a single Force of Production (FP) or of a single means of production (MP), for example, will affect outputs. The manager then tries to set the most efficient combination of inputs–outputs. This method is pertinent during the planning process and applies when changes are required during the transformation phase. It does not permit to identify the nature of POVs, but it may point at areas in the transformation phase where POVs are likely to surface. In particular, it is most advisable to identify walls (calendar-related activities) and ceilings (costs). The item with the greatest sensibility (e.g., a MP that has a greater impact on outputs vs. an FP, assuming MP and FP represent the same costs) is also the one most likely to become a POV once the project moves along its critical path. Hence, this method may help to predict the impact of POVs within the framework of a production taking place along the critical path.

7.4.8   Critical path analysis (CPA)

A critical path is the longest path that can be followed between inputs and outputs, given set parameters. This path must be analyzed in realistic terms: it may be possible that the way home for a citizen from Louisiana passes through a ravine; however, if the ravine hides rather unfriendly crocodiles, it would be recommended to cross the bridge. Similarly, choosing production paths that mean endangering management and/or the team members (Forces of Production) is counterintuitive, although it happens, especially in countries where work standards are low or where they are inefficiently applied.

The Function Analysis System Technique (FAST) rests on posing two questions, that is, not only "Why?" but also "How?" Using FAST, a feasibility analyst starts by determining a sequence of events along a critical path and then toys with a set of operational rules (making sure that the previous production step has been completed, and if not then asking why it has not been completed).

The best way to use the critical path method is, of course, to subtract all operational margins every step along the transformation phase, and to set an absolute pass or no-pass criteria, with no-pass meaning project failure. Answering the question "How?" is a way of identifying solutions to problems, while answering the question "Why?" simply leads to the root cause.

Technically, when working along a critical path, all elements of that path become POVs, because each one can make or break the project. Hence, the step that demands the most knotty set of solutions (the step for which the question "How?" is most difficult to answer) is indeed the one that represents this highest potential for being a POV.

7.4.9   Analysis of probability of risks

This tool serves to gauge the probability of a risk (as defined in this book) to evaluate the catastrophic potential of various scenarios. Recall that risks and POVs are different, yet they are related mathematically; this analysis was developed for risks, yet it makes a certain number of assumptions as far as POVs are concerned. Obviously, if risks are potent, then this means that the project is sensitive (thus vulnerable) to them.

Various authors have proposed various risk matrices[47] (see Table 7.5).

Table 7.5 Example of an analysis with probability

Probability of risk	Risk impact		
High	Medium	High	Low
Medium	Low	Medium	High
Low	Low	Low	Medium

The existence of risks supposes vulnerabilities, and this is what the feasibility analyst is after. Another example[48] is provided in Table 7.6.

Table 7.6 Example of probability analyses

Risk matrix	Low	Moderate	Medium	High	Very high
Prob.			Risk impact		
0.9	0.055	0.09	0.18	0.36	0.80
0.7	0.035	0.07	0.15	0.28	0.56
0.5	0.025	0.05	0.10	0.20	0.40
0.2	0.015	0.03	0.05	0.13	0.24
0.1	0.005	0.01	0.02	0.04	0.09

This method can help pinpoint the POVs that must be identified and dealt with.

7.4.10   Analysis by way of scenario simulation

Software such as Monte Carlo or even Excel can be used to test different scenarios by changing the settings of certain parameters. The objective is to see how changes in parameters impede on the triple constraints. It is useful to begin with parameters that affect each one of the three constraints separately. For costs, for example, the feasibility analyst would check the impact of adding inputs—Forces of Production, FP, Means of Production, MP, resources—and, in the case of resources, for example, extra machinery or inventory (R_n), hours of work (T) or tasks, necessary equipment, building, materials (R), and training sessions (T_h). Another way to use this method is by adapting it to each of the six strategic axes of project management analysis. For example, what would be the impact of adding a layer of management (Power in the four Ps)? Of a potent pessimistic scenario (PRO) (e.g., cold weather in the case of the Québec Multifunctional Amphitheatre (QMA) that delayed the pouring of concrete by several weeks)? Of changing a product that is part of the project (POW)? Of introducing a newly discovered POV? Of changing the Point of Equilibrium by modifying the Dominant strategy? Of enforcing commitment (work psychodynamics, PWP)?

All these questions can be tested using advanced simulation techniques.

7.4.11 Optimal path analysis

Optimal Path Analysis (OPA) requires the use of software to find the optimal path of the transformation process under scrutiny.

The main difference between the Critical Path Method (CPM) and the OPA is that not following the critical path leads projects to failure (zone of chaos), while not following the optimal path breeds unpredicted losses, but does not invariably cause failure. An optimal path may be chosen when the project is not in its critical path; when it is, the two are theoretically equal. "Optimal path" refers to finding the production processes that permit the completion of the project within the triple constraints, while minimizing POVs and while accepting realistic operating margins. One way of identifying potential POVs is, therefore, to set all production parameters at their maximum operational level and then one notch above, and guess which one is likely to flounder first. Then, continue with another notch above and see which fails next, and so forth. The ones that bend first are the ones that represent the most sensitive POVs. This tactic is frequently favored in chess; pressure is mounted against different parts of the opponent's game to see which one gives in.

7.4.12 Gantt, PERT

These methods are commonly employed in project management. They may help identify the relationships between various components of the project and thus clarify potential causal links.

7.4.13 Summary of types of analyses

The previously mentioned tools are available to the feasibility analyst and can be elected depending on the particular need for investigation and the situation being addressed. In a pretransformation phase, while doing a feasibility study, each of these methods has the potential to identify the POVs, their presence, and/or their importance.

Table 7.7 proposes some typical times during the project when the different methods can be used.[49]

I recommend that the feasibility analyst utilizes at least two contrasting types of analysis by stage (e.g., multicriteria and Gantt) in their quest to unveil the three **characteristics of POVs**: identity, presence, and

Table 7.7 Analytical arsenal and POVs

Stage	Proposed analytical tool	Helps finding POVs
Vision	Analysis by comparative tables	Identity
	Analysis using PRO scenarios	
	Analysis by decision tree	Presence
	Analysis of product tree	
	Analysis of probability of risks	Importance
	Multicriteria analysis	
Planning	Cause-to-effect analysis	Identity
	Gantt, PERT	Presence
	Critical path analysis	Importance
	Optimal path analysis	
	Sensitivity analysis	
Mobilization	Cause-to-effect analysis	Identity
	Critical path analysis	Importance
Deployment	Optimal path analysis	Importance
	Critical path analysis	
Completion and evaluation	Critical path analysis	Importance
	Cause-to-effect analysis	Identity

importance. In other words, no stage should be left unchecked and at least two analytical points of view must be adopted.

7.5   *Asymmetry of information*

Power is also about the capacity to decide and implement changes. This cannot occur without complete, useful, and valid information. I call this the **triple constraints of information** (TCI). It is a constraint because often, as an example, by the time one gathers the complete information, some of it has already become useless. In a more complicated model, some authors[50] express the idea that information must be consolidated, cumulative, easy to understand, in a simple format, meaningful, personalized to the recipient, relevant, and timely. It should also highlight problems and identify trends. The so-called SMART system favors data that is achievable, measurable, realistic, and temporally defined.

I prefer to summarize all of these attributes and subattributes in the triple constraints of information named TCI. Ultimately, the ability of a project leader to work within the requirements of the calendar, costs, and norms of quality implies that he has access to complete, useful, and valid information. Misuse of information can lead to confusion, deception, diversion, and silo thinking. The project leader is the guardian of the TCI (complete, useful, and valid information).

In many cases, there is what is called asymmetry of information between what he/she does and what the "situation knows," and this asymmetry is precisely **uncertainty**. What is not known by the project leader may, however, be known by some Forces of Production that, unintentionally or voluntarily, hide information; it may be also withheld by other stakeholders, such as sociocultural groups (external groups) that wish the project to abort.

The concept of an asymmetry of information applies particularly well when investors must decide whether to go ahead with funding a project as presented by the project promoter, and in the context of a project leader that must rely heavily on information they receive from other stakeholders, such as some managers and team members (Forces of Production) out in the field.

It is worth examining how asymmetry of information (or uncertainty) can affect a Dominant strategy (DS). To do this, I set up the following study.

7.5.1   A study on asymmetry of information

Let's set a hypothetical scenario involving two newspaper owners. The first newspaper owner is called J_1 and is referred to as "he" instead of "it" to avoid confusion in the upcoming explanations. J_1 is actually our project leader in search of day-to-day information in order to optimize his constrained operations. J_1 may also be an investor who is interested in funding a project. The second newspaper owner is called J_2 and represents the forces that play against J_1 (the project leader); J_2 is also referred to as "he" instead of "it." J_2 may also be viewed as another project investor who competes with J_1.

We assume that J_1—the project leader—wants to win against the forces that impede on his project, whether they are of natural or human cause; similarly, we assume J_1, as a project investor, wants to win against J_2. However, J_2—the opponent force—also wants to win. Each one is trying to build a competitive advantage.

We set a scenario that calls for a closed dynamic system. Newspaper 1 (J_1) is competing with Newspaper 2 (J_2) for the same customers; they both base their daily outputs on the same news. J_1 has a keen interest to access key information ("units of information" or news) before J_2; the reverse is true. We assume that the selling price of the newspapers, be they of J_1 or J_2, to remain at US$1 at all times. Should either player want to increase his price, customers will shy away. Should a player reduce his selling price, the other will start a price war. This potentiality for retaliatory actions forces a *status quo* on the selling price, on the assumption that each player targets the same customers.

When J_1 gets hot information, he can then publish groundbreaking news the following day, with no chance for J_2 to react until the third day. Being late, J_2 has sold fewer papers on the day the breaking news is released; his sales have subsided. J_1 has gained a competitive advantage over J_2. The reverse situation could happen, with J_2 being in the lead position.

Both J_1 and J_2 wish to maximize the quantity and quality of units of information. Customers will only spend money on the paper that is worth their while (costs). There is a race between J_1 and J_2; time is of the essence. Information should be sufficiently good (norms of quality). J_1 and J_2 are bound by the triple constraints and anything that delays publication, increases costs, or provides poor news is, truly, a utility drawback.

Should J_1 access unique units of information on day 1 and publish them on day 2, J_2 will work frantically to acquire the next newest units of information to publish by at least the third day. The interaction between J_1 and J_2 is thus dynamic, and there is no way out: there are no other players in the market. Consumers have a limited budget; they will only buy either newspaper J_1 or J_2, but they will never buy both of them on the same day.

Note that because both players try to minimize uncertainty, the cost of not knowing (not accessing units of information) is higher than the cost of knowing (the cost of acquiring units of information that are then sold to the public). We have seen how to represent a utility of the Dominant strategy, or U(DS) curve, as shown in Figure 7.5.

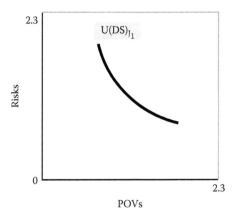

Figure 7.5 U(DS) for J_1.

J_1 operates in a portion of an equilateral hyperbola [Risks $= k$/POV]. As seen before, both J_1 and J_2 have a U(DS) curve that faces each other. This is rendered in Figure 7.6 by using an Edgeworth box.

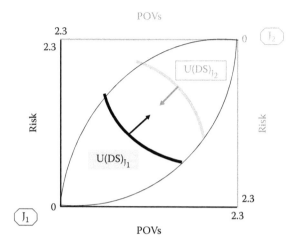

Figure 7.6 Edgeworth box for J_1 and J_2.

The U(DS) curve of J_1 meets or tends to meet that of J_2. Underneath each U(DS) curve is the market gained by either J_1 or J_2. J_1 and J_2 get more market share by moving toward the center, until they meet in the middle at the Point of Equilibrium (POE). Past this point, each one infringes on the market share of the other. This can be explained as follows: the surface underneath the U(DS) curve is equal to [(external risks | triple constraints × vulnerabilities)]. As J_1 expands toward J_2, he actually accepts to face more risks, which in this case are market risks (e.g., more customer complaints), and to expose himself to more vulnerabilities—what if the news that he published in a hurry is erroneous, thus damaging his reputation? The impact is much greater if he has a smaller market share than if he holds a large part of the market segment. Hence, it can be assumed that the surface underneath the U(DS) represents the market available to the player (J_1 in this case). When J_1 enters into the zone of J_2, past the Point of Equilibrium, this means that J_1 has eaten away at some of the market share of J_2, who for sure will retaliate sooner than later (recall our hostility formula—Equation 6.5).

Let's put this in the context of a project investor. He is competing for a project with another potential investor; he looks at the feasibility analysis and judges that the project is worth his investment. However, the other investor is equally interested. Each investor chases the most accurate information. The one who wins the bid will eat away not part, but most likely all of the market share (of the project), although it has been seen that competing firms accept to share responsibilities for a project in order to secure the business and capitalize on each firm's strengths.

Similarly, a project leader may long for the most up-to-date information on the cost of steel, as, for example, in the case of the Montréal Olympic
Stadium (MOS). Steel traders or manufacturers may withhold information about the likely cost of steel in the future. Should this price inflate
substantially, the project leader will face significant difficulties—his project will be compromised by well-informed steel traders who have taken
advantage of their monopolistic position.

On the other hand, recall that we have seen in Chapter 5 that
Collaboration and Trust are closely linked, with [Collaboration = 0.3 + 0.9
Trust - ε].[51] Pushing Trust to the limit of 2.3, that is, to Blind trust at
[$x = 2.3$], where the individual thinks that they are so safe they experience
no POVs, I am justified to draw an "arc" that takes into account how Blind
trust would operate over time, as shown in Figure 7.7.

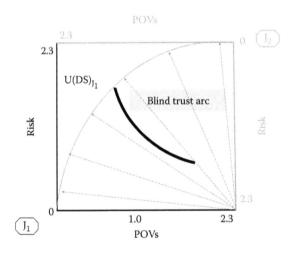

Figure 7.7 Blind trust arc (stylized).

For simplification purposes, I assume that the Collaboration curve
of J_1 equals that of J_2 with $\beta_0 = 0$ instead of 0.3. I assume that J_2 operates
at the maximum level of Trust—that is, at the Blind trust point—and
I keep in mind that J_2 wants to remain equal to himself at all times.
The arc that is generated represents the Collaboration-POVs limit of the
operating area.

In Figure 7.7, what happens to J_2 (he collaborates completely thinking
he has absolutely no POVs, thus expressing complete Blind trust) also happens for J_1 so that, *de facto*, there are two arcs being generated. The areas
outside the confines of the arcs represent chaos: it is impossible to go outside the ultimate level of Blind trust without entering another paradigm,
expressed by chaos. In these areas, there is a strong incentive for either

player to retaliate, that is, to seethe with anger (we have seen in Chapter 6 that [Hostility$_{t+1}$ = k Hostility$_t$ (1 – Hostility$_t$]).

Note that the U(DS) grows with time and achieves its maximum length at the dynamic Point of Equilibrium—see Figure 7.8.

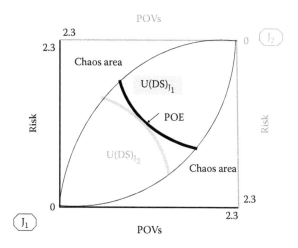

Figure 7.8 Dynamic Point of Equilibrium.

Recall that the dynamic Point of Equilibrium (POE) is the most efficient point in a project. As can be seen here, it is the most efficient point when, in the context set forth in this section, J$_1$ and J$_2$ interact closely. Some observations can be made.

First, at the dynamic Point of Equilibrium (POE), J$_1$ and J$_2$ have reached the maximum market penetration without intruding on the market share of the other player: in other words, both competing investors reach their optimal project tender at that point; both will be worse off past the Point of Equilibrium (POE). I interpret this as representing a Pareto-efficient stance for both of them. Second, at the POE, J$_1$, and J$_2$ exhibit the largest U(DS) curves, meaning they have all the experience (information) they need to reach a sound decision. Third, J$_1$ and J$_2$ can only go so far to the right or to the left along the U(DS) curves as otherwise they fall into the chaos area. Finally, each player has reached his best Dominant strategy: should J$_1$ invade J$_2$ territory, J$_2$ will retaliate so that each will torpedo his position. Given 40 buyers in the market, at the optimal POE, the stakes are as shown in Table 7.8.

J$_1$ and J$_2$ each sell 20 units (newspapers) to the market. Assuming their cost of producing each unit is US$0.50, they each earn US$10.00 [0.50 cents × 20 buyers each]. Should J$_1$ lower his price for a day, all 40 customers will buy J$_1$ newspaper that day, yet J$_2$ incurs a cost of US$10 since he produced 20 newspapers overnight in order to sell them first thing the

Table 7.8 Revenues and losses for J_1 and J_2

	J_1 keeps his current price (in US$)	J_1 lowers his selling price (in US$)
J_2 keeps his current price	20, 20	40, –10
J_2 lowers his selling price	–10, 40	0, 0

next morning, but in vain. When J_1 and J_2 enter into a price war, they both operate with zero margins and they eventually both go bankrupt. For two project investors, this means that they both lose the business opportunity, and in the case of a project leader facing negative forces, he ends up with the project being abandoned. Such a scenario occurs when management and a union cannot reach a deal, so that the former is eventually forced to close its business.

Recall that a project leader seeks rewards (satisfaction); this is true for J_1 and J_2. We have seen the equation for satisfaction as being $[S(q_m) = q_m + 2k/q_m - 2k^{1/2}]$ (Chapter 5). Before reaching the Point of Equilibrium (POE), J_1 is not completely happy with his potential—he could achieve more sales, or present a better tender, or else he could face adversity with more ingenuity. This third factor is what motivates J_1 to go for the maximum POE. The same holds true for J_2.

The question now is how does a change in J_1's position affect J_2's position? Put differently, how vulnerable is J_2 to the actions of J_1?

7.5.2   A short neurobiological study

To answer the question, I must revert back to the concept of "Apprehension." J_1 represents a threat for J_2 (J_2 feels vulnerable, otherwise he wouldn't perceive a threat in J_1). Likewise, it could be that J_2 represents a threat for J_1. For example, some negative force, like an environmental group, interferes with the project leader, or a competitor exercises some political influence to win the bid on a project. In all cases, there are POVs.

I devised an experiment whereby 20 participants aged between 18 and 25 had to navigate through a maze in order to escape a predator in a video-like game: a red triangle that is trying to catch them—it represents an unknown threat, or unknown predator–prey position (UPPP)—scenario 1; at times, random images of a snake appear—it represents a well-identified, or known threat, or known predator–prey position (KPPP)—scenario 2. Their brains were scanned by way of a functional magnetic resonance imaging (fMRI) scanner.

Using fMRI presents a number of advantages.

(1) fMRI permits interpretation of psychological processes in the brain as they are taking place; (2)

> fMRI enables the measurement of unconscious con-
> ditions and processes; (3) fMRI allows localization
> and differentiation of constructs that subjectively
> may seem similar but which are actually processed
> differently[52]

It is also noted that fMRI allows for measurements of the "simultaneous activation of two antithetical conditions and processes."

Another scenario (scenario 3) is that of the yellow circle acting as targets. When collecting yellow circles in the maze, participants earn money
(but how much was not disclosed to them ahead of time); however, when
being caught by the red triangle, they lose money (scenario 1); when being
caught by the red triangle while random images of snakes appear, they
also lose money (scenario 2). There were thus three scenarios: catching
yellow circles (which was the control position), and most importantly
running away from a red triangle (UPPP, or unknown predator–prey
position), or else running away from a red triangle while random images
of snakes appeared (KPPP, or known predator–prey position). This was
devised to test Apprehension.

The participants had, of course, to catch as many yellow circles as
possible in one scenario (scenario 3) and avoid being caught by the red
triangle in the other two scenarios (1 and 2). All participants had been
chosen because they had shown fear of snakes in the so-called SNAQ
questionnaire[53] they had filled out—this was their POV. Obviously, to
maximize the impact of the threatening images of snakes, I did not warn
participants that they would be flashed random snake images when performing their tasks. As well known by all of us, snakes induce fear[54] in
the majority of people. My sample mean age was 21, which falls within a
sensitive time for specific phobias of animals (which is approximately at
10 years of age[55]). Response time was approximately 800 milliseconds, not
enough time for cognitive functions to dictate the course of emotions;[54,57]
instantaneous fear of the threat was indeed what was measured.

Some of the most significant results[58] appear in Table 7.9. Note that
in order to read the results, one must take into account that: (1) the brain
image is a computed average of all the 20 brains that were scanned during
the test; (2) the active area is measured by subtracting the activity of one
of the three scenarios from the other; and (3) a red (dark gray) and yellow (moderately light gray) brain[59] area means an active brain area, with
the red (dark gray) being more active than the yellow (moderately light
gray). For example, in quadrant 1 of Table 7.9, the brain activity results
from subtracting the being chased condition (UPPP or red triangle) from
the chasing condition (attempting to catch a yellow circle in order to earn
money). In this case, there is more brain activity when the participant is
chased than when they are chasing. They perceive a threat: their basic

Table 7.9 Difference between predator and prey positions[145]

Doing the chasing versus being chased	UPPP versus KPPP
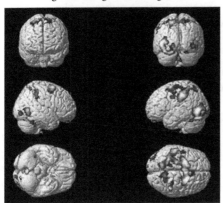	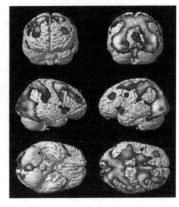
Being chased (being caught by the red triangle)—chasing (yellow circles): the brain is more active.	UPPP—KPPP: the brain also becomes very active.

motivation seems to be escaping danger.[60] In quadrant 2, the participants displayed on average more brain activity when they could not identify the predator (UPPP) than when being able to associate the perceived threat with an actual image, that of the snake (KPPP). This tends to indicate that the participants feel more threatened under uncertainty.

Table 7.9 shows what parts of the brain became active depending on the scenario presented to the participants.

Again, there are more brain areas being activated in the [UPPP—predator position] than in the [KPPP—predator position]: uncertainty (perceived threat) monopolizes decision-making resources. The fear of not knowing is worse than the fear of knowing the "enemy." My simple test provides results that are hardly debatable. I have evidence through a measuring instrument (the *f*MRI scanner) that the participants cannot conceivably fool, that perceived threat (or put differently, menacing uncertainty) negatively affects (I^-) the overall capacity to make decisions. In fact, I also found that participants hit roughly three times as many maze walls when facing threats by surprise (the snake appearing randomly) than when not being caught by surprise, as discussed in Section 7.8.

In Table 7.10, I use other means of analyzing the data. I elect what is named a "seed region": such an area of interest (AOI) serves as the central point out of which the computer analyzes the links that extend to other brain regions, without, however, making inferences on causality. When we choose the left hypothalamus (recall that the hypothalamus[61] is the center of hostile and defensive coping mechanisms), we see a negative connectivity with certain parts of the brain.

Table 7.10 Significant results from the *f*MRI study

Images	Seed region (0.005)	Negative connectivity
Left hypothalamus as a seed region 	Left hypothalamus	−44 −26 −14 Left superior temporal gyrus
Left amygdala as a seed region 	Left amygdala	−50 2 8 Left rolandic operculus and insular cortex 44 24 2 Right inferior frontal gyrus 6 −4 6 Right thalamus
Right amygdala as a seed region 	Right amygdala	54 26 26 Right inferior frontal gyrus (dorsolateral prefrontal cortex) 4 2 6 Basal ganglia

*p = 0.05; **p = 0.001.

I acknowledge that the hypothalamus is equipped with

> (...) displeasure centers, or negative-reinforcement sites [that] are located in more medial portions of the hypothalamus and lateral parts of the midbrain tegmental area. Stimulation at these locations may evoke a negative feeling such as fear.[62]

The same authors note that it is sensitive to stimuli that are induced by the presence of a predator.

My results allow me to state that, when in a predatory mood, functions indirectly related to language and emotions are negatively affected. A predator (or a person with a hostile strategy) speaks less and experiences fewer emotions (he/she is colder). When the seed region is the left amygdala[63] (knowing that the amygdala is closely associated with emotions and especially with fear or, put differently, perceived threat), we observe that the brain regions that are negatively affected indicate that as emotions increase in intensity (such as when seeing a threatening snake by surprise), reception of both internal and external information, language capacity, and capacity to take decisions (go/no go) perform more poorly. When we look at the right amygdala, we observe that as emotions become more intense due to a perceived threat, both cognitive capacities and the capacity to choose actions diminish (Table 7.10).

The predator becomes less sensitive to the pain that he/she is about to induce in his prey (he/she shows less empathy). The prey is temporarily numbed when realizing it is the target of the predator (freeze). The predator has a plan of attack. The prey is blindsided; it is purely reactive and has a hard time deciding what actions must be taken.

7.5.3   Discussion

Let us now assume that these kinds of results replicate many times over in the case of J_1 and J_2 facing each other every day of the year, with each competitor being a source of uncertainty. J_2 gains a competitive advantage by catching J_1 by surprise; during the time that J_1 has no valid information to print while J_2 has, J_2 conquers J_1 market share (100% of it), at least for the day the news is released.

When caught by surprise, the brain temporarily stops processing information. When numbed, J_1 (to pick this newspaper owner) is vulnerable. Graphically, this means that J_1's position moves to the right on his U(DS) curve as the level of his POV increases. J_1 is always trying to be equal to himself while he no longer processes messages from J_2, so that for a short while, he has moved along his U(DS) curve, to the right of it.

At the same time, his predator/prey (or Control/Transparency) ratio changes. He feels more on the defensive, so that his k'-value (ratio predator/prey) shies away from the standard value of $k = 1.3$. As such, J_1's U(DS) curve retreats toward the point of origin. In other words, he is giving up market shares if he is a newspaper owner, or his ability to propose a competitive bid if he is a project investor, or his capacity to face uncertainty if he is a project manager out in the field giving orders and supervising his workforce. The retreat means that the sheer length of his U(DS) curve diminishes—he has fewer options to base his actions upon than before.

Note that J_1's Satisfaction curve has also retreated, while that of J_2 has gained in strength.

Overall, J_1 is losing ground on five fronts due to uncertainty (or, put in economic terms, due to asymmetry of information): (1) he is off balance and to the right (more vulnerable) of his U(DS) curve; (2) his U(DS) curve has retreated; (3) he has fewer options (smaller U(DS) curve); (4) he has lost part of his power (market share/bid value); and (5) he is understandably less satisfied. This is summarized in Figure 7.9.

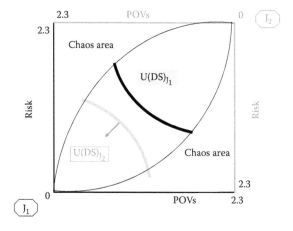

Figure 7.9 J_2's competitive advantage on J_1.

Based on my research and observations, I can tentatively claim that uncertainty, or being in a position of vulnerability, has a very large impact on the capacity to operate, decide, and succeed.

I can calculate by how much J_1 has been pushed along to his right limit along his U(DS) curve. He has moved his position to the limit of the right edge, just before the chaos area; he would like to retaliate, but cannot just yet as he does not have the necessary information to gain a competitive advantage of his own. He has performed an equal movement backward, toward the point of origin, to that of the move along the right of his U(DS) curve. Using simple trigonometry, I find that the move to the right of the U(DS) curve leads to a maximum point of $[x = 1.4]$ and $[y = 0.94]$. From there, I estimate an equal shift backward so that I can calculate a retreat along this curve for a final J_1 position of $[x = 0.45]$ and $[y = 0.63]$ (for simplification purposes, I assume that he adopted the Cooperation–Trust curve with a slope of 1, but I know that, in fact, the slope is 0.9).

J_2, symbolizing uncertainty or perceived threat, has caused J_1 to lose roughly 50% of his market if he is a newspaper owner, or equivalently, 50% of his bidding power if he is a project investor, or 50% of his decision capacity if he is a project leader. That's quite a predicament.

I thus have demonstrated that POVs can have a devastating impact on projects, in either the preliminary phase (when investors are sought for their money) or during the transformation phase (when project leaders are in full activity).

My demonstration rests on the concept of "uncertainty"; that is, on the lack of information that cripples the triple constraints of information, TCI (complete, useful, and valid information). Let me continue with a further analysis. I plot the cost of not knowing against the quantity of information using an elbow function, which I name "demand for information curve" in Figure 7.10.

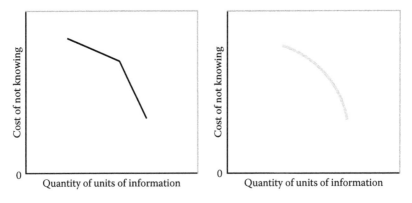

Figure 7.10 Demand for information curve for J_1: Cost of not knowing versus quantity of units of information

On the left side of Figure 7.10, I put a kink. Before the kink, which occurs at the POE, a small reduction in the cost of not knowing is beneficial: it allows J_1 to access a lot of units of information. J_1 does not want to stay in a state of not knowing. Past the kink, something different happens: the reduction in not knowing does not provide a huge amount of useful units of information. It is acceptable for J_1 to not know because whatever is to be known would bring little additional units of information that would have value to newspaper customers (or project leaders). This is true as J_2 has already published the additional news.

In reality, the kink point is not sharp. Both J_1 and J_2 are eager to find new, exciting information all the time; they interact constantly. They don't access units of information separately and sequentially. As a result, the Demand for information curve is in fact rather smooth, as illustrated on the right-hand side of Figure 7.10. I assume this curve to be an exact replica of J_2's U(DS) curve. The Demand for Information curve of J_1 is equal to the U(DS) curve of J_2. In a closed dynamic system such as a project, for J_1 not to eliminate uncertainty serves the interests of J_2.

J_1 does not need to know what J_2 does (and vice versa). Project bidders, for example, normally don't know what each other quote in order to obtain the contract. All J_1 has to do is to access units of information that have value in terms of quality, quantity (given timeliness), and at a cost that makes it worthwhile acquiring them.[64] In other words, a project leader must seek information that will help them deal with POVs, regardless of what shape uncertainty takes. Put differently, they are keenly interested in getting information regarding the calendar of activities, costs, and norms of quality; doing otherwise amounts to gambling with the project.

7.6 Biases

Project leaders are not perfect. They are under pressure and their ability to deal with threats and associated stress is at times challenged. External factors are not the only hurdle on a project leader's decision course. The fact is that they may well have their own mental or emotional POVs, which at times come in the form of biases. Biases are well studied in the area of decision-making, including in finance. In the end, the project's investor would be well advised to be aware of the biases that may lead them to make errors. Among the mishmash of biases found in the literature, let us cite a few: cognitive biases;[65] excessive optimism and overconfidence;[66] fear of missing out on an opportunity;[67] greater fool dynamic;[68] misconceptions given trends;[69] noise;[70] overconfident investors' divergence of opinion;[71] turnaround expectations;[72] short memory;[73] and finally an unscrupulous mentality[74] (if that can be considered a bias).

Based on literature review, I prepared[75] Table 7.11.

Project investors and project leaders are thus not short of biases! One bias that draws my attention and that is similar to overconfidence, is the so-called positivity bias. Many times, I have seen project proposals that claim to conquer the world at the snap of a finger while being peppered with calculation errors and blatantly short of complete, useful, and valid information. If I look at the lists of biases presented previously, I notice that two are directly related to a form of positivity bias, which, incidentally, can be studied scientifically: superfluous optimism/overconfidence and turnaround expectations. A few other biases are likely to be the result

Table 7.11 Biases discussed in the literature

Bias	Explanation	Four Ps
Groupthink	Team members developing a sense of invulnerability	People
Overconfidence	Belief that all is under control	Power and processes
Sunk cost "syndrome"	Refusal to accept sunk costs	Power and Plans
Overconservatism	Refusal to take new information into consideration	Power
Recency	Overemphasis on recent, most salient information	
Selective perception	Tunnel vision of sorts	

of a positivity bias: fear of missing out on an opportunity, groupthink, overconservatism, selective perception, and the sunk cost "syndrome."

In fact, I can relate the positivity bias to Blind trust, for which I have already established a mathematical formula and a visual representation as part of my six core competencies model: Blind trust is closing one's eyes on the presence and the imminence of danger (leading normally to Apprehension), reflecting incorrectly that everything will go well (positivity bias). Hence, from my perspective, the positivity bias must be examined as part of my wish to understanding Power in the field of project management. For all intents and purposes, I regroup all biases under a single one: the positivity bias.

7.6.1 Positivity bias

An interesting phenomenon with the positivity bias is that it is known to affect older people more intensely than younger people. Hence, it is possible to study it by comparing the behavior of different age groups. With age, mental capacities are assumed to decline,[76] thus generating more vulnerability[77] and suboptimal decision-making. In particular, older people turn a blind eye on the recollection of negative experiences and favor pleasant ones.[78] Indeed, numerous laboratory experiments have shown that older people tend to avoid looking at negative pictures and pay more attention to positive ones.[79,80] Put in the context of project management, whether as a potential investor, a promoter, or a project leader, this means that a positivity bias limits the perception of POVs.[81] Positivity biases in older people are generally attributed to two causes: either biological aging[82] or a cognitive phenomenon overruling emotions.

Let us take for granted that older people represent decision-makers that have more POVs than the average person, so that we can relate our observations to project management. Older people (vulnerable project-related decision-makers) display lower resistance and poorer adaptability to stressful environments;[83] emphasize choice repetition and purchase deferral;[84] have a harder time retaining and processing information;[85] stick to their habits and are adverse to change,[86] making the choosing of several complex options in minute detail[87]—a typical project situation—challenging, and prone to be filled with errors.

I have noted in the previous section that vulnerable people, when numbed by a threat (a red triangle/a snake in the experiment), do not receive all sensory information, probably because they would otherwise be overloaded and unable to react. A similar process seems to occur in older people.[88]

Some authors do not agree in full with biological aging, but rather point to so-called socioemotional selectivity.[89] The "strength and vulnerability integration theory" goes along the assumption that older people set more simple goals because their life horizon is shorter and they have access to lower amounts of resources.[90] Under the concept of socioemotional selectivity, older persons are seen as being more selective of the stimuli that elevate their well-being while limiting their exposure to negative ones.[91] They may ignore warning signs or show low interest in new information.[92] From my perspective, the result is the same whether biology or mental decline is to blame: poorer decision-making takes place.

Note, however, that older people are not necessarily constrained by POVs. Overall,[93] they regulate emotions better despite the reduction in their autonomic activity.[94] They deal more effectively with negative feelings such as anger and fear.[95,96] They put events in a life perspective.

7.6.2   A study on positivity bias

I conducted an experiment to determine if older people display a positivity bias. This does not answer the question of whether the project decision-makers suffer from a positivity bias, but at least it points in the right direction. Proving that we can somehow identify a positivity bias has its utility. We can also turn the conclusions around: if we can prove somehow that older people customarily have a positivity bias, knowing what such bias entails (phenomena such as eschewing negative feelings and having lower interest for valuable information and clues), we can then state that someone such as a project decision-maker who displays these behaviors is, most probably, subject to a positivity bias. This project decision-maker is displaying some level of Blind trust and is refusing to recognize the presence of danger (risks or POVs), which is not good news when managing a project. Thus, my study is a way of inferring that Blind trust in project decision-makers is, by itself, a POV.

I asked 807 people who participated in various projects to fill out my questionnaire aimed at evaluating their manager while being active in their project. I merged all the data after verifying that the participants were uniform to justify such procedure, and that the populations and residuals were statistically normal. I understand this step has its limits, but my effort is an exploratory one. I separated the entire group (named G123: 18–65+) into three age segments, as follows: G1: 398 participants aged 18–30; G2: 353 participants aged 31–64; G3: 56 participants aged 65+ (65 being the average age of retirement in Canada).[97] For all groups, 56% were female (G1: 56%; G2: 58%; G3: 33%) and 44% were male (G1: 44%; G2: 42%; G3: 66%).

7.6.2.1   Model's key variables

The six core competencies model's key variables according to the various age groups were measured, and the results are shown in Table 7.12.[98]

I note that what I call the "Hostile position/Defensive position ratio" (or equivalently Self-confidence, which, turned around and for all intents and purposes, is Apprehension) is functional for each group, with the strongest value being for G3 (the oldest group). The latter is also the group that feels the most defensive (at Defensive = 3.78 vs. G1 = 3.28 and G2 = 3.23), but it is also the group that has the highest score on the Hostility construct (4.92 vs. 3.82, 4.02). In other words, older people recognize they can be victims, yet they also know how to protect themselves.[99] In line with what I would have expected, the value of Trust is highest for the older group G3; also, the minimum level of Trust is much higher for the G3 group than for the two younger groups (47.62% vs. G1: 36.51%, G2: 42.86%) and the spread between minimum and maximum levels of Trust is also (consequently) much narrower for older than for younger people. This indicates that older people tend to be in the higher bracket of Trust level. The narrower view is in line with my previous comment: older people opt for fewer options. All of this points to a positivity bias.

7.6.2.2   Factorial analysis

I performed a factorial analysis with the results put in Table 7.13.

I observe that the same tendency occurs for each group, including for the overall group G123 (G1 + G2 + G3 compounded). When the value of the construct of Hostile position is higher than that of the Defensive position, Trust, Fairness, and Collaboration tend toward 1; when the value of Defensive position is greater than that of Hostile position, the values of Trust, Fairness, and Collaboration tend toward zero (0). This means that

Table 7.12 Key variables' scores ($n = 807$)

Group	Hostil.	Defen.	IP/DP	Trust	Fairn.	Coll.	Intent.	Dep.
	7-point Likert scale			In percentage				
Age 123								
Mean	3.99	3.30	1.44	78.70	75.20	75.76	79.71	64.02
St. D.	1.27	1.31	0.98	11.68	15.05	12.54	16.54	16.14
Min	1.00	1.00	0.14	36.51	14.29	21.43	14.29	14.26
Max	7.00	7.00	7.00	100.00	100.00	100.00	100.00	100.00
G1								
Mean	3.82	3.28	1.30	78.18	73.16	74.25	77.10	64.27
St. D.	1.08	1.10	0.68	11.42	14.45	12.02	17.76	15.68
Min	1.00	1.00	0.20	36.51	28.57	35.71	14.29	23.81
Max	7.00	7.00	7.00	100.00	100.00	100.00	100.00	100.00
G2								
Mean	4.02	3.23	1.54	78.63	77.05	76.60	81.64	63.65
St D.	1.36	1.45	1.11	11.53	14.92	12.60	15.13	15.90
Min	1.00	1.00	0.14	42.86	14.29	21.43	14.29	14.29
Max	7.00	7.00	7.00	100.00	100.00	100.00	100.00	100.00
G3 (Older Group)								
Mean	4.92	3.78	1.78	82.89	78.02	81.17	86.22	62.01
St. D.	1.51	1.69	1.51	13.64	18.13	13.98	12.29	27.59
Min	1.00	1.00	0.14	47.62	14.29	44.05	52.38	21.43
Max	7.00	7.00	7.00	100.00	100.00	100.00	100.00	100.00

Note: St. D. = Standard deviation.

people prefer to be in a position of control as opposed to being in the position of a victim. Put in managerial terms, it is much better to be in control than not.[100] This makes sense.

7.6.2.3 *Discriminant analysis*

Table 7.14 provides information from a discriminant analysis.

I arrived at three classes as the best solution. The values of Trust, Fairness, and Collaboration clearly fall into three different levels—the transactional (16% of the population) at 50%–65%, the relational (42%) at 70%–85%, and the interpersonal (41%) at 80%–85+%. This correlates well with the three interactional levels found in previous studies (see Appendix 5.3). Most of the participants (42 + 41 = 83%) are at ease with their manager. This is a climate conducive to the establishment of trust.

Table 7.13 Factorial analysis (*n* = 807)

			Group					
	Age 123		G1		G2		G3 (Older)	
Component	1	2	1	2	1	2	1	2
Trust	0.91	0.06	0.89	0.10	0.92	0.13	0.95	0.10
Fairness	0.87	−0.01	0.87	0.00	0.87	−0.04	0.89	0.12
Coll.	0.93	0.13	0.93	0.13	0.93	0.03	0.96	0.04
Hostile (IP)	0.39	0.71	0.36	0.72	0.42	0.71	0.24	0.71
Defens. (DP)	−0.16	0.87	−0.13	0.88	−0.21	0.87	−0.07	0.84

Table 7.14 Discriminant analysis (*n* = 807)

	Class		
	A	B	C
Construct	Transactional	Relational	Interpersonal
Trust	63.27	74.38	86.83
Fairness	53.37	66.45	80.59
Collaboration	55.74	68.48	82.55
Hostile (IP)	3.38	3.67	4.20
Defensive (DP)	3.31	3.34	3.20
Ratio IP/DP	1.02	1.10	1.31
Intention	48.59	72.94	85.79

7.6.2.4 *Perceptual map*

I produced a perceptual map with respect to Trust, as shown in Figure 7.11.

As can be seen, G3 differs from G1 and G2. The older group G3 gravitates around Trust levels of 10, 9, 4, and 3. By opposition, G1 and G2, the younger people, form one separate entity. The difference between the groups on the value of Trust appears in Table 7.15.

Indeed, G3 (older people) is different from both G1 and G2 (younger people); G1 and G2 are not significantly different from one another. Hence, we are justified to compare how G3 behaves in terms of Trust versus the grouping of G1 and G2 in order to detect a possible positivity bias.

7.6.2.5 *Predicted versus actual values*

I plotted the Trust levels (in increments of 10) for the three groups. Results show that the older group is more positive than the two other (younger) groups (Figure 7.12 and Table 7.16).

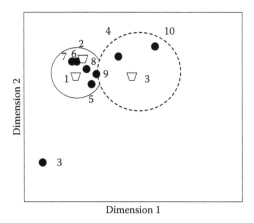

Figure 7.11 Perceptual map (*n* = 807).

As shown in Table 7.16, predicted values versus actual fit least for the older group assuming no positivity bias.

7.6.3 Discussion

To summarize, my research provides an indication that older people display a positivity bias in comparison to younger people. At the same time, a number of decision-making errors have been associated with old age, either due to cognitive decay or a growing social misfit, or both. The literature tends to correlate the two events: positivity bias is a way of minimizing the amount of information that can affect judgment. If we turn this observation around, we can tentatively recognize the fact that project leaders who operate on overoptimistic scenarios will likely make more decision-making errors than leaders who adopt a more realistic view. In other words, a positivity bias is a POV that can be associated with Power. A project feasibility analyst can detect such bias by examining the project plan that is submitted to him/her and deduce whether POVs will likely lead to errors and possibilities of failure.

7.7 Tunnel vision

As project leaders develop a false sense of invulnerability, they fuel their positivity bias, narrow down their receiving of TCI (complete, useful, and valid information), and limit their capacity to make sound decisions. This is especially true when they are under stress: concerns (apprehension) that the project may actually collapse may entice them to ignore negative data and to overemphasize an optimistic scenario. Positivity bias nourishes tunnel vision, which bolsters Blind trust. In a study on the subject,

Table 7.15 Analysis of differences (*n* = 807)

Trust		Levene test		t-test				Diff.		C.I. 95%	
		F	Sig.	t	df	S./b	Sup.	Inf.	Sup.	Inf.
G1G2	E.V.	0.06	0.81	-0.53	749.00	0.60	-0.44	0.84	-2.09	1.21
N.D.	U.V.	—	—	-0.53	736.62	0.60	-0.44	0.84	-2.09	1.21
G2G3	E.V.	2.02	0.16	-2.50	407.00	0.01	-4.26	1.70	-7.61	-0.91
N.D.	U.V.	—	—	-2.21	68.05	0.03	-4.26	1.92	-8.10	-0.42
G1G3	E.V.	2.44	0.12	-2.81	452.00	0.01	-4.70	1.67	-7.99	-1.41
Diff.	U.V.	—	—	-2.46	66.31	0.02	-4.70	1.91	-8.51	-0.89

Trust	Group stats.	Age123	N	Mean	SD	St. M error
G1 versus G2	No Diff.	1	398	78.18	11.42	0.57
		2	353	78.63	11.53	0.61
G2 versus G3	Difference	1	353	78.63	11.53	0.61
		2	56	82.89	13.64	1.82
G2 versus G3	Difference	1	398	78.18	11.42	0.57
		2	56	82.89	13.64	1.82

Note: G1G2: G1 vs. G2; N.D. = no difference; Diff. = difference; E.V. = equal variance; U.V. = unequal variance; S./b.: significance (bilateral); C.I. = confidence interval; Sup.= superior; Inf. = inferior; St. M. error = standard mean error.

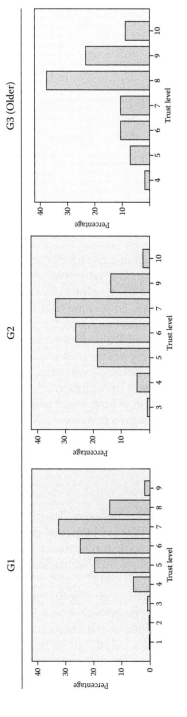

Figure 7.12 Positivity bias (*n* = 807).

Table 7.16 Differences by Trust levels ($n = 807$)

Tr. lv.	30	40	50	60	70	80	90	100	Total
				G1					
Actual	2	6	21	72	109	132	57	3	398
Pred.	1	3	19	71	105	132	58	9	398
				G2					
Actual	0	3	14	66	98	115	47	10	353
Pred.	1	3	17	63	93	117	51	8	353
				G3 (Older)					
Actual	0	1	4	6	6	21	13	5	56
Pred.	0	0	3	10	15	19	8	1	56

Note: Tr. lv. = trust level in %; Pred. = predicted.

some scholars[101] have found that overconfidence leads to a narrower focus (a landmark of fear as seen in my neurobiological experiment); some authors[102] express the idea that overconfidence reduces the perception of real risk. From this perspective, I can state that tunnel vision is a functional variable of an overreaction (overconfidence/overoptimism for the IP, overpessimism for the DP), which is itself an output of fear/apprehension (fear of not catching the prey or fear of being caught, both influencing (I^+) an overreaction)—overconfidence/overoptimism, of course, being to emotions what a positivity bias is to cognition.

I now examine tunnel vision in an attempt to detect whether it is present in the relationship between project leaders and their subordinates. To do this, I ran a study on 19 project groups, with the results discussed in Section 7.7.2.

7.7.1 A study on tunnel vision

I first examined the correlation between Trust and Collaboration for each of the 19 groups studied (834 participants) over a 4-year period; the bond between the two psychological constructs was statistically significant[103] (see Table 7.17). Participants had been asked to evaluate their project leader.

The shape of each correlation is elliptic, signifying that the data are well distributed along the regression line and the mean. Most data appear in the upper-right corner of the regression line, where levels of Trust and Collaboration are at a value[104] of above 3 over the Likert scale of 7 (put differently, above 40%). These results corroborate past studies.[105] This indicates that the groups represent, in all likelihood, an average project group.

Table 7.17 Pearson coefficient (R^2) for [Trust $\Rightarrow$ Collaboration] across the 19 groups ($n = 834$)

0.78/0.27/0.93

Year	R^2 for the 19 groups/alpha α/beta β						
2008	G1 = 35; 0.73/n.s./0.84	G2 = 28; 0.85/n.s./0.85	—	—	—	—	—
2009	G3 = 26; 0.47/n.s./0.79	G4 = 57; 0.59/1.73/0.67	G5 = 27; 0.68/n.s./1.12	G6 = 32; 0.69/n.s./0.98	G7 = 75; 0.68/0.78/0.85	G8 = 78; 0.64/0.80/0.83	G9 = 73; 0.85/ n.s./1.04
2010	G10 = 184; 0.86/.61/0.90	G11 = 13; 0.82/n.s./0.83	—	—	—	—	—
2011	G12 = 26; 0.62/1.71/0.61	G13 = 24; 0.81/n.s./0.74	G14 = 25; 0.88/n.s./1.05	G15 = 26; 0.73/n.s./0.90	G16 = 24; 0.82/n.s./0.84	G17 = 25; 0.80/ n.s./1.00	—
2012	G18 = 28; 0.79/n.s./0.96	G19 = 28; 0.54/n.s./0.80	—	—	—	—	—

Note: n.s. = not significant.

Using the values obtained for the 19 groups, I arrive at the following equation (which we have seen a few times before, e.g., in Equation 5.1), prepared with rounded-up numbers:

$$\text{Collaboration} = 0.3 + 0.9 \text{ Trust} - \varepsilon \qquad (7.1)$$

I tested the mediating role (I^+, I^+) of Fairness.[106] Indeed, in most cases, Fairness acts as a partial mediator, as predicted by my six core competencies model (Table 7.18).

Table 7.18 Verifying for the mediation role of Fairness ($n = 834$)

Comment	Regression	Significant at $p < 0.05$
Trust => Collaboration	Coop. = 0.272 + 0.925 Trust – ε	$F = 2{,}927.735$; $p = 0.000$
Trust => Fairness	Fairness = n.s. + 0.973 Trust – ε	$F = 1{,}425.433$; $p = 0.000$
Fairness => Collaboration	Coop. = 1.762 + 0.661 Fairness – ε	$F = 1{,}229.519$; $p = 0.000$
Trust + Fairness => Coop.	Coop. = 0.262 + 0.765 Trust + 0.165 Fairn. – ε	$F = 1{,}586.028$; $p = 0.000$; Durbin–Watson = 1.98 < 10
Trust coefficient change	Reduction from 0.925 to 0.765	—

Since the beta (β) coefficient for Trust has decreased with the introduction of the variable "Fairness," I conclude that Fairness acts as a partial mediator. Fairness balances the relationship between Trust and Collaboration.

7.7.1.1 *SEM and estimators for model fit*

A structural equation modeling (SEM) was run taking into accounts the structural variables of Trust and Cooperation, with results as shown in Figure 7.13 and Table 7.19.

The estimators for the structural variables of Trust and the functional variables of Collaboration, which are found to be significant, present similar values (approximately 0.80), except for Consideration.

For Trust, affinities and benevolence do not matter; for Collaboration, joint problem resolution does not account in the model. This is probably because these variables have little role in the context of project management. Put bluntly, who cares if a colleague relishes golf as much as his fellow worker (affinities) does, as long as the job is done? This is an indication of tunnel vision compared to other circumstances—in this case, it seems at least partly justified. The point is that team members rely on a small number of structural and functional variables that form Trust and that express Collaboration.

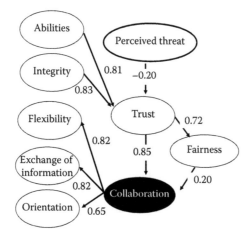

Figure 7.13 SEM model ($n = 834$).

Table 7.19 Fit indices ($n = 834$)

Adjustment indices (*fit*)	Target value	Actual values
Absolute index		
GFI	>0.9	0.937
TLI	>0.9	0.899
NFI	>0.9	0.934
Incremental index		
CFI	>0.9	0.937

7.7.1.2 *Factorial analysis*

When I run a factorial analysis using varimax rotation, the 19 groups produce the expected data supporting the law of Apprehension, as shown in Table 7.20.

When [Hostile position > Defensive position], meaning when the respondent is self-confident (does not perceive threats from their manager), they tend to trust and to collaborate, and they feel the relationship is fair. As soon as [Hostile position < Defensive position], the reverse occurs.

7.7.1.3 *Maps*

I produced graphics for the structural variables of Trust, shown in Figure 7.14.

I observe that most of the data occur within a narrow band. The band is systematically the same for each structural variable, with Self-confidence (the reverse of Apprehension), ranging from slightly over 1 at a minimum,

Table 7.20 Factorial analysis on the grouping of 834 participants

	Component			
	1	Tends toward	2	Tends toward
Trust	0.946	1	0.071	0
Fairness	0.906	1	−0.013	0
Collaboration	0.936	1	0.144	0
Hostile position (IP)	0.342	≈0	0.744	1
Defensive position (DP)	−0.150	0	0.865	1
Hostile vs. Defensive	>	—	<	—

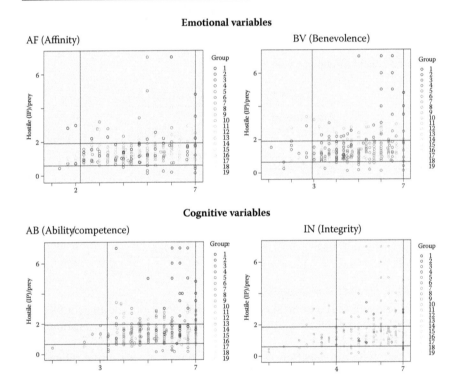

Figure 7.14 Self-confidence versus AF, BV, AB, and IN[146] ($n = 834$).

to a maximum 1.80. This band forms the **tolerance zone** whereby people are willing to be part of the project.

The tolerance zone varies, however, according to the structural variable under analysis: affinity (2–7); benevolence and ability (each 3–7); and integrity (4–7). This means that the team members are more accommodating in terms of affinity than in terms of their manager's integrity.

I can roughly derive four quadrants: first, the tolerance zone; second, the one above it where participants are overly confident and likely arrogant, if not hostile; third, the one to the left of the tolerance zone where participants lack self-confidence—they are likely to be of the avoiding type; and fourth, the area under the tolerance zone that covers the entire spectrum of the variable[107] from 1 to 7. These people vary widely in behavior, and are likely anxious. I illustrate my assumptions in Figure 7.15.

I performed the same kind of study for the construct of Collaboration, with results shown in Figure 7.16.

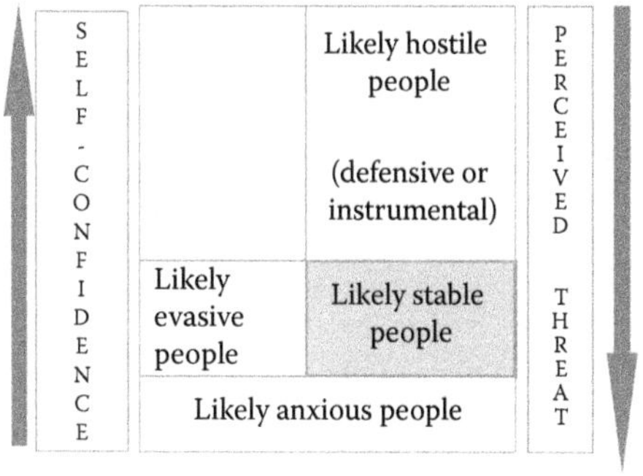

Figure 7.15 Summary of findings from the perceptual maps.

I note that one variable has a narrower spectrum (4–7 for exchange of information [EI]), two variables have a medium spectrum (Flexibility [FL] and Joint Problem Resolution [PR]: 3–7), and one has a wider spectrum (2–7 for Consideration [CD]). This means that exchanging information (EI) is more critical in the context of a project than anything else with respect to Collaboration. This makes sense.

I note also that as is the case with Trust, it is an emotional variable that has the widest spectrum (2–7 for Affinities in the case of Trust, and 2–7 for consideration in the case of Collaboration). As is also the case for Trust, it is a cognitive variable (Exchange of information, EI) that has the narrowest spectrum, and thus, that has more importance in the eyes of the participant. Put simply, Integrity and the Exchange of information are paramount features of the work culture (psychodynamics) between People and Power. Expressed differently, when it comes to day-to-day operations, management, and team members (the Forces of Production) narrow their vision of the essential: integrity on the part of people in a position of Power and exchange of information between them and people in a position of Power.

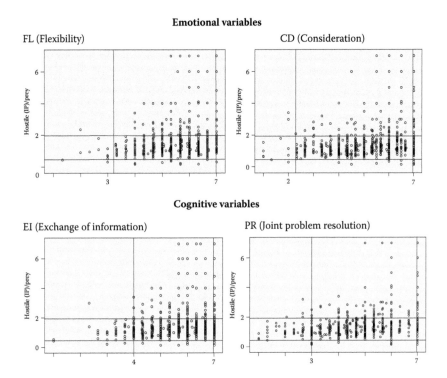

Figure 7.16 Self-confidence versus FL, EI, PR, and CD[147] ($n = 834$).

7.7.2 Discussion

I have identified the fact that tunnel vision may affect both People and Power. Chances are that if team members put emphasis on integrity and exchange of information, the same holds true for people in a position of authority. The glue that unites most People and Power is the structural component of Trust that is Integrity, and the functional component of Collaboration that is the Exchange of information. This is in line with the observations that I have made in previous chapters pointing to the importance of all stakeholders understanding what the project expectations are. It is also consistent with my comments on TCI—complete, useful, and valid information.

In certain circumstances, this shallow view may actually stall the development of stronger interpersonal relationships, which could be instrumental in deciding whether a project will succeed or not, especially in the face of adversity. Lack of "openmindness" may be, at times, a POV. I can probably establish that where there is little integrity (e.g., unruly

Unfits, FP_{nc}) and little exchange of information (e.g., poor managerial skills) lay the most damning POVs. POVs are related to Integrity taken individually, to Exchange of information (EI) taken individually, and to the mutual influence of Integrity and EI. In countries where corruption (a form of FP_{nc}) is rampant, it is unlikely that one will see projects being fully developed or else meet the triple constraints. The lesson here is that inasmuch as material infrastructures are a prerequisite for projects as seen in the introductory chapter ([Total strengths | time = Infrastructures/ Opportunity]), political infrastructures are equally needed.

I like to think that this observation can serve as a guide when evaluating a project, especially in countries where the political realm is questionable. Note that the importance of building infrastructures meets the objectives of the Keynesian economic theory: investing in infrastructures when the economy is in a dire state is a means to boost it, with the intent of encouraging growth opportunities in the near future. The Montréal International District (MID) case (case #6) is a perfect example.

7.8   *Errors in decision-making*

As previously mentioned, my neurobiological research has provided evidence that participants make more mistakes (by hitting the maze walls) when in a prey (Defensive, or vulnerable) position than when in a predator position[108]—almost twice as many. This is normal: they have more control in a predator position. I examined the number of yellow circle targets collected while playing the role of a predator (catching the yellow target to earn money) and the number of yellow circle targets collected while the participant knew ahead of time that they would soon become a prey[109] (being chased by a red triangle), either of a known predator (KPPP) or an unknown (UPPP) one, and, for all these scenarios, I looked at the number of times the maze walls were hit. When preparing to be a prey of either a known or unknown predator, participants caught more yellow circles; this is normal again, because they are more vigilant when in a potential prey position than when in a worry-free predator position. Also, participants make slightly more mistakes (hitting walls) when facing a known predator than when facing an unknown predator; again, the anticipation was probably heightened and corresponding vigilance levels rose when the participants were informed ahead of time that they would face a known versus an unknown predator. In the same vein, participants were slightly more efficient (efficiency being expressed by the number of yellow circles collected over the number of wall hits) when they were informed that they would face an unknown versus a known predator. The heightened level of vigilance seems to explain all of the results in that particular *f*MRI session.

During the research, I stood behind a window in a cubicle that allowed me to see the participants being scanned while playing the maze; I also had a view of the maze itself and could witness how the participants reacted counterintuitively by hitting walls when trying to escape the red triangle, sometimes even mistakenly driving toward it! It was quite surprising (and frankly, amusing) indeed, a little bit like if a lonely sheep, trying to escape a hungry tiger, ran straight into its path![110] All movements were registered in the mainframe computer for later analyses. Participants hit three times as many walls when under stress than not (Table 7.21).

A prey position is, of course, a manifestation of POVs.

To assess stress, I gathered cortisol samples from the participant's saliva before and after the experiment. Cortisol is a steroid hormone originating from the hypothalamus,[111] which is at the heart of predatory and prey positions.[112]

The number of participants was, of course, small; this makes the reaching of statistical conclusions a bit shaky (although, in this kind of neurobiological study, 20 participants is a common number).[113] Of the 20 participants, 3 (approximately 15%) showed significantly lower cortisol levels, while 6 out of the same 20 participants (approximately 30%) displayed significant higher levels of cortisol after the experiment (Table 7.22).

In all, 9 out of 20 participants (or nearly half the sample population) showed a change in cortisol levels, indicating some hypothalamic activity.

I found that participants exhibiting lower cortisol levels after the game (LCA) made more decision errors (wall hits) than participants whose cortisol levels were higher (HCA) or unchanged after the test; they were less vigilant. Undoubtedly, a certain amount of stress is helpful. Preparing to become a prey slightly intensifies the level of vigilance of the participants; that is, their capacity to maximize gains since their brain is more alert. The brain is more solicited by uncertainty (Figure 7.17).

Of note, there was no fatigue effect that could explain the results.[114] All participants remained quite alert during the entire experiment. Being in a prey position induces stress. Stress is a normal physiological response to a threatening stimulus; it provides the advantage of preparing the body and the mind to fight back. However, excessive stress has been known to reduce brain capacity. Indeed, stress affects brain structures in different ways depending on its intensity and how long it lasts. Causal relationships have been found between stress and cognitive abilities; toxic stress leads to the onset of diseases, with time lines having been established between the two in some cases (e.g., 11.5 years between the occurrence of child abuse and that of depression).[115] In short, a lower capacity to resort to contextual memory and fewer connections in the brain mean weaker mental capacities, and hence possibly more errors being made during decision-making processes. Additionally, my *f*MRI results show that one

Table 7.21 Results and assumed stress levels (n = 20)

Position	Description	Target collected	Wall hits	Ratio collected/ wall hits	Avg. efficiency
	**Scenario 3: Collecting yellow circles**
Predator	Predator position only (chasing the yellow circles)	5.85	3.35	1.77	1.80
	Low cortisol level after	5.67	3.54	1.60	
	High cortisol level after	6.00	2.90	2.07	
	No change in cortisol level	5.89	3.61	1.63	
	**Scenario 2: Known predator (snake image)**
Pre–KPPP	Preparing to be a prey to a known predator (KPPP)	5.99	3.14	1.92	1.90
	Low cortisol level after	6.03	3.52	1.71	
	High cortisol level after	5.78	3.00	1.93	
	No change in cortisol level	6.15	2.90	2.12	
Actual KPPP	Actually being a prey, to a known predator (KPPP)	—	7.48	—	—
	Low cortisol level after		9.13		
	High cortisol level after		6.83		
	No change in cortisol level		6.49		
	**Scenario 1: Unknown/faceless predator (red triangle only)**
Pre–UPPP	Preparing to be a prey to an unknown predator (UPPP)	6.19	3.19	1.95	1.95
	Low cortisol level after	6.06	3.46	1.75	
	High cortisol level after	6.20	3.20	1.94	
	No change in cortisol level	6.31	2.91	2.17	
Actual UPPP	Actually being a prey, to an unknown predator (UPPP)	—	7.33	—	—
	Low cortisol level after		8.92		
	High cortisol level after		6.29		
	No change in cortisol level		6.78		

Table 7.22 Cortisol levels ($n = 20$)

	Participants								
	Lower cortisol after			Higher cortisol level after					
Code	10	16	31	8	12	19	28	29	41
Sex	f	f	f	f	m	m	m	f	f
Before	0.246	0.268	0.415	0.059	0.017	0.009	0.101	0.067	0.067
After	0.094	0.130	0.267	0.350	0.191	0.243	0.201	0.178	0.131
Difference	0.152	0.138	1.148	0.291	0.174	0.234	0.00	0.148	0.064
% Diff.	−62	−51	−36	493	1024	2600	99	36	96

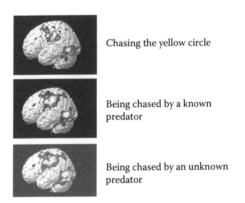

Chasing the yellow circle

Being chased by a known predator

Being chased by an unknown predator

Figure 7.17 Brain activation level.

of the brain regions involved in error detection (the anterior cingulate cortex or ACC[116]) is significantly activated when adopting a prey position.[117]

Overall, I believe that I have, with all due reservations, demonstrated that uncertainty (or Apprehension, or put differently, automatic self-assessment of one's own vulnerabilities) has a negative impact on decision-making after a certain level is reached, leading to delayed reactions, errors, and a feeble ability to use resources such as memory. Put differently, it affects the timing and quality of decisions; it increases its costs by way of greater monopolization of resources that, unfortunately, lead nevertheless to poor choices. In a project management perspective, this means that Apprehension or uncertainty affects the triple constraints (utility drawback). A manager must remain vigilant, but overextending himself leads to decision errors.

Knowing that uncertainty can be seen as risks measured as a function of one's own vulnerabilities (which is the U(DS)), as somehow exemplified by the experiment (the prey position is such because the participant becomes vulnerable), it thus can be said that the U(DS) is shifted up and down and left to right in my graphical model (e.g., Figure 4.12) by uncertainty—that is, by poor TCI (incomplete, useless, and erroneous information). A Total strength strategy can therefore be seen as the ability to manage the U(DS) given TCI. This seems normal: the project leader is in fact the guardian of strengths, contingencies, and information. This is a far cry from the policy of *laissez-faire*.[118] In fact, project management is all about the opposite philosophy. Don't let the lonely sheep, hungry tigers, Unfits, and bad apples run loose, unless you want carnage!

7.8.1   Overoptimism and overpessimism

I have recognized the fact that project leaders and investors may suffer from biases and tunnel vision, not to mention that their hands are occasionally tied by a certain amount of asymmetry of information.

They may also navigate in various forms of conflicts of interest or embezzlement (their own or that of others), engage in speculative behaviors or else they may display complacency; they may thrive on lack of accountability, due diligence, enforcement, and supervision. Because projects have an ephemeral time line, they may just have at heart short-term interests—this is something I see, for example, in the case of some CEOs of organizations who know their stay at the helm is only temporary—they may be bluntly unscrupulous.

In other words, project leaders and investors may simply be overoptimistic. After all, a number of large international projects financed by altruistic organizations end up grounded—or else six feet underground! Haiti is a prime example: barring a few undertakings, little has been achieved after the devastating 2005 earthquake, despite billions of dollars in aid having been promised. Perhaps some projects are doomed because of underlying overoptimistic behaviors—to that effect, there is currently an investigation in the case of the Montréal Hospital whereby the chief project manager has been accused of a fraud of C$1.3 billion.[119] Some scholars point to unbridled greed along with other features, such as a general culture consisting of abuse of people's vulnerabilities, lack of transparency, misrepresentations, and voluntary oversight.[120]

On the other hand, growth of uncertainty, pressures from stakeholders, and rapidly changing circumstances may actually sway the project management group toward overpessimism.

In either case, overoptimism or overpessimism, the core raw emotion is fear: fear (resulting from apprehension or as an expression of apprehension) of missing out on an opportunity to invest or fear of missing out on an opportunity to exit. Exit can take the form of using shortcuts in a project plan in order to gain financially, should the project be completed ahead of schedule; indeed, a provision that is nowadays commonly found in project contracts rewards project leaders for completion of projects ahead of time. Given enough rewards (quick-earned money, status, etc.), the project leader has an incentive to turn a blind eye on his/her formal duties and on commitment to quality. Similarly, a project investor may accept a project in order to please his/her employer: this could be the case of a government agency that supplies public funds to would-be entrepreneurs where the project evaluator decides in favor of projects in order to meet government quotas set for the agency.

In those cases, there is a high potential for POVs (overoptimism, overpessimism) to taint a project. This potential is not related to uncertainty coming from the outside environment, but rather stems from the internal psychological volatility of key stakeholders. In short, the project is not robust, a term I defined in the introductory chapter (pp. 20–21).

7.9   Hungry tiger personality

I have touched on the fact that project leaders and investors are not exempt from flaws; in some cases, they may actually be the main source of POVs due to their questionable practices. In economics terms, this is referred to as "moral hazard." Some authors describe moral hazard as "the failure of either to behave diligently or in good faith at any point in the exchange."[121] Characteristics of the industries in which the project takes place, market structure, and the regulatory environments are sometimes blamed for it. Some authors[122] point to macro- as well as to micro-level factors. The latter includes the behavioral profile of the key stakeholders.

Some experts, especially in regard to financial decisions, speak of "irrationality." I separate irrationality into six different increasing levels of POVs, as follows:

1. Stressed and uninformed[123]
2. Imperfectly rational[124]
3. Inconsistent[125]
4. Irrational[126]
5. Delinquent[127]
6. Pathological[128]

Many scholars associate pathological behavior with all or some of the four following key characteristics:[129] (1) a distorted view of reality; (2) a lack of empathy for the feeling or fate of others/remorseless behaviors; (3) selfish interests;[130] and (4) using a calculated approach to serve egoistic agendas.

In short, what is described here are the structural components of a Hostile position. This is to say that project leaders and investors are not exempt from hostile behaviors, profiles, or personalities.

7.10   Decision to invest in a project

I now present a mathematical model that attempts to explain how a decision to invest (DI) in a project can be subject to biases.

Let σ_m be the external risk (market risk) and σ_j be the project bidder's POVs. For the present analysis, let's agree that both σ_m and σ_j range in value from 0 to 3 and are deemed to be standard deviations from the norm, having positive values only. At a value of [$\sigma_m = 0$], the market of bidders behaves normally; as σ_m rises toward 3, corruption in the bidding process takes place. At a value of [$\sigma_j = 0$], J_1 behaves normally; as σ_j increases toward 3, J_1 thinks he/she is invulnerable, that is, they become completely aloof to his POVs. Both sigmas (σ), when at a value of 3, call for trouble.

Ready to invest?

When [$\sigma_j = 3$] and σ_m tends toward 0, J_1 self-confidence is boosted to a maximum (and we know disproportionate confidence amounts to arrogant if not destructive behaviors).

J_1 is a bidder who is keenly interested in undertaking a government-sponsored mega-infrastructure project, but who faces the competition of numerous other bidders (all compiled under the term J_2) that form the overall market of bidders. Recall that J_1 represents the project leader or the project investor depending on circumstances; likewise, J_2 represents negative forces or a competitor (or group of competitors). The point is that the present scenarios apply to project managers as well: they too face competing forces (J_2s) and work conditions (market) composed of multiple J_2s wanting to take their slice of the project.

J_1 is eager to win the bid; he/she needs as much TCI (complete, useful, and valid information) as possible. Their tender must come in within a certain period of time, past which they will not be eligible anymore, so that J_2 will likely win the bid. As we know, market risks (σ_m) and J's POVs (σ_j) are correlated, given the project's triple constraints.

The Utility U(DS) curve expresses the fact that investor J_1 finds a certain utility in presenting their bid to the project promoter despite market risks and their own vulnerabilities. Such a Utility U(DS) curve has five core characteristics: (1) an ordinal measure is used; (2) more is always preferred to less up to the dynamic POE (dynamic Point of Equilibrium); (3) no quantities of risk or POVs are negative; (4) these curves can be associated with a budget constraint; and (5) they do not cross each other. As seen previously, before the area where Blind trust causes turmoil, J_1 is active on a small portion of the Utility U(DS) curve. They are naturally bound by chaos (see Figure 7.8): should they play outside the tender rules, they will be expelled from the process.

I can express the U(DS) curve as follows:

$$\sigma_m = \frac{k}{\sigma_j} \tag{7.2}$$

Because time (calendar) is part of the triple constraints, the derivative of this function corresponds to the velocity of the trade between the market's risks and J_1's vulnerabilities due to asymmetry of information. The first derivative is thus:[131]

$$\frac{\partial \sigma_m}{\partial \sigma_j} = \frac{-k}{\sigma_j^2} \tag{7.3}$$

The covariance (Cov) is

$$\text{Cov}\left(\sigma_m, \sigma_j\right) = \text{Corr}\left(\sigma_m, \sigma_j\right) \sigma_m'\left(\sigma_m\right) \sigma_j'\left(\sigma_j\right)$$

or

$$\text{Cov}\left(\sigma_m, \sigma_j\right) = k' \sigma_m'\left(\sigma_m\right) \sigma_j'\left(\sigma_j\right) \tag{7.4}$$

where σ' is the standard deviation of risk σ (which is itself a standard deviation) and where the correlation value (Corr) is equal to k'. There is a point along the Utility U(DS) curve where the function $[\sigma_m = k/\sigma_j]$ equals the function $[\sigma_m = k\,\sigma_j]$; the value of $[\sigma_j = 1]$ and that of $[\sigma_m = 1.3]$; this is the Point of Equilibrium.

Of course, investor J_1 has not entered into a bidding war just for pleasure: he/she is eager to earn a profit. Resorting to the well-known financial CAPM formula,[132] we have[133]

$$E\left(r_a\right) = r_f + \beta_a \left[E\left(r_m\right) - r_f\right] \tag{7.5}$$

The expected return $[E(r_a)]$ is a function of the current cost of money the investor is willing to invest (the current market rate of return or risk-free rate r_f) and of expected market return $[E(r_m)]$.

The beta (β) for the project being bid on is

$$\beta_a = \frac{k' \sigma_m'\left(\sigma_m\right) \sigma_j'\left(\sigma_j\right)}{\sigma_m^2} \tag{7.6}$$

As seen before, the bidder experiences the fear of missing out on an opportunity to *enter* the market on time (Apprehension), that is to make a bid (fear of not entering or F_{in}) while they may experience the fear of missing out on an opportunity to *exit* the bid (or the market) on time once they have won the bid and see that conditions change for the worse (fear of not exiting, or F_{out}).[134] Changes could stem from disparate causes: new conditions imposed on the bid, such as penalties for delays; political agenda being modified; or unexpected cost increases of raw material. It may be the case that a bidder has won the bid and has started to finance the project, but that they then prefer to opt out, after realizing that it will not likely materialize. Apprehension leading to fear of not entering (J_2 gets the business) and fear of not exiting the market (J_1 will suffer from the changes causing the market to implode) positions are at the heart of the bidder's behavior.[135]

The decision to invest (DI)[136] in the project (amounting to presenting the bid with the hope of winning) is

$$DI = E(r_a)\psi = f(\text{fear of not entering or not exiting on time}) \quad (7.7)$$

Psi (ψ) refers to the positivity bias that I examined previously in this chapter. ψ ranges from 0 to 1. A high positivity bias (near 1) implies a little Apprehension while a low positivity bias (near 0) suggests a high level of Apprehension, and thus, a feigned desire to invest. Blind trust is rendered by $\psi = 1$. The decision to invest is based on the expected return gained once the project is completed and J_1's positive outlook on market risks and personal vulnerabilities (σ_m and σ_j).

I propose henceforth a stylized CAPM model, as follows:[137]

$$E(r_a) = r_f + \beta_a\left[E(r_m) - r_f\right] - \varepsilon \quad (7.8)$$

The term "risk-free rate" (r_f) amounts to initial trust (Pretrust); "β_a $[E(r_m)-r_f]$" represents a risk premium where "β_a" is the systemic risk that affects the anticipated gain. "$E(r_m)-r_f$" is the market risk premium, as threatened by the potentiality of chaos, ε—chaos being the extreme case of brazen behaviors with all bidders retaliating against each other. Equations 7.8 and 7.9 suggest that the decision to invest depends on the expectations of rewards ("$r_f + \beta_a [E(r_m)-r_f]$") as subjected to the possibility of retaliation from other bidders (ε) by way of harassment, lawsuits, political interference, pressure groups, sabotage, and so forth.

In an optimistic market, the decision to invest becomes more imperative as the fear of not entering the bid on time grows—overoptimism may even kick in. In a context where things go sour, the bidders may become fretful and wish to quickly exit the process, even if they are already engaged in it, for fear of losing everything. These two possibilities are represented in Figure 7.20.

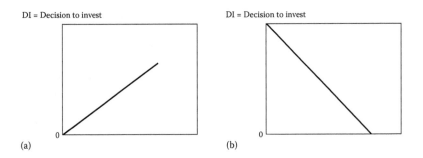

(a)   (b)

Figure 7.18 Fear of missing out on the opportunity to enter or exit the bid. (a) F_{in} or overoptimism curve: The fear of missing out on the opportunity to *enter* the project (to bid). (b) F_{out} or overpessimism curve: The fear of missing out on the opportunity to *exit* the project

In an optimistic outlook:

$$\Psi_{in} > \Psi_{out} \tag{7.9}$$

Note that the slope of the F_{out} curve (overpessimism curve) is steeper than that of the F_{in} curve.[138] Figure 7.19 joins the two curves—overoptimism and overpessimism.

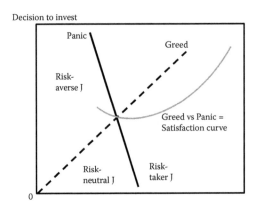

Figure 7.19 Fear of missing out on the opportunity to enter and to exit the project.

The area to the left of the meeting point between the two curves represents a risk-averse J_1 (quadrants II and III): $[F_{out} > F_{in}]$ J_1 is anxious or avoidant. At the meeting point between the two curves, J_1 is balancing the options of not exiting and not entering the market—he/she is stable. To the right of the meeting point, J_1 is an active bidder/investor (quadrants

I and IV) and they may even become increasingly hostile (IP) as the possibilities of profits lure them into believing they are invulnerable.

7.10.1   Overoptimism and overpessimism in action

As mentioned, because the U(DS) develops as a function of time (time
being one of the three constraints of projects), its derivative is an expression of velocity. We have seen that robustness is what a project feasibility analyst seeks when evaluating a project: both high levels of risks and
POVs as well as high velocity betray acute problems.

In working with the derivative of the U(DS) curve, $[d\sigma_m/d\sigma_j = -k/\sigma_j^2]$, I
replace $[k = 1.3]$ by $[k = 1]$: I thus create a point with a value of 0, which will
be my reference point. I then set the values generated by the derivative
into absolute values, a step that generates a curve that I then transform
into a natural log function ("ln") to make it as linear as possible. This
mathematical feat allows us to put the Point of Equilibrium at a minimum
value of zero (0), so that I have forged a reference point. I set all data into
absolute values once more so that comparisons can be made given that
behaviors cannot be negative *per se*—they can only be above the reference
point. Figure 7.20 illustrates the ensuing result.

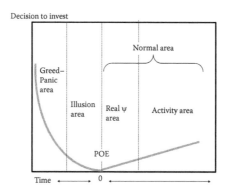

Figure 7.20 Overoptimism and overpessimism in action.

Point $[y = 0]$ corresponds to the Point of Equilibrium on J_1's U(DS) curve.
I observe a potential for growth to the left and to the right of it. The left part
is composed of two sections. I name the first area the **Illusion area** the curve
portion mimics a normal growth curve such as the one found on the right
of the Point of Equilibrium. It has a slope of 6%, while the slope on the right
side of that point being of 5%.[139] To the left of the Point of Equilibrium, J_1 is
lured into thinking that the project will be profitable and will succeed. It is
a dreadful positivity bias ψ, however; as J_1 continues to engage to the left,
things rapidly grow out of control (quasi-exponential growth signifying a

fear factor); moving along to the left (overoptimism) and retreating rapidly (overpessimistic) are the two behavioral options past the Illusion area.

To the left, J_1 dismisses past mishaps, ignores some valuable information TCI, and ignores market risk σ_m and his/her own vulnerabilities. He/she is convinced the odds are on his/her side. He/she finds justification in past successes to prove that his/her strategy is working.[140] He/she cannot afford to adjust to reality because he/she wants to remain equal to him/herself (this being a key tenet of our model). This kind of behavior is commonly found in gambling. In the Illusion area, J_1 is convinced that, overall, the probability of cashing in is greater than the probability of market risks and POVs beating him/her.

J_1 assumes that the market is homogeneous, that is, that the market is composed of bidders that act just like him/her, all fearing to miss out on the opportunity to enter a particular bid. This nourishes an urge that renders the bidder market the more volatile, so that σ_m keeps increasing. The market for project bidders becomes inefficient because quoted prices do not reflect intrinsic, fundamental values associated with the project,[141] but rather an inflated positivity bias ψ.

A different scenario develops to the right of the Point of Equilibrium. The **realistic ψ area** expresses the fact that J_1 is reasonably justified to be optimistic—otherwise, he/she would not bid at all. He/she is alert and rational rather than fooled. The upward outlook with a 5% slope does not accelerate; it is smooth (proving control and lack of fear). I posit that this is where my proposed stylized CAPM formula operates.

7.10.2   Greed

Let me now introduce the fear factor (fear of missing out on the opportunity to bid), using the Greek letter φ, into our equation. We have[142]

$$DI = E(r_a)\psi^\varphi \text{ where } E(r_a) = r_f + \beta_a^\varphi \left[E(r_m) - r_f \right] - \varepsilon \qquad (7.10)$$

The positivity bias has a base value of 1 (at 1, it does not impact the decision to invest DI): there is complete (Blind) trust in the market. As greed φ kicks (going from 0 to 1; 1 being the highest fear of missing out on the opportunity to tender), the positivity bias is affected. The more the positivity bias bulges, the more likely the decision to invest (or to tender) materializes; the more fear there is that a great opportunity to earn a profit is being missed (that the project will be a success), the more this fear fuels the positivity bias.

I can estimate the value of ψ by reasoning as follows: when σ_j is equal to 3 standard deviations, J_1 believes he/she is statistically better than 99% of the rest of the bidders—J_2 (all the Js interested in the project forming the market of bidders). He/she thus takes excessive risks because he/she is overconfident. The lower and upper limits of σ_j are zero (0) when J_1 is equal to all persons bidding and 3 when J_1 thinks he/she is virtually God in the market of bidders. It follows that

$$\psi = \frac{\sigma_j}{3} \qquad (7.11)$$

Thus, ψ ranges between 0 and 1.

On the other hand, the higher J_1's standard deviation is versus that of the market of bidders, the more firmly he/she believes he/she can beat his competitors, the greedier he/she becomes. Fear of missing out on the opportunity to enter the market (to bid) results from calculating the chances of success attributed to oneself versus the chances of success of the competitors; more precisely, fear results from calculating the chances of failure[143] (expressed as a function of the number of standard deviations) versus the chances that all the other bidders might fail. Thus:

$$\varphi = \frac{\sigma_j}{\sigma_m} \qquad (7.12)$$

When J_1 thinks he/she cannot win the bid ($\sigma_j = 0$), he/she experiences no greed, so that ψ equals 1: that's the best he/she can do. When he/she thinks he/she can win the bid at all times with no chances whatsoever for the other bidders (when greed kicks in), even to enter into a joint venture with him/her ($\sigma_m = 0$), then $\varphi = \infty$. We have

$$\psi^\varphi = \left(\frac{\sigma_j}{3}\right)^{\frac{\sigma_j}{\sigma_m}} \qquad (7.13)$$

My demonstration shows how a decision to invest in a project (to enter a bid or actually to put money into a project) can quickly and truly become irrational.

In the end, I can summarize the decision to invest as follows:

$$DI = E(r_a)\psi^\varphi \text{ where } E(r_a) = r_f + \beta_a^\varphi \left[E(r_m) - r_f\right] - \varepsilon$$

and where:

$$\psi^\varphi = \left(\frac{\sigma_j}{3}\right)^{\frac{\sigma_j}{\sigma_m}} \qquad (7.14)$$

The decision to invest is a function of expected returns and of a positivity bias influenced by greed. It invariably has a cognitive component (expected returns, bias) and an emotional component (fear of missing out on an opportunity or put differently, greed).

Lesson learned? The project feasibility analyst will want to measure overconfidence and fear (apprehension and greed) ahead of time. This can be done by examining the PRO scenarios put forth by the project promoters and by questioning them on their deepest motivations.

The logic I have applied in this section can be used not only for speculators bidding on projects and promoters but also for project leaders. As I have discussed in the case of the Québec Multifunctional Amphitheatre, QMA (see the introductory chapter), a project's plans are constantly changing despite the best efforts to set them straight; hence, project leaders are regularly obliged to make decisions that can have little or large impacts on the future of the project. From a POV perspective, it is crucial to ensure that these decisions are as sound-proof as possible.

7.11   Conclusion to Chapter 7

I have seen that while people in Power can be a source of POV, they can use a variety of tools to assess themselves or to judge the project's organization.

Six strategic tools, which I refer to as the six strategic Ps, are available: POV, PWP, POE, PRO, four Ps (SVOR), and POW. POVs are at the heart of a feasibility analysis and I have listed a few tools in this chapter that can be utilized, one of which is the cause-to-effect analysis. PWP is my psychodynamics model that underscores the notion of Trust.

The Point of Equilibrium (POE) is an abstract frame that can assist in understanding how all the different parts of a project and the bid work together, with the six core competencies model being present in the background. As I have just discussed, looking at PRO can tell a lot about the presence of a positivity bias and the inherent motivation (in particular, about the underlying fear) of stakeholders. POW allows the analyst to go to the heart of the project in very practical terms. The six strategic Ps are structural variables of the project and we have seen that all its aspects can be brought down to them.

From this perspective, as mentioned before, management has three fundamental preoccupations—managing and minimizing:

1. Risks (related to calendar)
2. POVs, including a positivity bias (related to costs)
3. The error term ε (related to norms of quality)

In other words, my vision of a feasibility study revolves around the double triple constraints, so to speak: risks–calendar, POVs–costs; and error terms–norms of quality. A large portion of this book has been dedicated, of course, to POVs, which have seldom been addressed in the project management literature.

7.12   What we have learned about POVs: Chapter 7

POVs …

1. Can be enhanced by the lack of balance between Control and Transparency.
2. Can be addressed by real power.
3. Affect both perceived and added value.
4. Are numerous (everything that can go wrong).
5. Are unexposed due to the lack of proper definition of the key attributes of the project.
6. Must be measured with the proper instruments.
7. Must be measured for their description, presence, and importance.
8. Are fostered by complexity.
9. May be highlighted by comparing contrasting cases/projects.
10. Have a greater chance of occurring when there is excessive weight given to a particular criterion versus the others.
11. Obstruct the chances of success as the project is less and less rooted in reality.
12. Can be partially assessed by looking at walls, ceilings, and floors.
13. Can be identified by setting all production parameters at their maximum operational level and then one notch above, and then by estimating which one will likely fail first.
14. Can be highlighted by using various types of analyses (e.g., multicriteria analysis).
15. Are exemplified by biases such as the positivity bias (overoptimism), Blind trust, narrow-mindedness (tunnel vision), and prey positions.
16. Are related to Integrity taken individually, to Exchange of information (EI) taken individually, and to the mutual influence of Integrity and EI.
17. Are, in the best of scenarios, acknowledged by project managers.
18. Are part of the triple core functions of a project manager (managing and minimizing: risks—related to calendar; POVs—related to costs; and the error term—related to norms of quality).

7.13   Key managerial considerations: Chapter 7

1. The *real power*, from my point of view, is the capacity to address POVs.
2. Strategic project management is looking at projects from six angles: Plans–Processes–People–Power (four Ps); Pessimistic–Realistic–Optimistic (PRO) scenarios; POW; PWP (six core competencies model); points of vulnerability (POVs) using SVOR; and dynamic Point of Equilibrium (POE).
3. Aim to simplify complexity.
4. When negotiating, shy away from asymmetry of information.
5. Decisions are influenced by a positivity bias, of which the project feasibility analyst should be wary.
6. A project manager has three fundamental preoccupations: (1) risks, (2) POVs, and (3) the error term ε.

7.14   Case study Chapter 7: BB's highs and lows

There has been plenty of press coverage over the development of two aircrafts (or lines of aircrafts) by two major worldwide competitors in the airplane manufacturing industry: Boeing in the USA (revenues of US$96 billion in 2015) and Bombardier in Canada (C$19 billion).

The 100-year-old Boeing Corporation announced that its new Dreamline airplane would be delivered in 2011. The 200-seat-plane project was initiated in the early 2000s but officially launched in 2004; it went on to experience a 40-month delay while the development cost rose to twice what was initially estimated, at a whopping US$40 billion. It had originally been scheduled for a 2007 maiden flight, with deliveries expected in 2008. The planes were grounded in 2013 by the US Federal Aviation Administration after some faulty batteries were discovered.

In all, the development totaled eight major delays imputed to reasons such as engine failure, incomplete software, shortage of fasteners, and structural flaws. A number of root causes have been identified: the complexity of the product, the heavy reliance on external suppliers, and the novelty of many assembly/production and management processes.

Through the course of the project development, a handful of modifications were made, including improvement of the fuselage joints and a management shake-up. Other challenges grounded the project, one of which was an acrimonious relationship with the machinists union.

Bombardier, headquartered in Montréal (Canada), is known worldwide as a manufacturer of planes and trains; it employs roughly 71,000 people. The company was founded by French-Canadian Joseph Armand Bombardier, who invented the snowmobile in the 1930s.

The Bombardier CSeries is a family of narrow-body, twin-engine, medium-range airplanes. The CS100 can seat 100 people and the CS300 can seat 135. These airplanes are meant to compete with the European Airbus A319 NEO, the Boeing 737 MAX 7, and the Brazilian Embraer E195-E2.

The CSeries project was announced in July 2004. The plane was to be fuel-efficient thanks in part to the use of novel composite materials in its fuselage. Early in January 2006, Bombardier cancelled its plans and decided that it would not develop the CSeries after all. The following year, Bombardier changed its mind once again. It officially launched the CSeries program in 2008 with the expected first flight scheduled to take place in 2012, and the first deliveries in 2013. Letters of interest signed from various customers kept building up.

The maiden flight for the CS100 took place in September 2013. However, the airplane development suffered a number of setbacks thereafter, which led to some cancellation of prebooked orders. The maiden flight for the CS300 took place in February 2015.

The difficulties Bombardier encountered forced it to seek a range of financial arrangements, most notably with the Québec government. However, as of 2016, things started to fall into place, with test flights being convincing and orders firming up.

7.14.1   *Questions related to Case 7: Boeing and Bombardier*

The reader is invited to do some research and to answer the following questions:

1. For each of the two projects (Boeing and Bombardier), please determine if the characteristics of a project are met and discuss how they apply given the circumstances (Table Case 7.A).
2. Please provide a short SVOR analysis for each of the two companies—Boeing and Bombardier (Table Case 7.B).
   Do not forget to include, in the negative forces, the roles of FP_{nc} and think about the dreadful combination.
3. Based on what is written in the present case and your own research, please discuss the level of vulnerability for each of the two projects (Table Case 7.C).
4. For each of the two projects, the US and the Canadian one, please highlight from the list presented in Table Case 7.D what problems were encountered. You may do additional research to substantiate your answer.
5. In hindsight, please answer the questions presented in Table Case 7.E, for each of the two projects.

Table Case 7.A BB's characteristics

Characteristics	Comment
Inputs (structural elements: All are *sine qua non* conditions)	
Plans	Is unique
	Incurs some costs
	Has a set time frame
	Respects preset norms of quality
Processes	Is innovative
	Offers challenges
People	Responds to an opportunity
Power	Has a defined line of authority

Outputs (at least two out of three must be identified)
Provides deliverables
Generates formalized knowledge
Has impacts (+ and −)

Table Case 7.B BB's SVOR

	Internal	External
Positive forces	Total strengths	Opportunities
Negative forces	Vulnerabilities	Risks

Table Case 7.C BB's vulnerabilities

Low-vulnerability (internal) context	High-vulnerability (internal) context	Four Ps
Appropriate resource commitment	Poor resource commitment	Plans
Carefully planned changes, solid results	Rapid changes, quick results	
Realistic expectations	Unrealistic expectations	
Set responsibilities	Ambiguous roles	
Operational changes	Strategic changes	Processes
Shared vision	Conflicting perceptions	People
Supportive top management	Uncommitted top management	Power

Table Case 7.D BB's problems

For outputs items	For time	For costs	For norms of quality
Forms processed	Cycle time	Budget variances	Complaints
Items assembled	Equipment downtime	Contingency costs	Defects
Productivity	Late reporting	Cost by account	Error rates
Sales	Overtime	Delay costs	Number of accidents
Units produced	Response time to complaints	Overhead costs	Rejects
Work backlog	Repair time	Penalties/fines	Rework
—	Work stoppage	Unit costs	Scrap
Others	—	Variable costs	Waste

Table Case 7.E BB's potentiality

List	Four Ps
Scope: Was the plan realistic?	Plans
Well-articulated plan: Was it so?	
Change processes: Were they adequate?	Processes
Measures of success: Were they well established?	
Group of managers: Had the best managers been chosen in the circumstances?	Power
Monitoring system: Had the necessary safeguards been put in place to prevent derailment?	

6. Try to draw a growth curve for each of the projects, taking into account walls, ceilings, and floors.
7. Determine the KSF and KFF for each of the two projects (Table Case 7.F).

    If all the KSF were actually met, would it be likely that there were POVs? Which ones?
8. Substantiating your response with additional research, please determine if the POW seems to be in marching order for each of the two projects (Table Case 7.G).

    If indeed the POWs were well established, is it possible that there were indiscernible POVs? Which ones?
9. Did the development of each of the airplanes include straight linear processes or did it have diagonal processes as well? Please explain.
10. Some major adjustments and rework were needed in both cases. What causal bonds were most likely implicated, and how critical were the POVs (Table Case 7.H)?

Table Case 7.F   BB's KSF and KFF

KSF	
Dominant strategy (DS)	

Plans

Well-defined charter, objectives, impacts, and roles	Clarity
Access to financial resources	Efficacy
Articulate budget	
Establishment of norms	
Realistic calendar	

Processes

Formal methodology	Efficiency
Solid infrastructures	

Power

Experienced project management	Competencies

Contingency strategy (CS)
Preliminary risk management plan
Dreadful combination
Divergent stakeholders' interests
Management blindness and overoptimism

Table Case 7.G   BB's POWs

Item	POW
Key performance indicators (KPIs) were identified	PBS
Deliverables were well defined	
Management of resources were well planned	OBS
DS and CS were established	
Budget was completed	
Tasks were defined	WBS
Calendar of tasks/activities were established	
Forecasted work was done	

Table Case 7.H BB's Causal links

Type of causal link	Critical level of POVs
Not ending the previous activity $(C^+)\rightarrow$ not starting the next activity	Low because no costs are associated with the next activity
Not ending the previous activity $(C^+)\rightarrow$ not ending the current activity	Moderate because both activities are assumed to be near their end
Start of previous activity $(C^+)\rightarrow$ start of current activity	Serious because the process will not work unless both activities (which both incur costs) are started
Start of previous activity $(C^+)\rightarrow$ end of current activity	Critical because there is intense pressure once the entire process has begun.

11. For each of the two projects, classify the POVs you have identified depending on whether they relate to errors (e.g., manufacturing errors) or adherence of outputs to the initial plan (Table Case 7.I).

Table Case 7.I BB's efficiency and efficacy

Inputs (beginning of and during the transformation process)	Outputs (end of transformation process)
Efficiency	Efficacy
POVs linked to ε	POVs linked to outputs

12. In the case of Boeing, given the outstanding record of the company over the decades, is it probable that there was an element of positivity bias involved in the decision to go ahead with the project? Given Bombardier's relatively short history in the airplane manufacturing business, was there some level of apprehension in engaging funds in the projects? You are invited to do the necessary research.
13. Review the two projects with respect to the notion of complexity (Table Case 7.J).

Endnotes

1. PMI Today (July 2015, p. 3).
2. Harris (1993) in an interview with Peter F. Drucker.
3. Lj et al. (2015).
4. Graham and Englund (2003).
5. See Divjak and Kukec (2008).
6. Hence the importance for project leaders to establish a strong legitimacy. See Hooge and Dalmasso (2015).
7. See Reid et al. (2004).
8. Zwikael and Smyrk (2015).

Table Case 7.J BB's complexity

Elements of complexity	Boeing	Bombardier
Interdependence between project's components		
Change factors		
Number of elements in project's system		
Uncertainty		
Project's elements dynamism		
Closed system		
Overall complexity		

9. In this book, I do not discuss governance. For initial comments on the subject, see PMBOK 5 (p. 34): "Examples of the elements of a project governance framework include: project success and deliverable acceptance criteria (...); project organization chart that identifies project roles (...)" I assume the word "governance' has an ethical connotation attached to it in this particular case.
10. Russell and Stone (2002).
11. See Bstieler and Hemmert (2010, p. 486).
12. Berg and Karlsen (2007).
13. Deveau, J.P. Interviewed March 1, 2016.
14. Meredith and Mantel (2009).
15. PM network (July 2015, p. 21).
16. Brière et al. (2015).
17. See, for example, Blake and Mouton (1964).
18. These could potentially be Key Performance Indicators (KPIs) during the course of the project.
19. We know by now that the optimization of satisfaction given a project context occurs at or near the POE.
20. Actually, it is not a recycling process but rather a process whereby the cardboard's pulp is recovered through a chemical mean.
21. PMBOK 3 (2004, p. 39).
22. OPM3 (Organizational Project Management Maturity Model, 2013).
23. The Standard for Portfolio Management (PMI, 2013, p. 36).
24. Mantel et al. (2011, p. 244).
25. Nunnally (1970, p. 133).
26. See Plouffe et al. (2009).
27. Oliver (1980).
28. This is why I have been careful not to determine that trust is the exact opposite of distrust when discussing the construct of Trust.
29. This is why I do not consider distrust to be the opposite of trust.
30. I must refer the readers to my past works on psychological measurements, as this book's goal is not to expand on that subject for lack of space.
31. Sawyer and Peter (1983) did a study and concluded that 33% of articles published in scientific, peer-reviewed journals were biased toward the initial hypotheses.

32. Nunnally and Bernstein (1994).
33. Coviello and Brodie (2001, p. 391) write: "Following minor modifications to structure and wording, the instrument was pretested with a set of executive students similar to those ultimately targeted to participate in the research. The results suggested the instrument was understandable, interpreted appropriately, and captured the characteristics of marketing practice of interest in this investigation."
34. This is done to boost Cronbach's alpha and shows a lack of rigor in creating the questionnaire.
35. See Morris and Pinto (2004), as well as the OPM3 (2008, p. 26).
36. Morris and Pinto (2004).
37. Qureshi and Kang (2015).
38. Bstieler and Hemmert (2010) state that uncertainty can generate misunderstanding of the project's mission and create psychological barriers leading to poor collaboration.
39. Steffey and Anantatmula (2011).
40. For more information on the radial map associated with my model, see my previous work, notably Mesly (2010).
41. Note the direction of the arrows. I am referring to descriptive variables and not consequent arrows. Flight can be described as light or heavy for example; the thinking is not that flight leads to light or heavy (left to right). This may appear counterintuitive to some readers but the trick is to differentiate between descriptive and consequent arrows.
42. See Marzouk et al. (2013).
43. There are 13 different configurations by which one or more of the elements of the triple constraints can affect one or more of the other elements. See Sporns (2011, p. 109).
44. Note than I use consequent arrows and parallelograms, as this process is one that implies action and not merely description, which would entail the use of S and F variables. Clearly, we are dealing here with a diagonal flow because nothing is purely linear.
45. When the time comes to make a choice between two mutually exclusive projects it is the decision based on the NPV and not the IRR that must be made.
46. Also named Fishbone or Ishikawa diagrams.
47. Such as authors Wang and Yang (2012, p. 825).
48. Inspired by The Stationary Office (2009, p. 84).
49. Mulcahy (2013, p. 302) considers that there are seven basic quality measurement tools: cause-and-effect diagrams, flowcharts, check sheets, Pareto diagrams, histograms, control charts, and scatter diagrams. In effect, some are used when auditing the project and others are results of data sorting.
50. Graham and Englund (2003).
51. Note that when Trust is pushed to the limit of 2.3, Collaboration = 2.37, which is not possible since the maximum value of Collaboration is 2.3. Thus, the error term is equal, in this case, to [2.37 − 2.3 = 0.07], which is [2 − k].
52. Reimann et al. (2011, p. 609).
53. Developed by Klorman et al. (1974).
54. Fredrikson et al. (1996); Kindt and Brosschot (1997); Kandel et al. (2000); Nili et al. (2010).

55. See Marks and Gelder (1966); Öst (1987); Craske et al. (1995).
56. Mobbs et al. (2009) experiment served as a source of inspiration. E-Prime 2.0 (Psychology Software Tools, Inc; www.pstnet.com). Snake images were meticulously chosen from the IAPS database (Lang et al., 2008) and presented on-screen to the participants for 2000 milliseconds. Brain imaging was obtained on a 1.5 Tesla Siemens Magnetom Symphony MR scanner. A gradient echo localizer was used to prescribe a subsequent 3-D FLASH (Fast Low Angle Shot, a spoiled gradient sequence) (TR/TE 22/9.2 ms, flip angle 30°, field of view [FOV] 256 – 256 mm scan). Whole brain echo planar fMRI, based on the blood oxygen level dependent (BOLD) effect, was conducted. I used a gradient echo pulse sequence (TR/TE 3000/40 ms, flip angle 90°, FOV 250 – 187.5 mm, 64 – 64 matrix, slice thickness 5 mm, 27 axial slices, bandwidth 2430 Hz/pixel). Statistical Parametric Mapping 8 (SPM8) was used to conduct the statistical analyses. Images (574 images per participant) were realigned to correct for motion (Friston et al., 1995) and spatially normalized to match the echo planar imaging (EPI) template provided in SPM8. Images were smoothed with a 10 mm full-width at half-maximum Gaussian filter given the slice thickness of 5 mm. Regions of interests (ROIs) were chosen for psychophysiological interactions (PPI) analysis and seed regions. In my preliminary analyses, elected seeds regions were the amygdala ($x, y, z = 24 -2 -20$ and $-24 -2 -20$), the hypothalamus ($x, y, z = 8 -2 -7$ and $-8 -2 -7$), and the dorsolateral prefrontal cortex ($x, y, z = -32\ 42\ 18$ and $-32\ 42\ 18$).
57. Côté and Bouchard (2005).
58. I report the most significant results related to the present book.
59. No relation to the yellow circles and the red triangles.
60. Family-wise error (FWE) = 0.001, voxels = 100. FEW is a term used in neuroimaging to identify Type-I errors, known as the probability of reaching one or more false conclusions.
61. As most parts of the brain, it is divided in left and right sections.
62. Bear et al. (1996, p. 452).
63. Again, almost all of the brain regions are divided in either left or right (e.g., amygdala, hypothalamus), or anterior and posterior (e.g., the pituitary gland) areas.
64. We see here that the triple constraints are present.
65. Kahneman and Tversky (1979); Lam et al. (2010).
66. Campello and Graham (2013); Scheinkman and Xiong (2003); Shiller (2005).
67. See Lux (1995).
68. Keynes (2006).
69. Soros (2008).
70. De Long et al. (1990).
71. Scheinkman and Xiong (2003).
72. Gjerstad and Smith (2009).
73. Galbraith (1987, 1992, 1993, 1995).
74. Ferguson (2008).
75. Inspired by Shore (2008).
76. Phillips et al. (2003) state that the following brain areas are being affected with age: the anterior cingulate gyrus (decision; reward anticipation) and prefrontal cortex (cognitive), the hippocampus (memory), the insula, the

ventral regions—the amygdala (emotions), and the ventral striatum. The hippocampus is known to be particularly affected by both age and stress. Some authors assert that the prefrontal cortex (responsible for advanced cognitive efforts) of older people is actually better utilized (Monk et al., 2006) and that their amygdala (responsible for emotions such as fear) is fully functional, depending on the type of stimulus—images or words—(Leclerc and Kensinger, 2011). The anterior cingulate cortex (ACC)—a structure involved in reward anticipation, decision-making and error detection (Cardinal et al., 2002)—has been found to be used more by older adults than by their younger counterparts (Brassen et al., 2011).

77. Moschis et al. (2011).
78. Mather and Carstensen (2005); Zebrowitz et al. (2013).
79. It must be noted that most of the research has been conducted in laboratory conditions, not in real-life situations (Tomaszczyk and Fernandes, 2013). Some of my research was performed in real-life conditions.
80. Isaacowitz et al. (2009); Brassen et al. (2011).
81. See Rozin and Royzman (2001); Isaacowitz et al. (2009).
82. Foster et al. (2013).
83. Lambert-Pandraud et al. (2005).
84. East et al. (2013).
85. Raz et al. (2004).
86. Sudbury and Simcock (2009).
87. Tun et al. (1998); Koutstaal (2003).
88. Weierich et al. (2011).
89. Carstensen (2006); Scheibe and Carstensen (2010); Gross (2013).
90. Charles and Piazza (2009); see also Mather and Carstensen (2005); Kryla-Lighthall and Mather (2009); Mather (2012).
91. East et al. (2013).
92. Lambert-Pandraud et al. (2005).
93. Mather (2012).
94. Feder et al. (2009); see also Kunzmann et al. (2014).
95. Mather and Knight (2005) argue that older people display higher cognitive (prefrontal cortex) control over emotions (amygdala) compared with their younger counterparts.
96. Angie et al. (2011).
97. The relatively small number of older people by itself as well as *versus* the size of the two other groups forces us to moderate my conclusions.
98. Instrumentally hostile position (IP) and Defensive position (DP) are measured on a 7-point Likert scale and reported as such. Other constructs use the same scale but are reported on a percentage basis. Past research shows that this method makes it easier to relate to the constructs by the reader. Also, the ratio predator–prey (or hostile/defensive), which refers to self-confidence, is also better understood than its reverse (prey/predator, or defensive/hostile, which expresses perceived threat). See the various works of Rubicik outlined in the "references" section, including the doctoral thesis (Rubicik, 2010).
99. In line with results by Hedden and Park (2003).
100. I have also referred to this as the law of perceived threat in some other writings, which does not apply to projects alone, but also to the general human realm.

101. De Bondt (1993).
102. Odean (1998).
103. All of my analyses use a 95% confidence level; with $p < 0.05$.
104. A 7-point Likert scale is used in my questionnaire. The value 3 thus corresponds roughly to $3/7 \approx 40\%$.
105. Anderson and Narus 1990s study has an estimator is of 0.73; Palmatier et al. (2006).
106. Using Baron and Kenny's 1986 method.
107. Remembering this refers to the 7-point Likert scale, with 1 being "do not agree at all" and 7 being "completely agree."
108. See Starcke and Brand (2012).
109. For this test, I let the participants know in advance that they would soon become a prey, without specifying if it would be of an unknown predator (red triangle only) or from a known predator (red triangle and random images of a snake).
110. I also ran a similar test using the software I developed in virtual reality setting through the cyberpsychology laboratory to which I belong. I obtained similar behavioral discrepancies, with the participants getting at times completely confused when facing a surprised predator (a mean-looking virtual creature I designed for the purpose of the experiment).
111. The stimuli provided in the experiment were not deemed strong enough to generate *f*MRI images of the hypothalamus. I scanning results confirmed this statement. Perhaps, however, the medial prefrontal activity found in my results indicate some potential level of ventral tegmental area (VTA) and lateral hypothalamus activity as orexin neurons (some of which are instrumental in reward-seeking) interconnect these three brain areas (Aston-Jones et al., 2010).
112. As mentioned, the hypothalamus is actively involved in all major homeostasis-related activities (Squire et al., 2003; McCullough et al., 2007; Siegel and Sapru, 2011) as well as in information processing and behavioral, autonomic, endocrine, and motor activities under predator–prey conditions (Berthoud and Münzberg, 2011; Hinds et al., 2010; Squire et al., 2003). Homeostasis activities include body temperature, heart rate, sexual drive, sleep, thirst, and so on.
113. I could not find a difference between male and female participants.
114. I computed the results by dividing them in two sections; a first half of the experiment and a second half, in order to verify whether fatigue was a factor or not. I did not find significant differences between the two halves.
115. The brain center for emotion and anxiety—the amygdala—increases in activity under stress, while the hippocampus (linked to learning and memory) sees a reduction in size. Meanwhile, the prefrontal cortex, responsible for decision-making, suffers from dendritic reduction and retraction (Lupien et al., 2009).
116. See Cardinal et al. (2002).
117. Family-wise error (FEW) = 0.001, voxels at 100.
118. Such policy is a landmark of American policies. Some view it as what nurtured the creation and exploitation of predatory mortgages that rigged the market during the year 2000–2008, leading to the grave consequences that we all know, in the United States as well as in numerous countries around the world (e.g., Iceland and Spain) (see Gayraud, 2011).

119. CBC News (2013).
120. Sama and Shoaf (2005).
121. Ericson and Doyle (2003, p. 11).
122. Such as Aguilera and Vadera (2007).
123. Barlevy and Veronesi (2003).
124. Cohen and Kudryavtsev (2012).
125. Smith et al. (1988).
126. Shiller (2005).
127. Danis and Pennington-Cross (2008).
128. Kamihigashi (2008).
129. Christie and Geis (1970); Babiak and Hare (2006).
130. Petrick (2011).
131. The derivative will be used further in this section.
132. Sharpe (1970).
133. Note that r_m must be greater than r_f at all times, otherwise the tender is not worth it.
134. See Lux (1995).
135. Three profiles are readily identifiable for all bidders: risk-averse, risk-neutral, and risk-taker.
136. Of course, the decision to withdraw one's investments is related to the fear of not exiting the project on time.
137. The subtraction of the factor ε in is in line with the build-up method used to estimate the required return of an asset (Pinto et al., 2010).
138. The exact slope of both F_{out} and F_{in} curves can be calculated since we know the shape of the Satisfaction curve and since we know that the lowest point for the overpessimistic curve sits at $[y = 0]$.
139. A thought: The ln of 5% is ln $0.05 = -3$, which is $-10 * (k\ 1)$! This is worth celebrating!
140. De Bondt and Thaler (1985) speak of the "winner–loser" effect. It is not the risk σ, but rather ψ that explains such phenomenon.
141. See Fama (1965).
142. The stylized Sharpe ratio hence becomes $[(R_p - r_f)\ \psi^\varphi]/\sigma_p$.
143. "Chances of failure" is tantamount to perceived threat or (in the case of my neurobiological experiment) to the events of being caught by a menacing red triangle.
144. Process by which the cardboard pulp is extracted.
145. FWE: family-wise error. FWE = 0.001, voxels at 100.
146. These maps are not read as explanatory variables (e.g., Affinity) of the ratio IP/DP, but rather as a correlation between say Affinity and IP/DP.
147. Again, these maps reflect the correlation between CD, for example, and IP/DP.

chapter eight

General conclusion

So it is, said the Wise Eagle, that things never turn out the way you thought they would, which means you didn't think right in the first place.

Darloz

Projects are much like hurricanes. They have a beginning and, fortunately, an end. They have a ceiling (they can go up to the troposphere) and a floor (the central pressure level, which is a key measurement of their strength). They are the results of causes, some of which are undisputed: sea surface temperature and vertical wind shear to name just two (the human influence and its participation in global warming is still being debated). They are unique and innovative in their own right. They are composed of recognizable characteristics: rotation, size, strength, vapor content, and so forth. You can infer them from their functional elements: (1) a storm surge (easily exceeding 20 feet of water pouring over the land for Category Five); (2) high winds, some recorded at over 170 miles/hour; (3) massive loads of rain water; and finally (4) tornadoes spawned from their release of energy on land. They are bound by seasonal variations, that is, by the duration and timing (e.g., the Atlantic seasonal hurricane season) resulting from the Earth's movements, by their structure (e.g., what they can gather in terms of dust, heat, vapor, wind, etc.), as well as by their energetic qualities (e.g., an initial independent disturbance that eventually gets them churning out their power, their origin, their proximity to the equator and to land, etc.).

Hurricanes have, of course, impacts on human lives, nature, and infrastructure. As Hurricane Katrina has proved in the most cruel manner, they can hit where vulnerable points are (the inadequate levees in the case of New Orleans), causing havoc and suffering in their trail. They are part of a feedback system: they alter the course of climate, and climate changes, in turn, shape them somehow. That's because they belong to a closed dynamic system: the Earth.

What if nature, in fact, never evolved based on an assumed random effect but rather on somewhat coordinated attempts to reorganize itself? After all, hurricanes displace massive amounts of heat from the tropics to the North Pole, for example, and in the process stir the sea, so that it produces, in the end, submarine conveyors that transport sediments and food for the various species away from that pole. What if nature operated on the basis of projects? We humans do just that, after all.

The difference between human projects and hurricanes is, evidently, that the former are aimed at curving points of vulnerability, be they a need or a problem. Levees in New Orleans are an engineering response to the threats of deadly hurricanes relentlessly taunting the Gulf of Mexico. Unfortunately, that response had its own points of vulnerability, as tragically demonstrated by the passage of Katrina.

Hence, there is no sensible solution other than to make sure that the project that is being envisioned is feasible, that is, that it will realistically address the point of vulnerability it seeks to combat, rather than being dotted with more points of vulnerability itself.

The present book took us on a journey into the realm of project feasibility and points of vulnerability. While scientific in essence, each chapter and case read much like a series of short novels that sought to solve the mystery of points of vulnerability. The hero has been the feasibility analyst and the numerous villains included stakeholders who, maliciously or not, were found to impede on the successful realization of the project.

Projects are everywhere. They even exist in the underground economy. The mob and the police have their own projects, which may increase my readership! In the 1970s, to take this example, Project Alpha was a covert operation with an initial time line of six months, which was thereafter extended. Because of budget constraints, the Federal Bureau of Investigation (FBI) and the New Jersey (USA) State Police had to cooperate in order to share costs. Norms of quality had to be met: the infiltrators had to look credible. The dragnet that was designed to catch the Mafia, which was occupied in particular with gambling, intimidation, sharking, and theft, was all based on mutual trust—between the FBI and the state police agents and between the members of the Mafia and their servicemen. Suspicion (read: apprehension) ran high on both sides of the equation. Gangsters looked for signs everywhere: the way certain cars were parked (police cars usually park outside coffee shops with the front of the car facing the exit, for quick pursuit) or the way "customers" sitting in a bar unwittingly looked at people walking in from head to toe (or, more matter-of-factly, from head to shoes!). The FBI and the police had, evidently, a hidden agenda. The setup relied on one initial infrastructure: a mock trucking company.

Whether in Africa, Canada, Haiti, or the United States, whether in an open or a closed, a legal, corrupt or black market economy, projects invade every aspect of our lives. Once a project is envisioned, the key question is: "Is it feasible?"

I believe that the present book has offered the reader a fair bit of information that can help with the management of projects (hopefully legal ones!), and in particular in preparing an ironclad feasibility study by taking into account points of vulnerability (POVs). I also suppose that at least some of the ideas and the results I discussed could be of use in the regular operations of small, medium, and large companies.

I had fun. I enjoyed a little zoological tour, being acquainted with bees, brown trout, cats, chimpanzees, crocodiles, donkeys, eagles, headless chickens, herring (certainly not red ones!), horses, gazelles, hungry tigers, lonely sheep, mice, minks, pernicious jellyfish, pigs, rare long-peak birds, snakes, some academics (!), squirrels, and white elephants! I analyzed a number of cases pertaining to a plethora of fields: Recycl'Art, Sea Crest Fisheries, the African and Haitian situation, the AFT boat-building company, the Boeing–Bombardier predicament, the cat litter box, the Champlain bridge, the Italian Floorlite improvement project, the marriage example, the mean jellyfish, the lavender manufacturing farm, the Maine East Pharmacy (MEP), the Mervel Farm, the Montréal International District (MID), the Montréal Olympic Stadium (MOS), the NSTP-Environ projects, the oil filter, the Québec Multifunctional Amphitheatre (QMA), the Stradivarius concert, and the TAPI project (see Brainteaser 8). Some of these projects were great successes, others catastrophic, and yet others potential successes or failures. I performed a number of field and laboratory studies, including one using functional magnetic resonance imaging—*f*MRI) and some tests in a cyberpsychology laboratory with which I work. I provided a large number of drawings, figures, images, pictures, and tables in order to render the content as concrete as possible. The objective was to make the reader feel immersed in the reality of projects, so to speak. Using the methodology called 'data percolation', I dug into all kinds of sources, including literature reviews and contemporary interviews. I provided tools, such as

- A questionnaire (Appendix 5.2)[1]
- Prefeasibility and feasibility templates (Appendices 2.2 and 3.1)
- Reference tables (e.g., Appendix 5.3; Tables 1.5, 2.8, 2.13, 2.14, 3.4, 4.1, 4.4, 4.6, 4.8, 5.3, 5.7, 6.1, and 6.32)

- SVOR (Table I.12)
- The summative triangle (Figures 3.11 and 3.12)
- Three core models (Figures 1.6, 2.3, and 5.4)

I came across a great number of passionate people, including Tutankhamen (he didn't talk much though)!

I certainly favored a multidisciplinary approach. I touched in particular on project management, of course, but also on chess and football strategy, economics, ethology, marketing, meteorology, neuropsychology, organizational behavior, psychology, sociology, and statistics. I proposed some mathematical formulae in an attempt to describe how behaviors actually emerge, develop, and become extinct within the confines of the closed dynamic system of a project. Are these formulae the ultimate truth? Hardly. However, they are an honest attempt at modeling the reality of project feasibility.

I readily acknowledge that the content of this book has its limitations. I expressed a number of hypotheses, and I worked around my core model in order to focus on the constant that keeps emerging through my data set, $[k=1.3]$. My effort should be viewed as an exploratory one; much research remains to be done and much constructive debate can ensue that will allow for the betterment of my approach. I tried to stick to conventional wisdom, and to express project management phenomena in a manner that can be generalized. In fact, many project managers have told me that I had put down, in a formalized way, what they were experiencing but never had a chance to put into words, for lack of existing relevant vocabulary.

Being both a business analyst and professor, I had, at heart, to communicate without adopting a heavy academic style; rather, I opted for imagery such as hungry tigers, and for humor, such as the third eye on the back of Caesar's head, which Cleopatra may not have found very attractive! In fact, I improved the perceived value of the present book by merging two outstanding features: first, it is a book of knowledge in itself, and second, it can serve as a sleeping pill if you read it all at once!

I thought I would summarize the present book by going over, once more, the novel ideas I presented; the key components of a feasibility study as outlined in the previous chapters; the six laws of project feasibility as we discovered them; the managerial recommendations I made with great humility and without any intention of imposing my viewpoints; the lessons learned about the POVs; and finally the existing theories that I challenged.

8.1 Novel ideas

I presented a number of novel ideas and tools, such as (1) the concepts of walls (related to calendars), ceilings (related to costs), and floors (related to norms of quality); (2) the five frames of a prefeasibility analysis; (3) the four Ps of project feasibility analysis; (4) the g-spread; (5) the summative triangle; (6) the notion of 'bad apples'; (7) the PCI Index; (8) the six Ps of strategic project management, which includes the SVOR analysis (with the mathematical links between opportunity, risks, strengths, and vulnerabilities); (9) the triangle of value (perceived, added, and residual) as it is linked to the triple constraints; and (10) the barometer of Trust.

8.2 Key components of a project feasibility analysis

I found key components of a project feasibility analysis, including: (1) the role of the constant k; (2) the role of stars and hungry tigers; (3) the six core competencies model; and (4) the role of numerous mathematical functions such as the Satisfaction curve.

8.3 Six laws of a project feasibility assessment

I identified six 'laws' of a project feasibility[2] assessment that can serve as a general guide for double-checking a feasibility study. The first law of project feasibility (the law of positive and negative forces) states that a project is not feasible if the positive forces (which play in favor of the project, i.e., which maintain a functional g-spread) are smaller than the negative forces (which play against the project, such as risks). The second law of project feasibility on depen-

Window of opportunity.

dencies notes that the higher the dependencies between the project's tasks, the more vulnerable the project is. High task interdependence engenders high potential vulnerability. The third law of project feasibility (the law of points of vulnerability) stipulates that the higher the total vulnerability is (the sum of all g-spreads along each of the transformation stages) and the weaker the remedial actions are, the less the project is feasible. The fourth law of project feasibility (the law on the Forces of Production) posits that the more $FP_{nc} > FP_c$, the more the probability of collapse augments. Notably, I referred to a deadly interplay between FP_{nc} and poor planning. The fifth law of project feasibility (law on conflicts) states that the more the wrangling among stakeholders is intense, frequent, and covers critical issues, the less likely the feasibility of the project. The sixth and final law of project feasibility (law of complexity) assumes that the more complex a project is, the riskier it is and the more POVs it contains, and the more probabilities of failure exist.

I also described how the law of Apprehension, which applies to other fields (such as behavioral finance), is present in every aspect of life.

8.4 *Managerial considerations*

To address the challenges of contemporary projects and feasibility analyses, I humbly proposed a number of key managerial recommendations. In the general introduction of this book, I concluded that (1) POVs cannot be ignored; (2) POVs exist for each of the four Ps—Plans, Processes, People, and Power; (3) managers must strive for a Total strength strategy (Dominant + Contingency+Short strategies); and (4) risks cannot be assessed without evaluating POVs at the same time.

In Chapter 1 on Plans, I suggested the following: (1) ensure full and accurate definition of the project; (2) understand inputs and outputs; (3) identify the three types of values: perceived, added, and residual; and (4) examine functionality, design, and costs.

In Chapter 2, also on Plans but with an emphasis on prefeasibility, I concluded that (1) all four Ps have to be evaluated in a prefeasibility study; (2) poor planning along with the presence of Unfits (uncontrolled Forces of Production) is to be avoided; (3) all five frames of analysis as a way of shedding light to uncover dark spots have to be used; and (4) an aim for a robust strategy is to be favored.

In Chapter 3 on Plans and feasibility analysis, I showed that it was advisable to (1) clearly define the products, the organization, and the work to be done; (2) laser focus on what can go wrong in search of POVs; (3) face risks and reduce vulnerabilities appropriately, and build on strengths while answering adequately to opportunities; (4) use both a detailed and a global picture approach; (5) evaluate the transformation stages in terms of calendar of tasks and activities, costs, and norms of quality; (6) actively scrutinize the causal links between the tasks; and (7) use the summative triangle to render the concept in a simple format that everyone can comprehend.

Chapter 4 on Processes led me to the following recommendations: (1) reduce all process flows to causal relationships; (2) seek equilibrium; (3) measure variance punctually and longitudinally; (4) check for risks, POVs, and errors; (5) use modeling and modeling language to extirpate hidden POVs; and (6) zoom in on the three fundamental preoccupations, which are the managing and minimizing of risks, POVs, and ε, the error term.

In Chapter 5 on People, I said that (1) four kinds of stakeholders must be taken into account—customers, suppliers, regulators, and 'bad apples'; (2) the manager should do all he can to minimize the influence of Unfits (FP_{nc}) and bad apples; (3) the manager should observe, observe, and observe again; and (4) the manager must master the project from a parametric, conceptual, and process point of view and seek to be in control, transparent, and fair. He/she should expect from all personnel at all the levels of the project to display trust, collaboration, and commitment. I noted that hostility (especially instrumental hostility) destroys a work atmosphere and that a dynamic Point of Equilibrium is more important to achieve than everyone's satisfaction.

In Chapter 6 on People in action, I concluded that (1) equilibrium in project management is a dynamic event; it is constantly challenged by circumstances and people; (2) three simple questions may be posed to rapidly assess a project: (a) "Are adequate managerial skills in place?"; (b) "Do all the stakeholders trust the project and each other?"; and (c) "Are all the necessary direct and indirect resources available ($|R$ and $|T$)?"; (3) top performers must be quickly identified and encouraged for what they bring to the team; (4) a manager should scour instrumental hostility and deal with elusive agendas and dysfunctional Forces of Production; and (5) he/she must address problems and squabbles right away so as not to allow things to become rotten (they inevitably will).

Chapter 7 on Power can be summarized by the following: (1) the real power is the capacity of people in a position of authority to address POVs; (2) strategic project management is looking at projects from six angles: Plans–Processes–People–Power (four Ps), pessimistic–realistic–optimistic scenarios (PRO), product–organization–WBS (POW), present work psychodynamics (PWP) (six core competencies model), points of vulnerability (POVs) using SVOR, and dynamic Point of Equilibrium (POE); (3) aiming to simplify complexity; (4) when negotiating, shying away from asymmetry of information; (5) realizing that decisions are influenced by a positivity bias that the project feasibility analyst should be weary of; and finally (6) having three fundamental preoccupations: (a) risks, (b) POVs, and (c) the error term ε.

8.5 *Lessons learned about POVs*

Regarding what we have learned about POVs, this amounts to the following. In the preface, we saw that POVs make for predator–prey dynamics—they are costly, they are tantamount to poor loss management, and they are created by humans. In the general introduction, I suggested that POVs that are not managed impair managers' Dominant and Contingency strategies; POVs that are managed can help a manager achieve success; POVs spell trouble down the road; they are present in any project at a minimum level of 4%–7%; they exist within each one of the four Ps; POVs have degrees of gravity; they are sensitive to the triple constraints, but are mostly associated with costs; they exist in both national and international projects; POVs can be somewhat tricky; and they are directly connected to risks.

In the first chapter, I concluded that POVs are seldom discussed in the scientific literature and in feasibility studies; when internally controlled, they express a Contingency Strategy (CS); they are exemplified by such an element as the Unfits (uncontrolled Forces of Production, FP_{nc}); they can be hidden; if deciphered, they can present a dual advantage: the possibility of reducing costs (by way of finding inventive ways to rethink a particular process, for example) and to prevent the rise of costs due to processing mishaps; POVs can sneak in along the process chain inputs–transformation–outputs; they can compromise a project after its completion; they foster uncertainty; they are associated with ambiguity and unknown variables; they are hosted or fostered by intangibles; they require an analysis of the motivation (the opportunity) behind the project from the point of view of each important stakeholder; POVs can be spotted by a large difference between perceived value, added value, and residual value; they typically occur when a latent need and the corresponding innovation don't match; and they are also associated with a short strategy (when tackled, they end up in the Contingency strategy area).

In the second chapter, I showed that POVs can appear in any of the eight different contextual risk areas; they may include a nonrealistic plan and/or a poorly articulated plan, diffuse measures of success, an inadequate change process, incompetent managers, and a lack of safeguards to prevent derailment; POVs may be uncovered by resorting to qualitative measurements; they may also be unmasked through proper modeling; they exist because people do not think of them, do not wish to admit that they exist, or else ignore them; they often result from poor planning and an inadequate workforce; they can be subdued in a feasibility study by way of: (plan) a well-defined charter, objectives, impacts, and roles; access to financial resources; detailed resource requirements; the establishment of norms; a realistic calendar; an articulate budget; (processes) a formal methodology; the use of proven standards, procedures, and technology; a reliance on solid infrastructures; (power) resorting to experienced project managers; and adequate control procedures.

In the third chapter, I showed that POVs affect costs most directly; they are minimized by great managers, thus allowing them to better handle the triple constraints; they have a tendency to throw the project outside the 'football' formed by the three constraints, in an area that I call 'chaos'; they are actuated by the presence of risks; they can be detected by examining everything that can go wrong (PRO system); they are highlighted by the fact that a promoter cannot measure his/her endeavor or else that he/she doesn't care enough about finding ways to measure them other than financially or operationally; they develop because there are no measurements; they are theoretical in nature at the planning stage; they start being active once the [effort | time] curve adopts the linear ascending shape; they represent a source of clear and present danger during the stages of mobilization and deployment; they are sometimes discovered after the project is completed; they are temporally spaced along the calendar of tasks; they are related to proper *description* during the planning stage, their *influence* during mobilization, their level of *causality* during implementation, and their *timeliness* during completion of the project; they can compromise POW; and they are closely associated with cost control management.

In Chapter 4, I showed that POVs become obvious when the direct linear process starts going awry; they can be awoken by uncontrollable factors; they can be detected when the project starts to be both ineffective and inefficient; they are brought under control when straight, direct, and diagonal processes work in tandem to achieve a result that is close enough to the intended output; they are hidden in a variety of entry points, end points, or nondirect flows, and will likely remain undetected until after they cause damage to the project; they always affect one or more than one of the three constraints: time, costs, and/or norms of quality; they wreak havoc in the transformation phase of the project when the most critical linkage level is reached; they have four levels of criticality (low, moderate, serious, critical) depending on the nature of the causal bond in the process; they can be outlined by the causal links appearing in the process (especially in a critical path assessment); they are more dangerous on the critical path than on paths where there are mediating options; they cripple any process system; they can be awoken by the catalyst effect of a moderating variable; they are sensitive to the type of process bond (D, T, I, C) in place; they grow like mushrooms (in dark areas!); they can be anticipated to a certain degree; they can be identified through my modeling technique; they are to be included in a manager's list of duties, which also includes: controlling the Unfits (FP_{nc}), eliminating errors, harmonizing the four Ps, managing risks, and relying on proper infrastructures; they have two faces: one with respect to the error term ε (linked to efficiency), and one with respect to the final outputs (linked to efficacy); they are temporal: they have to be measured in regard to inputs, outputs, and to the error term ε; they are empowered by the lack of forecasting of outputs; they are linked to the Blindness curve; they are linked to utility (a zero difference between beginning and end POVs = high utility); they are not classified as magic moments; they can be somewhat reduced by having some excesses R_n; they are at the highest level when far from the POE; they can hardly be measured directly because they are often buried; they can be uncovered by analyzing the POE; they can somewhat be represented by the g-spread, and can be viewed punctually and longitudinally.

Chapter 5 taught us that POVs: are inherent in the case of competing interests among stakeholders; they increase as $FP_{nc} > FP_c$; they go hand in hand with Trust in a psychodynamics context; they can be estimated by looking at curriculums and answers to particular psychometric tests; they are closely associated with the barometer to Trust; they have an impact on a project outcome along with risks; they are part of Apprehension when internalized; they drive people's behaviors; they are often known to people, consciously or not (gut feeling); they weaken a Dominant strategy;

they are linked to Trust given positive intentions; they influence commitment; they can reach a value of 2.3, at which point the project system crashes; they can have a value of zero which means zero Apprehension; they are fostered by lack of due diligence/vigilance; they come to light when people sit on their laurels; they are minimized at the dynamic POE; they are related to compliance/resistance to change (the more POVs, the more resistance); they are sensitive to cultural contexts; and they are an integral part of human behavior.

In Chapter 6, I outlined the fact that POVs may play against People's own welfare; they diminish with People's commitment; they are related to the constant k; they are best tackled by top performers; they are linked to nefarious behaviors; at their peak, they point to the fact that there are people wanting to intentionally disrupt the project (Instrumentally hostile individuals or hungry tigers); they are related to $|R$ and $|T$; they foster the normalization of conflicts; they affect trust, which, being intimately bonded with a sense of fairness and collaboration, impacts conflict levels; they foster blind management when left unchecked; they are in part unpredictable; and they demand employee training and the supply of $|R$.

Finally, in Chapter 7, I showed that POVs can be enhanced by the lack of balance between control and transparency; they can be addressed by real power; they affect both perceived and added value; they are numerous (everything that can go wrong); they are masked by the lack of proper definition of the key attributes of the project; they must be measured with the proper instruments; they must be measured for their description, presence, and importance; they are fostered by convoluted plans; they may be highlighted by comparing contrasting cases/projects; they have a greater chance of occurring when there is disproportionate weight given to a particular criterion versus the others; they obstruct the chances of success as the project is less and less rooted in reality; they can be somewhat assessed by looking at walls, ceilings, and floors; they can be identified by setting all production parameters at their maximum operational level and then one notch above and by seeing which one will likely fail first; they can be highlighted using diverse types of analyses (e.g., multicriteria analysis); they are exemplified by biases such as the positivity bias (overoptimism), Blind trust, narrow-mindedness (tunnel vision), and prey positions; they are related to integrity taken individually, to exchange of information (EI) taken individually, and to the mutual influence of integrity and EI; they are, in the best of scenarios, acknowledged by project managers; and they are part of the triple core functions of a project manager (managing and minimizing: risks—related to calendar, POVs—related to costs, and the error term—related to norms of quality).

Put differently, POVs (with each subsection presented in the chronological order the features appear in the book):

Characteristics

Description

Identity

1. Are an integral part of human behavior.
2. Are not classified as magic moments.
3. Are part of Apprehension, when internalized.
4. Are part of the triple core functions of a project manager (managing and minimizing: risks—related to calendar, POVs—related to costs, and the error term—related to norms of quality).
5. Can be somewhat tricky.
6. Are linked to Trust given positive intentions.
7. Go hand in hand with Trust in a psychodynamics (work culture) context.
8. Are numerous (everything that can go wrong).
9. Grow like mushrooms (in dark areas).
10. Have three characteristics: identity, presence, and importance.
11. Are temporal: They have to be measured in regard to inputs, outputs, and the error term ε.
12. Represent a source of clear and present danger during the stages of mobilization and deployment.
13. Are related to proper *description* during the planning stage; their *influence* during mobilization; their level of *causality* during implementation; and their *timeliness* during completion of the project.

Presence

1. Exist in both national and international projects.
2. Exist within each one of the four Ps.
3. Are often known to people, consciously or not (gut feeling).
4. Can be represented by the *g*-spread.
5. Are seldom discussed in scientific literature and in feasibility studies.
6. Are sometimes discovered after the project is completed.
7. Are temporally spaced along the calendar of tasks.
8. Can appear in any of the eight different contextual risk areas.
9. Can be viewed punctually and longitudinally.
10. Can sneak in along the process chain inputs–transformation–outputs.

Importance

1. Drive people's behaviors.
2. Are more dangerous on the critical path than on paths where mediating options exist.
3. Are present in any project at a minimum level of 4%–7%.
4. Are theoretical in nature at the planning stage.
5. Are, at their peak, the fact of people wanting to intentionally disrupt the project (hungry tigers).
6. Can reach a value of 2.3, at which point the project system crashes.
7. Have degrees of gravity.

(Continued)

Characteristics (Continued)

8. Have four levels of criticality (low, moderate, serious, critical) depending on the nature of the causal bond in the process.
9. Increase as $FP_{nc} > FP_c$.
10. Start being active once the [effort | time] curve adopts the linear ascending shape.
11. Come to light when people sit on their laurels.

Causes and Triggers

1. Are activated by the presence of risks.
2. Are empowered by the lack of forecasting of outputs.
3. Are fostered by complexity.
4. Are fostered by lack of due diligence/vigilance.
5. Can be awoken by the catalyst effect of a moderating variable.
6. Can be awoken by uncontrollable factors.
7. Develop because there are no measurements.
8. Exist because people do not think of them, do not wish to admit that they exist, or else ignore them.
9. May be caused by a nonrealistic plan; a poorly articulated plan; diffuse measures of success; an inadequate change process; incompetent managers; and a lack of safeguards to prevent derailment.

Links

1. Have two faces: one with respect to the error term ε (linked to efficiency), and one with respect to the final outputs (linked to efficacy).
2. Are associated with a Short strategy (when not tackled).
3. Are associated with ambiguity and unknown variables.
4. Are closely associated with cost control management.
5. Are closely associated with the barometer to Trust.
6. Are directly connected to risks.
7. Are linked to hostile behaviors.
8. Are linked to the Blindness curve.
9. Are linked to utility (a zero difference between beginning and end POVS = high utility).
10. Are related to $|R$ and $|T$.
11. Are related to compliance/resistance to change (the more POVs, the more resistance).
12. Are related to Integrity taken individually, to EI taken individually, and to the mutual influence of Integrity and EI.
13. Are related to the constant k.

Influences Sustained

1. Are sensitive to cultural contexts.
2. Are sensitive to the triple constraint, but are mostly associated with costs.
3. Are sensitive to the type of process bond (D, T, I, C).

Detection

1. Are at the highest level when far from the POE.

Characteristics (Continued)

2. Are exemplified by biases such as the positivity bias (overoptimism), blind trust, narrow-mindedness (tunnel vision), and prey positions.
3. Are exemplified by such element as Unfits (uncontrolled Forces of Production, FP_{nc}).
4. Become obvious when the direct linear process starts going awry.
5. Can be detected when a latent need and the corresponding innovation don't match.
6. Can be detected when the project is both ineffective and inefficient.
7. Can be detected when there is disproportionate weight given to a particular criterion versus others.
8. Can be hidden.
9. Can be highlighted using various types of analyses (e.g., multicriteria analysis, SVOR).
10. Can be identified by setting all production parameters at their maximum operational level and then one notch above and see which one will likely fail first.
11. Can be identified through my modeling technique.
12. Can be outlined by the causal links appearing in the process (especially in a critical path assessment).
13. Can be somewhat assessed by looking at walls, ceilings, and floors.
14. Can be somewhat guessed by looking at curriculums and answers to particular psychometric tests.
15. Can be spotted by a large difference between perceived value, added value, and residual value.
16. Can be uncovered by analyzing the POE.
17. Can be uncovered by examining everything that can go wrong (PRO system).
18. Can hardly be measured directly because they are often hidden.
19. May be uncovered by resorting to qualitative measurements.
20. May be uncovered through proper modeling.

Strategies

1. Are best tackled by top performers (the Stars).
2. Are brought under control when straight, direct, and diagonal processes work in tandem to achieve a result that is close enough to the intended output.
3. Are minimized at the dynamic POE.
4. Are minimized by great managers who know how to handle the triple constraints.
5. Are to be considered by a manager along with controlling the FP_{nc}, controlling risks, harmonizing the four Ps, relying on proper infrastructures, and eliminating errors.
6. Are, in the best of scenarios, acknowledged by project managers.
7. Can be addressed by real power.
8. Can be anticipated to a certain degree.

(Continued)

Characteristics (Continued)

9. Can be somewhat reduced by having some excess (R_n).
10. Can be subdued by the proper balance between control and transparency.
11. Can be subdued, in a feasibility study, by way of: (Plans) well-defined charter, objectives, impacts, and roles; access to financial resources; detailed resource requirements; the establishment of norms; a realistic calendar; an articulate budget; (Processes) a formal methodology; the use of proven standards and procedures; the use of proven technology; a reliance on solid infrastructures; (Power) resorting to experienced project managers; and adequate control procedures.
12. Demand employee training and the supply of $|R$.
13. May be highlighted by comparing contrasting cases/projects.
14. Must be measured for their description, presence, and importance.
15. Must be measured with proper instruments.
16. Require an analysis of the motivation (the opportunity) behind the project from the point of view of each important stakeholder.
17. Can help a manager achieving success when well managed.
18. Can impair managers' Dominant and Contingency positions when not managed.
19. Express a Contingency Strategy (CS) when internally controlled.

Impacts

1. Affect both perceived and added value (and, ultimately, the residual value).
2. Affect the costs more directly.
3. Affect Trust, which, being intimately bonded with a sense of fairness and collaboration, impacts conflict levels.
4. Always affect one or more than one of the three constraints: time, costs, and/or norms of quality.
5. Can compromise POW.
6. Can jeopardize a project *after* its completion.
7. Wreak havoc in the transformation phase of the project when the most critical linkage level is reached.
8. Foster uncertainty.
9. Diminish with People's commitment.
10. Foster blind management when left unchecked.
11. Foster the normalization of conflicts.
12. Have a tendency to throw the project outside the 'football' formed by the three constraints, in an area called 'chaos'.
13. Have an impact on a project outcome along with risks.
14. Influence commitment.
15. If tackled, can present a dual advantage: the possibility of reducing costs (by way of finding inventive ways to rethink a particular process, for example) and to prevent the rise of costs due to processing mishaps.
16. May play against People's own welfare.
17. Obstruct the chances of success as the project is less and less rooted in reality.

Characteristics (Continued)
18. Often result from poor planning and an inadequate workforce (dreadful combination).
19. Spell trouble down the road.
20. That have a value of zero lead to zero Apprehension.
21. That are hidden in a variety of entry points, end points, or non-direct flows will likely remain undetected until after they cause damage to the project.
22. Weaken a Dominant strategy.
23. Weaken any process system.

8.6   Theories challenged

I challenged some of the theories or methods put forth by other authors or sources, such as the *Project Management Body of Knowledge* (PMBOK). Examples include: (1) the notion of groups of processes in PMBOK; (2) the accuracy of the 10 domains of knowledge in PMBOK; and (3) the views on constraints in PMBOK 5.[3]

In some other domains, I opened some research avenues by proposing to reorganize the DSM-V nomenclature to fit the six core competencies model, and I postulated that oxytocin had to be considered along with cortisol, testosterone, and serotonin in order to fully comprehend the four basic coping mechanisms anchored in the hypothalamus. On this point, I hypothesized that in fact all four responses are ultimately defensive in nature, considering that Instrumentally hostile behaviors had one differentiating particularity: they are a delayed response that involves higher cognitive inputs. All four coping measures are meant to deal with POVs.

My comments and critiques were meant to be constructive and by no means must be deemed denigrating; it is by way of scrupulous analyses that knowledge progresses.

8.7   Cases and brain teasers

In the sections labeled 'Case' at the end of each chapter, I presented some real project cases as well as some challenging questions. Analyses of the cases and answers can be made available by the author upon request by the reader. I invite the reader to go through the cases and instinctively apply to them the knowledge that he/she has acquired in the introduction, the seven chapters and the conclusion of this book. Project management professionals (PMPs) will also find that they can apply PMBOK theory to these cases, as they are constructed to match some of PMBOK 5 content. I also added some "brain teasers" at the end of the present book to somewhat entertain the reader.

8.8   Humor

I have tried to make the reading both entertaining and serious. My use of dry wit should not be misinterpreted to infer that my discussion was not stamped with scientific rigor. Quite the contrary! I have researched extensively and looked at POVs in all kinds of scientific ways. Yet, I believe communication has to be lively and that it should raise interest. Communication is important. The author of this book, who stutters, confides:

> "Communication and use of proper wording is crucial. You have to make sure your crew understands you loud and clear. To take a personal example, I told my wife on our honeymoon that I would I give her an orchid every week (she loves flowers). She misunderstood the word and has held me up to it ever since!"

8.9   The future

There is no doubt that a fair amount of future work is advisable in order to develop the field of project feasibility analysis from the angle of POVs. We have learned throughout the present book that people commit to projects, but more particularly, that they commit to meeting the deadlines, to the tasks and activities (WBS), to respecting set cost limits, and to abiding by the norms of quality. In order to guarantee efficient commitment, there has to be collaboration among team members. However, collaboration is unlikely if there is no mutual trust and a sense of fairness in the general work culture (psychodynamics) of the project. Trust is negatively affected by apprehension, so that it is most important to get an unflappable grip on hostility and to be transparent within reason in order to make the project plan as limpid as possible for all stakeholders. I have tried to show that confrontational management encourages process and product flaws, which, of course, means that it leads to a reduction in quality. People have all kinds of ways of sabotaging their work—by conveying the wrong information, delaying outputs, procrastinating, spending time for personal reasons, and so forth. Goodwill, an intangible asset that I discussed previously, builds on the psychological core of trust, fairness, and collaboration.

I hope that this book provides an initial spark to encourage the effort of understanding the link between human behavior and relationships and quality outputs. Team efforts permeate every aspect of projects. Again, I do not pretend to hold the truth, or that this book is perfect. If it could merely serve the project feasibility analysts in one way or another, I would be pleased and would assume that I have achieved my modest goal. I hope that some of my imaginative ways of expressing some of my ideas—such as the "I gotta go" people, to take only this example—have made the reading enjoyable and memorable.

In summary, a project feasibility analyst looks for Dominant and Contingency strategies; he/she detects hostile and defensive positions that can jeopardize the project. He/she is aware of the need for Short strategies when difficulties surface. An analyst aims for **robust** project management, one that is able to deal with POVs ahead of time, whether this entails being agile on occasions or else standing firm in front of adversity. Think of a project as a football game: the team has to come together. It establishes a strategy to get to the finish line the best way possible, as pressured by time, and given physical energy (cost), quality of the plays, of the players, and of the Fits (you don't want to put Joe Montana in a tight end position, do you?). You face adversity (the opponent in the football example) that you must respect and outsmart with the proper skills. When you score, everyone is proud and celebrates, possibly by the thousands. There are random setbacks, but you keep going forward, to the very last second. Your plays are realistic; you try to find a gap in the opponent's actions and reactions, while managing and minimizing risks, POVs, and errors.

I invite the reader to apply the knowledge expanded in this book in real-life situations—there is no better way to integrate it and make improvements on it.[4]

Thank you for having taken the time to read this book and remember, despite the best efforts at due diligence, there will always be Fits and Unfits, hungry tigers and lonely sheep, stars and average team members, yet try not to hire any extraterrestrial creatures: they do great work but then they vanish without warning and without leaving any contact information!

8.10   *Main hypothesized behavioral mathematical functions*

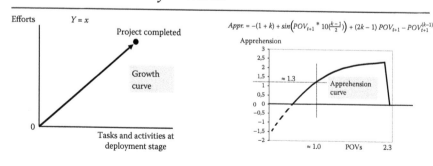

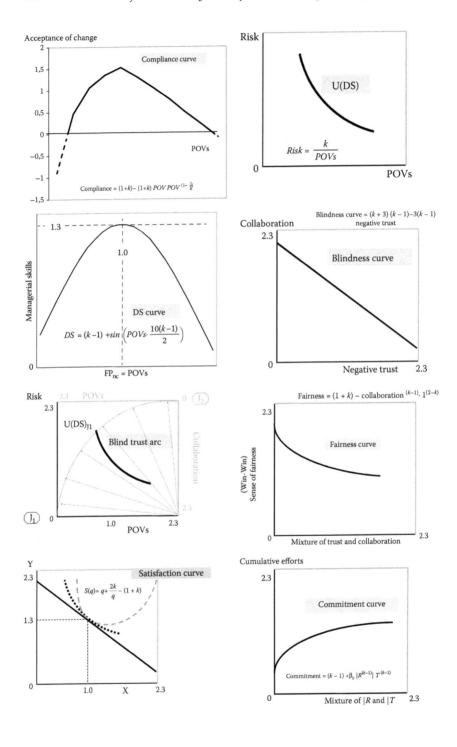

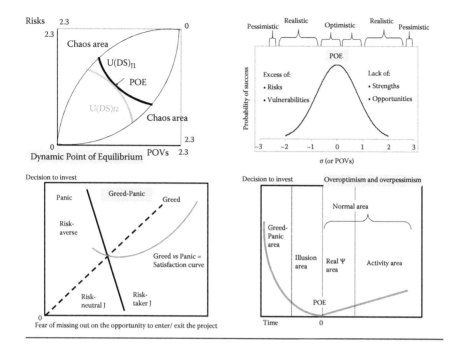

8.11   Brain teasers

8.11.1   Brain teaser 1: The Bermuda Triangle

Comment on the following:

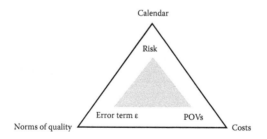

8.11.2   Brain teaser 2: PMBOK's 10 domains

Comment on the following (see PMBOK 5, p. 99):

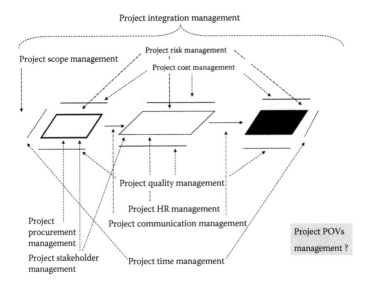

8.11.3   Brain teaser 3: PMBOK groups of processes

Going back to Chapter 2, about the idea of an oil filter for cars and trucks, please comment using the knowledge acquired in this book:

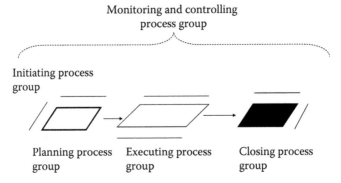

8.11.4   Brain teaser 4: Top performers

An analysis of top performers, regular performers, and below average performers for nearly 2000 team members, reveals the following graph, where the x-axis is the participants and the y-axis the project capability index (PCI), on a scale of 0 (poor) to 10 (excellent):

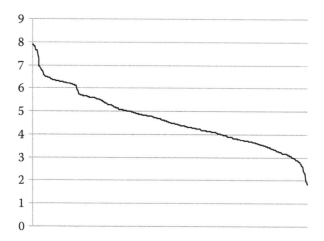

In addition, the distribution around the mean is as follows (with the *x*-axis being the standard deviation and the *y*-axis being the number of participants in percentage of the overall group):

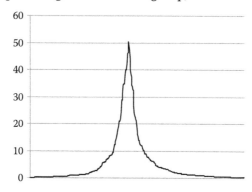

Please comment on the observations you make referring back to Chapter 6, including by identifying the top performers.

8.11.5   Brain teaser 5: Behavior

A large multinational firm ran a study across its offices and a number of statistical outputs were produced. Please answer the following questions:

1. Based on the following graph, is it fair to say that Instrumentally hostile behavior is not linearly related to defensive behavior (*x*-axis = Defensive position; *y*-axis = Instrumentally hostile position; $R^2 = .094$)[5]?

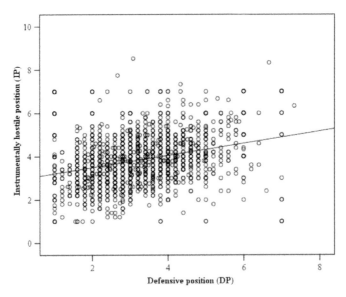

2. For 1717 respondents, the average *k*-value and the levels of some of the key constructs are as follows:

	k' value	Trust	Fairness	Collaboration	Depend.	Commit.
Valid *n*	1717	1717	1717	1717	840	1638
Missing	0	0	0	0	877	799
Average	1.36	78	74	74	59	75

Please comment.

3. From the following table (and with the same group as above), can you infer the law of Apprehension?

	Component 1	Component 2
Control	0.311	0.849
Transparency	− 0.174	0.750
Trust	0.879	0.106
Fairness	0.864	0.042
Collaboration	0.882	0.149
Commitment	0.654	− 0.073

4. Using the same database and looking at the following graph (*x*-axis = Trust, *y*-axis = Collaboration; $R^2 = 0.575$), what can you conclude[6]?

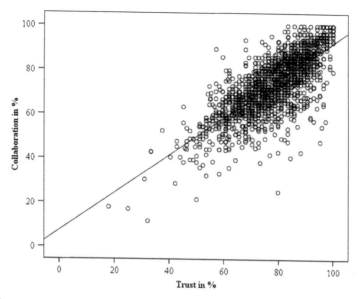

5. When management plots Fairness (on the *x*-axis) against Collaboration (*y*-axis), it notices the following, on which please draw a conclusion ($R^2 = 0.531$)[7]:

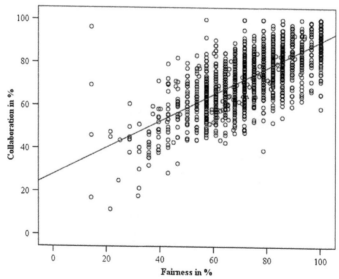

6. When management plots each of the four core types of coping mechanisms—hostile-offensive (instrumentally hostile), anxious, resisting, and fleeing—it obtains the following graph, about which please comment with respect to the *k*-value[8]:

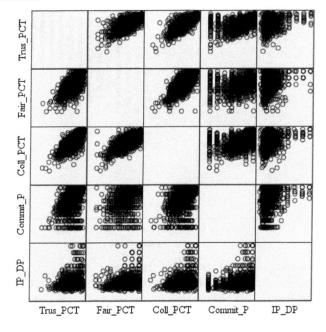

7. Based on the following matrix, can management conclude that trust, fairness, and collaboration display an identical pattern (the behavioral core), while IP/DP is different, just like Commitment is also different?

8. Based on the following 3-D graph, is it fair for management to conclude that Trust, Fairness, and Collaboration form a single behavioral core?

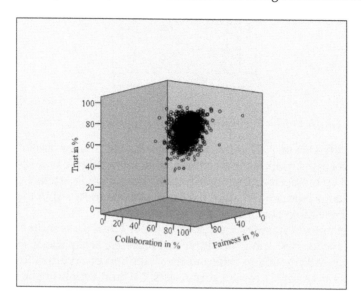

8.11.6   Brain teaser 6: PMI talent triangle

1. The Project Management Institute (PMI) promotes the following concept, called the 'PMI talent triangle':

In particular, ask yourself: is "strategic management" not a component of overall business management? Can "strategic and business management" be associated with Plans, "technical project management" with Processes, and "leadership" with Power? Is "People" not missing from the model?

2. Please compare the PMI model with the following model, inspired from the MID (The *Quartier international de Montréal*) case:

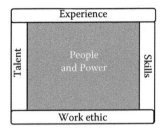

8.11.7   Brain teaser 7: Organizational process assets

PMBOK 5th edition, p. 27, refers to "organizational process assets" as "(…) plans, processes, policies, procedures, and knowledge bases specific to and used by the performing organization. They include artefact, practice, or knowledge from any or all of the organizations involved in the project that can be used to perform or govern the project."

   Assume for a moment that plans refer to Plans as seen in the present book, processes, and procedures to Processes, and policies to Power, Would it be admissible to reformat the definition of organizational process assets as simply "project's core Assets"? Would People not be missing from the equation? In other words, could we regroup the four Ps of project feasibility analyses under the heading "project's core assets," as follows (stylized model)?

8.11.8   Brain teaser 8: In sync or sink

In a project feasibility analysis, as we have seen throughout the present book, the experts try to determine ahead of time whether the four Ps are synchronized. Is the Plan in line with what can be realistically achieved? Are the processes, linear and diagonal, adapted to the Plan in a way that will make the project efficient and efficacious? Are the four groups of People (customers, suppliers, regulators, and bad apples) acting on the same understanding and with the same beneficial intentions, without hidden agendas? Is Power legitimate and focused on controlling hostility and on minimizing dissension? Failing proper alignment of means and ends, the project is likely to fizzle.

In the realm of project feasibility, the motto is "in sync or sink."

A project is currently in the making whereby a $10 billion pipeline will be built to channel natural gas from the former Soviet state of Turkmenistan on a 1118-mile journey, going through Afghanistan and Pakistan to end in Fazilka, India. The so-called TAPI project would be operational by the end of 2019 and would be capable of transporting 24 billion gallons of gas a day.

After doing the necessary research and going through the four Ps, the POVs as well as the PRO scenarios, what would be your take on the "in sync or sink" point of view?

8.11.9 Brain teaser 9: Hostility and failure

Please comment on the following two models, based on what you have learned in this book (hint: find the type of bonds that exist between the constructs):

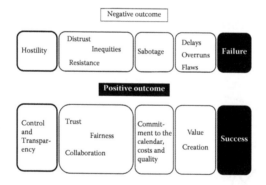

8.11.10 Brain teaser 10: A losing hand

Please adapt the following to a project you know:

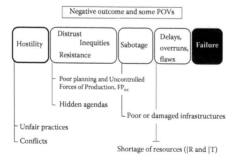

8.11.11 Brain teaser 11: A winning hand

Please adapt the following to a project you know:

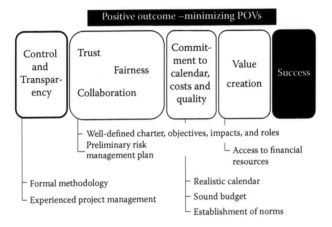

8.11.12 Brain teaser 12: Cleopatra's necklace

Have fun with the following necklace that Caesar offered Cleopatra in an attempt to help her build a few pyramids:

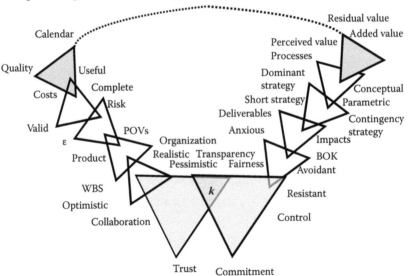

8.11.13   Brain teaser 13: Who's at work

Please look at the following and see if you can associate the various kinds of team members with people you know within your project endeavor:

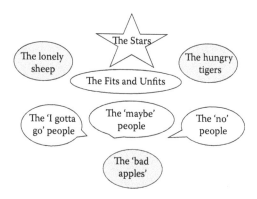

8.11.14   Brain teaser 14: POVs

Try to make a list of everything you remember about POVs and compare it with the list found in the general conclusion.

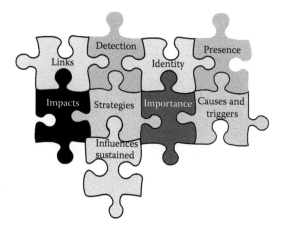

Endnotes

1. As mentioned, the questionnaire on psychological constructs can be made available by contacting the author and sending me, for example, a pre-paid Porsche Cayenne (marine blue please)!
2. Not to mention the law of Apprehension (perceived threat), which is used in other venues such as psychology and behavioral finance.
3. And of Mulcahy (2013).

4. The "most effective learning is working in real-life situations" (Divjak and Kukec, 2008, p. 251).
5. Residuals show a very normal distribution with no noticeable tendencies.
6. Residuals display a normal behavior. F-value $= 2{,}320.454$, $p = 0.000$ (at 95%).
7. Residuals display a normal behavior. F-value $= 1{,}938.727$; $p = 0.000$ (at 95%).
8. Hint: In what quadrants/combinations are most of the respondents sitting?

Glossary

Absolute time is simply the normal passage of time as set by nature.

Added value is the actual total cost of the project, calculated by adding the cost of each material and service unit along the transformation stages.

Anxious The "maybe" people.

Apprehension Sentiment or sensation that one experiences when in a situation of perceived threat.

Asymmetry of information When one person has more and/or better information than another.

Avoidant The "I gotta go" people.

Barometer of Trust A scale that classifies trust, from Blind trust to perceived predation.

Beginning wall $|\uparrow$, associated with a calendar.

Bermuda Triangle Figurative expression for the triple constraints.

Blindness curve Occurs when the promoter underestimates the risks (a typical case when an overly enthusiastic project promoter seeks funding) or else overestimates his own capacity (thus suffering from some kind of inflated ego) = The reverse of Trust => Collaboration curve.

Calendar of tasks and activities Calendar of the project with specifically related tasks and activities, which is at the core of the process of the project.

Causal (C) (+ or –): C^+nens; C^+nene; C^+ss; C^+se.

Ceiling $\uparrow$ relates to costs. The maximum amount of money the promoter wants to dedicate to inputs, transformation and outputs.

Characteristics of POVs: Identity, presence, and importance POVs can be measured by way of their characteristics, whether or not they are present (if detected), and the strength of their action in the project.

Closed dynamic systems A system that is constantly active and that is bounded by definite borders.

Collaboration–trust operating area Confined area set by the arcs out of which chaos takes place. Inside it, trust and collaboration are assumed to take place.

Consequent parallelograms Process elements that imply a time factor. Expressed in three forms (influence: I, longitudinal: T, or causal: C).

Contingency strategy Contains strengths that allow the project manager to minimize the effect of risks and vulnerabilities on his project.

Coping mechanisms Instrumentally hostile and defensive positions. They are awakened in a situation of apprehension (perceived threat).

Critical causal thinking Most people, consciously or not, bring events to their critical path, that is, to their causal dynamics.

Curve types Composed of six basic types.

Defensive hostile behaviors are temporary in nature. There are three types: anxious, avoidant, and resistant. Also referred to as **lonely sheep behaviors**.

Defensive position (DP) A coping mechanism that does not imply planning but rather high impulsivity. A behavioral position that develops when the manager is caught by surprise; he loses control and feels he is outsmarted by uncontrollable human-generated or naturally occurring behaviors and/or events.

Descriptive Parallelograms D = S *and* F One process element is linked to another. Can be expressed through structural (S): binary or continuous, or functional (F): binary or continuous arrows. No time factor.

Diagonal Flow of a project that is not linear because it involves changes in the direction of what would have been a direct flow from point A to point B.

Diagonal process flows Show deviations from the planned output.

Dominant strategy Behavioral strategy that allows the project manager to complete all sets of tasks, drive the organization to the level that is assigned to him, and deliver the final product within the specific constraints of time, costs, and norms of quality.

Dreadful combination Combination between the FP_{nc} and poor planning. A major cause for project failure.

Dynamic point of equilibrium or dynamic POE At this particular point (with $x = 1$ and $y = 1.3$), all active psychological forces that affect the project are aligned and, presumably, productivity is optimized.

End wall $|↓$, concerns the calendar.

Feasibility study A project feasibility study is a comprehensive study that examines in detail the five frames of analysis of a given project in consideration of the four Ps, its POVs, and its constraints (costs, calendars, and norms of quality) in order to determine whether it should go ahead, be reconsidered, or else totally abandoned.

Fits People who fit right into the project without being stars, hungry tigers, or lonely sheep.

Floor ↓: relates to norms of quality. The minimum quality required to ensure that the process is functional.

fMRI Functional magnetic resonance imaging, a brain imaging process.

Functional arrows (F) A functional arrow expressing the fact that the subprocess element is not essential to the definition of the key process element, but that it is a key manifestation of it. The main process element can be inferred from the functional process elements. Minimum: 2 sub process elements. No time factor.

Groupthink phenomenon This occurs when members of a group narrow down their decision-making options with the feeling that they hold the ultimate truth: they have blind trust in themselves.

Growth curve of the project The linear ascending curve.

g-spread A temporal representation of the variance σ around the project's dynamic point of equilibrium.

Hidden truths What people know and refuse to admit, or don't know, but that still exists and colors people's motives.

Hostile behaviors Impede on projects.

I, T, or C The three fundamental ways of expressing the nature of links between process elements with a time factor.

Indirect variable Moderating (±) or mediating (+ or –).

Influence (I) Direct (+ or –).

Initial value proposition of a project Project layout based on functionality, design, and costs.

Instrumentally hostile position (IP) A relatively enduring behavioral stance, by which one takes advantage of another person's vulnerability for one's own benefit, causing the other harm (e.g., loss), by surprise. Referred to as a **"hungry tiger's behaviors."**

Intangibles Elements of a project that cannot be measured and that are not necessarily part of the project.

Key performance indicators (KPI) The main parameters that are used to measure how well a project is doing.

KPPP In my *fMRI* study, the situation whereby the participant faces a known predator (the snake).

Latent needs Needs have not been formalized or universally recognized yet; they are dormant, or put differently, unconscious.

Longitudinal (T) or (t) No polarity.

Observables Imply a change in behavior that can be observed. Are general (in general, happy people laugh or smile) or a contextual framework (e.g., according to Japanese tradition, one doesn't shake hands but bow).

Parallelograms A process element.

Perceived threat A psychological construct formed by the ratio of *defensive* (one's sense of vulnerability) over *predatory* (one's ability to attack) positions.

Perceived value is the total value attributed to the project, including what it produces in terms of intangibles, by the users/clients.

Phase Each step within the transformation chain. Inputs → Transformation → Outputs.

Point of autonomy Point reached when the whole production/transformation process could theoretically operate without supervision.

Point of no return Point in the production process from which the process cannot be stopped or reversed.

Points of vulnerability (POVs) Temporal and physical points along the transformation stage of a project that impede on the cost, calendars, or quality of the project when faced with adverse conditions (negative forces), whether these conditions are under human control or not.

Prefeasibility study A study that follows an initial value proposition of the project and that offers a general view of said project using five analytical frameworks and that allows the feasibility expert to make a recommendation on the suitability to conduct a feasibility study.

Pretrust The often automatic or subconscious trust that one allocates to another before even engaging in a relationship.

Psychological core (Trust + Fairness + Cooperation)/3.

Punctual cost of changes Cost of changes without considering the historical costs of change.

Real power The capacity to address POVs.

Red flags Signal poor work psychodynamics; examples include dismissal, fallacy, the normalization of deviance, and overoptimism.

Relative cost of changes is the cost of change relative to the budget that has been expended so far during the project.

Relative time is the path that is designed for the project (the full transformation chain), and that can become critical as the concert date gets closer—it is created by humans by way of calendars of activities and tasks.

Residual value Related to norms of quality as quality persists over time.

Resistant The "no" people.

Robustness Capacity to deal with points of vulnerability, whether this entails being agile at times or else standing firm in front of adversity.

Rule of seven Rule by which one establishes that any set of seven or more points appearing consecutively, either in the lower or upper range (on one side of the mean), is considered not to be a random occurrence, suggesting that the process is actually out of control.

Saturation point Achieved at the dynamic point of equilibrium, when nothing can be added to improve the project or a person's well-being.

Stage Each step within the life cycle of the project (vision, planning, mobilization, deployment, and completion/evaluation).

Stars People who excel at their job.

Straight direct Flow of a project that is linear, going from A to point B without changes of direction.

Straight direct flow Inputs are not diverted; they follow a linear logic that sees them create a predictable output.

Structural arrows (S) Structural arrow that express the fact that there is a *sine qua non* condition for the definition of the main process element; this condition is expressed by at least two sub process elements. Minimum: 2. No time factor.

Summative triangle Figure that merges the three axes of the triple constraints with the five stages of the project's life cycle.

SVOR Strengths, vulnerabilities, opportunities, and risks. Replaces the SWOT system in feasibility analysis, and, as an analytical tool, serves to identify POVs.

Transformation chain Stages and phases combined.

Transparency Defined as the sharing of useful information with others to facilitate their work and nurture their motivation.

Triple constraints Refers to the quasi-mutually exclusive requirements to meet a particular deadline, to respect forecasted costs, and to meet quality criteria (sometimes referred to as the Bermuda Triangle).

Triple constraints of information (TCI) Valid, useful, and complete information.

UPPP In my *f*MRI experiment, a situation whereby participants face an unidentified predator (a red triangle).

Utility drawback Occurs when management has to use time, incur costs and/or sacrifice quality (perhaps only temporarily) in order to achieve its goal.

Utility of the dominant strategy or U(DS) A curve that expresses the fact that given the triple constraints, the strongest managerial strategy is achieved at [Risks = 1.3/POVs].

Vulnerability Condition experienced in any one of the four Ps of a project that makes it susceptible to failure.

Walls There are two walls, one at the beginning of the process and one at the end of the process—these relate to the calendar (beginning and end).

Work psychodynamics (PWP) Internal culture and social interactions between stakeholders. Put differently, it refers to how team members use trust to achieve the ultimate goal: completing the project on time, within costs, while meeting preset norms of quality. Maximized when DS is at its peak.

References

Ackoff, R. (1957). *The Design of Social Research*. Chicago, IL: University of Chicago Press.

Adams, D.B. (2006). Brain mechanisms of aggressive behavior: An updated review. *Neuroscience and Biobehavioral Reviews 30*(3), 304–318.

Adenfelt, M. (2010). Exploring the performance of transnational projects: Shared knowledge, coordination and communication. *International Journal of Project Management 28*(6), 529–538.

Afitep. (2000). *Le management de projet, principes et pratiques*. Paris: Ed. Afnor.

Aguilera, R.V., and Vadera, A.K. (2008). The dark side of authority: Antecedents, mechanisms, and outcomes of organizational corruption. *Journal of Business Ethics 77*, 431–449.

Ainamo, A., Artto, K., Levitt, R.E., and Orr, R.J. (2010). Collaboratory for research on global projects. *Scandinavian Journal of Management 26*(4), 343–351.

Albescu, F., Pugna, I., and Paraschiv, D. (2009). Cross-cultural knowledge management. *Informatica Economică 13*(4), 39–50.

Al-Maghraby, R. (2008). Project human resources management; itSMF Egypt chapter establishment project case study. 2008 ITI 6th International Conference on Information and Communications Technology, Published in *Information & Communications Technology*, 2008, Egypt, Dec., 37–42.

Alvarez, R.P., Biggs, A., Chen, G., Pine, D.S. and Grillon, C. (2008). Contextual fear conditioning in humans: Cortical-hippocampal and amygdala contributions. *The Journal of Neuroscience 11*(June), 6211–6219.

Amaral, D.G. (2002). The primate amygdala and the neurobiology of social behavior: Implications for understanding social anxiety. *Biological Psychiatry 51*, 11–17.

Amit, R., and Schoemaker, P.J. (1993). Strategic assets and organizational rent. *Strategic Management Journal 14*, 33–46.

Amsel, A. (1992). *Frustration Theory: An Analysis of Dispositional Learning and Memory*. Cambridge, UK: Cambridge University Press.

Anderson, J.C., and Gerbing, D.W. (1988). Structural equation modeling in practice: A review and recommended two-step approach. *Psychological Bulletin 103*(3), 411–423.

Anderson, J.C., and Narus, J.A. (1990). A model of distributor firm and manufacturer firm working partnerships. *Journal of Marketing 54*(1), 42–58.

Anderson, P.F. (1983). Marketing, scientific progress, and scientific method. *Journal of Marketing 47*, 18–31.

Ang, J.S., Cole, R.A., and Lin, J.W. (2000). Agency costs and ownership structure. *Journal of Finance LV*(1), 81–106.

Angie, A.D., Connelly, S., Waples, E.P., and Kligyte, V. (2011). The influence of discrete emotions on judgement and decision-making: A meta-analytic review. *Cognition and Emotion 25*(8), 1393–1422.

Anvuur, A.M., Kumaraswamy, M.M., and Mahesh, G. (2011). Building "relationally integrated value networks" (RIVANS). *Engineering, Construction and Architectural Management 18*(1), 102–120.

Appelbaum, S.H., Shapiro, B., and Elbaz, D. (1998). The management of multicultural group conflict. *Team Performance Management 4*(5), 211–234.

Asquin, A., Falcoz, C., and Picq, T. (2005). *Ce que manager par projet veut dire*. Paris: Ed. d'Organisations.

Assaf, S.A., and Al-Hejji, S. (2006). Causes of delay in large construction projects. *International Journal of Project Management 24*, 349–357.

Aston-Jones, G., Smith, R.J., Sartor, G.C., Moorman, D.E., Massi, L., Tahsili-Fahadan, P., and Richardson, K.A. (2010). Lateral hypothalamic orexin/hypocretin neurons: A role in reward-seeking and addiction. *Brain Research 1314*, 74–90.

Austin, R., and Luecke, R. (2011). *L'essentiel pour manager un projet*. Paris: Ed. Les Echos.

Babiak, P., and Hare, R.D. (2006). *Snakes in Suits When Psychopaths Go to Work*, 1st edn. New York: HarperCollins.

Bagazonzya, H.K, Devkota, K., Mahul, O., and Stutley, C.N. (2009). Agricultural insurance feasibility study; Report no: 46521. http://documents.worldbank.org/curated/en/2009/01/16388163/nepal-agricultural-insurance-feasibility-study.

Barbin, L.F.J., and de Oliveira Moraes, R. (2006). Managing conflicts in IT projects in Brazilian companies. *2006 Technology Management for the Global Future: PICMET 2006 Conference, July, 5*, 2322–2329.

Barlevy, G., and Veronesi, P. (2003). Rational panics and stock market crashes. *Journal of Economic Theory 110*(2), 234–263.

Baron, R.M., and Kenny, D.A. (1986). The moderator-mediator variable distinction in social psychological research: Conceptual, strategic, and statistical considerations. *Journal of Personality and Social Psychology 51*(6), 1173–1182.

Barrick, M., Mount, M., and Judge, T. (2001). Personality and performance at the beginning of the new millennium: What do we know and where do we go next? *Personality and Performance 9*, 9–30.

Bear, M.F., Connors, B.W., and Paradiso, M.A. (1996). *Neuroscience: Exploring the Brain*. Philadelphia, PA: Williams and Wilkins.

Bechara, M., and Gupta, A.K. (1999). Trust within the organization: Integrating the trust literature with agency theory and transaction cost economics. *Public Administration Quarterly 23*(2), 177–203.

Bejarano, C., Fuzzell, L., Clay, C., Leonard, S., Shirley, E., and Wysocki, T. (2015). Shared decision making in pediatrics: A pilot and feasibility project. *Clinical Practice in Pediatric Psychology 3*(1), 187–192.

Bell, G.G., Oppenheimer, R.J., and Bastien, A. (2002). Trust deterioration in an international buyer-supplier relationship. *Journal of Business Ethics 36*(1/2), 65–78.

Belout, A., and Gauvreau, C. (2004). Factors influencing project success: The impact of human resource management. *International Journal of Project Management 22*(1), 1–11.

Berg, M.E., and Karlsen, J.T. (2007). Mental models in project management coaching. *Engineering Management Journal 19*(3), 3–13.

Berssaneti, F.T., and Carvalho, M.M. (2015). Identification of variables that impact project success in Brazilian companies. *International Journal of Project Management 33*, 638–649.

Berthoud, H.R., and Münzberg, H. (2011). The lateral hypothalamus as integrator of metabolic and environmental needs: From electrical self-stimulation to opto-genetics. *Physiology and Behavior 104*, 29–39.

Besanko, D., Doraszelski, U., and Kryukov, Y. (2011). The economics of predation: That drives pricing when there is learning-by-doing? *Twelfth CEPR/JIE Conference on Applied Industrial Organization.* Tel Aviv; 24–27 May.

Birch, D.G.W., and McEvoy, N.A. (1992). Risk analysis for information systems. *Journal of Information Technology 7*, 44–53.

Birnholtz, J., Dixon, G., and Hancock, J. (2012). Distance, ambiguity and appropriation: Structures affording impression management in a collocated organization. *Computers in Human Behavior 28*(3), 1028–1035.

Biró, M., Deák, C., Ivanyos, J., and Messnarz, R. (2006). From compliance to business success: Improving outsourcing service controls by adopting external regulatory requirements. *Software Process Improvement and Practice 11*(3), 239–249.

Black, C., Akintoye, A., and Fitzgerald, E. (2000). Analysis of success factors and benefits of partnering in construction. *International Journal of Project Management 18*(6), 423–434.

Blaikie, P.T., Cannon, I.D., and Wisner, B. (1994*). At Risk: Natural Hazards, People's Vulnerability, and Disasters*. London: Routledge.

Blake, R.R., and Mouton, J.S. (1964). *The Managerial Grid: The Key to Leadership Excellence*. Houston, TX: Gulf Publishing.

Blanchard, R.J., and Blanchard, D.C. (1989). Antipredator and defensive behaviors in a visible burrow system. *Journal of Comparative Psychology 103*, 70–82.

Bogataj, D., and Bogataj, M. (2007). Measuring the supply chain risk and vulnerability in frequency space. *International Journal of Production Economics 108*(1/2), 291–301.

Bollen, K., and Lennox, R. (1991). Conventional wisdom on measurement: A structural equation perspective. *Psychological Bulletin 110*(2), 305–314.

Bolz, U. (2012). *Guide pour les partenariats public-privé (PPP) dans le domaine de la cyberadministration et des TIC*. Editor: Direction opérationnelle E-Government Suisse, Unité de pilotage informatique de la Confédération UPIC. Switzerland.

Botham, M.S., Kerfoot, C.J., Louca, V., and Krause, J. (2006). The effects of different predator species on antipredator behavior in the Trinidadian guppy, Poecilia reticulata. *Naturwissenschaftern 93*, 431–439.

Bourgoin, D. *Manager, Project Management Office/Energy, Mining and Environment Portfolio*. National Research Council Canada/Government of Canada. Interviewed February 2016.

Bourne, L. (2015). *Making Project Work*. Boca Raton, FL: CRC Press.

Bowlby, J. (1973). *Attachment and Loss: Vol. 2. Separation: Anxiety and Anger*. New York: Basic Books.

Bowles, J.T. (2000). Sex, kings, and serial killers and group-selected human traits. *Medical Hypotheses 54*(6), 864–894.

Brannen, J. (ed) (1992). *Mixed Methods: Qualitative and Quantitative Research*. Aldershot, UK: Awebury.

Brassen, S., Gamer, M., and Büchel, C. (2011). Anterior cingulate activation is related to a positivity bias and emotional stability in successful aging. *Biological Psychiatry, 70,* 131–137.

Brett, J., Behfar, K., and Kern, M.C. (2006). Managing multicultural teams. *Harvard Business Review,* November, 85–97.

Brewer, J., and Hunter, A. (1989). *Multimethod Research: A Synthesis of Styles.* Newbury Park, CA: Sage Publications.

Brière, S., and Proulx, D. (2013). The success of an international development project: Lessons drawn from a case between Morocco and Canada. *International Review of Administrative Sciences 79*(1), 165–186.

Brière, S., Proulx, D., Flores, O.N., and Laporte, M. (2015). Competencies of project managers in international NGO's: Perceptions of practitioners. *International Journal of Project Management 33,* 116–125.

Brière, S., Proulx, D., Hondeghem, A., and Steen, T. (2013). The success of an international development project: Lessons drawn from a case between Morocco and Canada. *International Review of Administrative Sciences 79*(1), 165–186.

Briscoe, G., Dainty, A.R.J., and Millett, S. (2001). Construction supply chain partnerships: Skills, knowledge and attitudinal requirements. *European Journal of Purchasing and Supply Management 7*(2), 243–255.

Brock, J.K.U., Johnson, J.E., and Zhou, J.Y. (2011). Does distance matter for internationally-oriented small firms? *Industrial Marketing Management 40*(3), 384–394.

Brooks, B.W., and Rose, R.L. (2008). The influences of matched versus mismatched negotiation orientations on negotiating processes and outcomes. *Journal of Marketing Theory and Practice 16*(3), 199–217.

Brucks, W., Reips, U., and Ryf, B. (2007). Group norms, physical distance, and ecological efficiency in common pool resource management. *Social Influence 2*(2), 112–135.

Bstieler, L. (2006). Trust formation in collaborative new product development. *Journal of Production and Innovation Management 23,* 56–72.

Bstieler, L., and Hemmert, M. (2010). Increasing learning and time efficiency in interorganizational new product development teams. *Journal of Product Innovation Management 27*(4), 485–499.

Buchanan, D.A. (1991). Beyond content and control: Project vulnerability and the process agenda. *International Journal of Project Management 9*(4), 233–239.

Buchanan, D.A., and Bryman, A. (2007). Contextualizing methods choice in organizational research. *Organizational Research Methods 10*(3), 483–501.

Campbell, M.C., and Kirmani, A. (2000). Consumers' use of persuasion knowledge: The effects of accessibility and cognitive capacity on perceptions of an influence agent. *Journal of Consumer Research 27,* 69–83.

Campello, M., and Graham, J.R. (2013). Do stock prices influence corporate decisions? Evidence from the technology bubble. *Journal of Financial Economics 107*(1), 89–110.

Cardinal, R.N., Parkinson, J.A., Hall, J., and Everitt, B.J. (2002). Emotion and motivation: The role of the amygdala, ventral striatum, and prefrontal cortex. *Neuroscience Biobehavioral Review, 26(3),* 321–52.

Carmel, E., and Agarwal, R. (2001). Tactical approaches for alleviating distance in global software development. *IEEE Software 18*(2), 22–29.

Carney, D.R., Cuddy, A.J.C., and Yap, A.J. (2010). Power posing: Brief nonverbal displays affect neuroendocrine levels and risk tolerance. *Psychological Science 20*(10), 1–6.

Carstensen, L.L. (2006). The influence of a sense of time on human development. *Science 312*(5782), 1913–1915.

Cartlidge, D.P. (2006). *Public Private Partnerships in Construction*. London, New York: Taylor & Francis.

CBC News (2013). Ex-McGill hospital boss Arthur Porter arrested in Panama. CBC News Montreal. May 27. http://www.cbc.ca/news/canada/Montréal/ex-mcgill-hospital-boss-arthur-porter-arrested-in-panama-1.1302252.

Chambers, R. (1983). *Rural Development: Putting the Last First*. Essex, UK: Longmans Scientific and Technical Publishers; New York: Wiley.

Charles, S.T., and Piazza, J.R. (2009). Age differences in affective well-being: Context matters. *Social and Personality Psychology Compass 3*(5), 711–724.

Chen, H.L. (2015). Performance measurement and the prediction of capital project failure. *International Journal of Project Management 33*, 1393–1404.

Chen, I.J., and Popovich, K. (2003). Understanding customer relationship management (CRM); People, process and technology. *Business Process Management Journal 9*(5), 672–688.

Chen, P., and Partington, D. (2004). An interpretive comparison of Chinese and Western conceptions of relationships in construction project management work. *International Journal of Project Management 22*(5), 397–406.

Cheung, S.O., Ng, T.S.T., Wong, S.P., and Suen, H.C.H. (2003). Behavioral aspects in construction partnering. *International Journal of Project Management 21*(5), 333–343.

Chevrier, S. (2003). Cross-cultural management in multinational project groups. *Journal of World Business 38*(2), 141–149.

Christie, R., and Geis, F.L. (1970). *Studies in Machiavellianism*. New York: Academic Press.

Chua, R.Y.J., Morris, M.W., and Mor, S. (2012). Collaborating across cultures: Cultural metacognition and affect-based trust in creative collaboration. *Organizational Behavior and Human Decision Processes 118*(2), 116–131.

Clark, K.B., and Wheelwright, S.C. (1992). Organizing and leading "heavyweight" development teams. *California Management Review 34*(3), 9–28.

Clarke, A. (1999). A practical use of key success factors to improve the effectiveness of project management. *International Journal of Project Management 17*(3), 139–145.

Cleland, D.I. (1999). *Project Management: Strategic Design and Implementation*. New York; Montréal: McGraw-Hill.

Cleland, D.I., and Ireland, L.R. (2006). *Project Management: Strategic Design and Implementation* (5th ed.). New York: McGraw-Hill.

Cleland, D.I., and Kerzner, H. (1986). *Engineering Team Management*. New York: Van Nostrand Rheinhold.

Cleland, D.I., and King, W.R. (eds). (1988). *Project Management Handbook*, 2nd edn. New York: Van Nostrand Reinhold.

Cohen, G., and Kudryavtsev, A. (2012). Investor rationality and financial decisions. *Journal of Behavioral Finance 13*(1), 11–16.

Collier, J.E., and Bienstock, C.C. (2009). Model misspecification: Contrasting formative and reflective indicators for a model of e-service quality. *Journal of Marketing Theory and Practice 17*(3), 283–293.

Collins, N.L., and Feeney, B.C. (2004). Working models of attachment shape perceptions of social support: Evidence from experimental and observational studies. *Journal of Personality and Social Psychology 87*(3), 363–383.

Collins, N.L., and Miller, L.C. (1994). Self-disclosure and liking: A meta-analytic review. *Psychological Bulletin 116*(3), 457–475.

Consoli, G.G.S. (2006). Conflict and managing consortia in private prison projects in Australia: Private prison operator responses. *International Journal of Project Management 24*(1), 75–82.

Cooke, H.S., and Tate, T. (2010). *The McGraw-Hill 36-Hour Project Management Course.* New York: McGraw-Hill.

Cooke-Davies, T. (2002). The "real" success factors on projects. *International Journal of Project Management 20*(3), 185–190.

Corriveau, G. (2007). *Exceller dans la gestion de projet.* Québec: Ed. Transcontinental et Fondation de l'entrepreneurship.

Cossette, P., and Lapointe, A. (1997). A mapping approach to conceptual models: The case of macroeconomic theory. *Revue Canadienne des Sciences de l'Administration 14*(1), 41–51.

Costa, P.T. Jr., and McRae, R.R. (1992). Revised NEO Personality Inventory (NEO-PI-R) and NEO Five-Factor Inventory (NEO-FFI) professional manual. Odessa, FL: Psychological Assessment Resources.

Côté, S., and Bouchard, S. (2005). Documenting the efficacy of virtual reality exposure with psychophysiological and information processing measures. *Applied Psychophysiology and Biofeedback 30*(3), 217–232.

Courtot, H. (1998). *La gestion des risques dans les projets.* Paris: Ed. Economica.

Coviello, N.E., and Brodie, R.J. (2001). Contemporary marketing practices of consumer and business-to-business firms: How different are they? *Journal of Business and Industrial Marketing 16*(5), 382–400.

Cowan, C.E. (1991). Strategy for partnering in the public sector. In *Preparing for Construction in the 21st Century.* Edited by L.M. Chang, 721–726. Cambridge, MA: ASCE.

Craske, M.G., Mohlman, J., Yi, J., Glover, D., and Valeri, S. (1995). Treatment of claustrophobias and snake/spider phobias: Fear of arousal and fear of context. *Behavior Research and Therapy 33*(2), 197–203.

Creasy, T., and Anantatmula, V.S. (2013). From every direction: How personality traits and dimensions of project managers can conceptually affect project success. *Project Management Journal 44*(6), 36–51.

Creswell, J.W. (1994). *Research Design: Qualitative and Quantitative Approach.* Thousand Oaks, CA: Sage Publications.

Crosby, L.A., Evans, K.R., and Cowles, D. (1990). Relationship quality in services selling: An interpersonal influence perspective. *Journal of Marketing 54*, 68–80.

Crosby, P. (1989). *Let's Talk Quality.* New York: McGraw-Hill.

Dainty, A.R.J., Bryman, A., Price, A.D.F., Greasley, K., Soetanto, R., and King, N. (2005). Project affinity: The role of emotional attachment in construction projects. *Construction Management and Economics 23*(3), 241–244.

Dalal, A.F. (2012). *The 12 Pillars of Project Excellence: A Lean Approach to Improving Project Excellence.* Boca Raton, FL: CRC Press.

Danis, M.A., and Pennington-Cross, A. (2008). The delinquency of subprime mortgages. *Journal of Economics and Business 60*(1), 67–90.

Davies, A., and Brady, T. (2000). Organisational capabilities and learning in complex product systems: Towards repeatable solutions. *Research Policy 29*(7–8), 931–953.

De Bondt, W.F.M., and Thaler, R. (1985). Does the stock market overreact? *Journal of Finance 15*(3), 793–805.

De Bondt, W.P.M. (1993). Betting on trends: Intuitive forecasts of financial risk and return. *International Journal of Forecasting 9*(3), 355–371.

De Brentani, U., and Kleinschmidt, E.J. (2004). Corporate culture and commitment: Impact on performance of international new product development programs. *Journal of Product Innovation Management 21*(5), 309–333.

De Fraja, G. (2009). The origin of utility: Sexual selection and conspicuous consumption. *Journal of Economic Behavior and Organization.* doi:10.1016/j.jebo.2009.05.019

De Long, J.B., Shleifer, A., Summers, L., and Waldmann, R.J. (1990). Positive feedback investment strategies and destabilizing rational speculation. *Journal of Finance 45*(2), 375–395.

del Caño, A. (1992). Continuous project feasibility study and continuous project risk assessment. *International Journal of Project Management 10*(3), 165–170.

Delahaye, J.P. (2015). Comment jouer parfaitement au poker. *Pour la science 453,* 78–83.

Deming, W.E. (n.d.). https://www.deming.org/.

Deng, X., Pheng, L.S., and Zhao, X. (2014). Project system vulnerability to political risks in international construction projects: The case of Chinese contractors. *Project Management Journal 45*(2), 20–33.

Déry, É. (B.A.A.). Deputy Vice-president, Business Development, BDC, Gatineau. Interviewed March 2015.

Devaux, S.A. (2015). *Total Project Control: A Practioner's Guide to Managing Projects as Investments,* 2nd edn. Boca Raton, FL: CRC Press.

Deveau, J.-P. President *Les Algues acadiennes.* Interviewed March 1st, 2016.

Diallo, A., and Thuillier, D. (2005). The success of international development projects, trust and communication: An African perspective. *International Journal of Project Management 23*(3), 237–252.

Diamantopoulos, A., Riefler, P., and Roth, K.P. (2008). Advancing formative measurement models. *Journal of Business Research 61*(12), 1203–1218.

Diamantopoulos, A., and Siguaw, J.A. (2006). Formative versus reflective indicators in organizational measure development: A comparison and empirical illustration. *British Journal of Management 17*(4), 263–282.

Diamantopoulos, A., and Winklhofer, H.M. (2001). Index construction with formative indicators: An alternative to scale development. *Journal of Marketing Research 38*(2), 269–277.

Dillon, S., and Taylor, H. (2015). Employing grounded theory to uncover behavioral competencies of information technology project managers. *Project Management Journal 46*(4), 90–104.

Dingle, J. (1985). Project feasibility and manageability. *Project Management 3*(2), 94–103.

Di Scala, G., Schmitt, P., and Karli, P. (1984). Flight induced by infusion of bicuculline methiodide into periventricular structures. *Brain Research 309*(2), 199–208.

Divjak, B., and Kukec, S.K. (2008). Teaching methods for international project management. *International Journal of Project Management 26,* 251–257.

Doloi, H.K. (2011). Understanding stakeholders' perspective of cost estimation in project management. *International Journal of Project Management 29*(5), 622–636.

Donaldson, T., and Preston, L.E. (1995). The stakeholder theory of the corporation: Concepts, evidence, and implications. *Academy of Management Review 20,* 65–91.

Doney, P.M., and Cannon, J.P. (1997). An examination of the nature of trust in buyer-seller relationships. *Journal of Marketing 61*(2), 35–51.

Dong, L., and Glaister, K.W. (2006). Motives and partner selection criteria in international strategic alliances: Perspectives of Chinese firms. *International Business Review 15*(6), 577–600.

Dontenwill, E., and Reynaud, E. (2005). Le rôle de l'approche par les parties prenantes dans l'initiation d'une politique de développement durable: étude de cas d'une entreprise de transport urbain. *Journée du développement durable, AIMS*. France: IAE-Aix en Provence, May 11th.

Douhou, S. and Magnus, J.R. (2012). Peer reporting and the perception of fairness. *De Economist 160*, 289–310.

Dulaimi, M.F., Ling, F.Y.Y., and Bajracharya, A. (2003). Organizational motivation and inter-organizational interaction in construction innovation in Singapore. *Construction Management Economy 21*(3), 307–318.

Dulewicz, V., and Herbert, P. (1999). Predicting advancement to senior management from competencies and personality data: A seven-year follow-up study. *British Journal of Management 10*(1), 13–22.

Duncan, T., and Moriarty, S.E. (1998). A communication-based marketing model for managing relationships. *Journal of Marketing 62*(2), 1–13.

East, R., Uncles, M.D., and Lomax, W. (2013). Hear nothing, do nothing: The role of word of mouth in the decision-making of older consumers. *Journal of Marketing Management 30*(7–8), 786–801.

Eibl-Eibesfeldt, I. (1971). *Love and Hate*. New York: Holt, Rinehart and Winston.

Eibl-Eibesfeldt, I., and Sütterlin, C. (1990). Fear, defense and aggression in animals and man: Some ethological perspectives. In *Fear and Defense*. Edited by P.F. Brain, S. Parmigiani, R.J. Blanchard, and R. Mainardi, 381–408. London: Harwood Academic.

Ein-Dor, T., Mikulincer, M., and Shaver, P.R. (2011). Attachment insecurities and the processing of threat-related information: Studying the schemas involved in insecure people's coping strategies. *Journal of Personality and Social Psychology 101*(1), 78–93.

Ekman, P. (1999). Basic emotions. In *Handbook of Cognition and Emotion*. Edited by T. Dalgleish and M. Power, 45–60. New York: Wiley.

Ericson, R.V., and Doyle, A. (2003). The moral risks of private justice: The case of insurance fraud. In *Risk and Morality*, 317–362. Toronto: University of Toronto Press.

Esteves, J., and Pastor, J. (2001). Analysis of critical success factors relevance along SAP implementation phases. In *Proceedings Seventh Americas Conference on Information Systems*. Boston, MA: Association for Information Systems.

Fama, E. (1965). The behavior of stock market prices. *Journal of Business 38*(1), 34–105.

Farrell, L.M. (1995). Feasability analysis of artistic and cultural production projects. *International Journal of Project Management, 13*(6), 395–401.

Feder, A., Nestler, E.J., and Charney, D. (2009). Psychobiology and molecular genetics of resilience. *Natural Review of Neuroscience 10*(6), 446–457.

Fehr, E., Fishbacher, U., and Kosfeld, M. (2005). Neuroeconomics foundations of trust and social preferences: Initial evidence. *Neuroscientific Foundations of Economic Decision-Making 95*(2), 346–351.

Ferguson, N. (2008). *The Ascent of Money: A Financial History of the World*. New York: Penguins Books.

Ferrer, M., Santa, R., and Hyland, P.W. (2010). Relational factors that explain supply chain relationships. *Asia Pacific Journal of Marketing and Logistics 22*(3), 419–440.

Fetchenhauer, D., and Dunning, D. (2009). Do people trust too much or too little? *Journal of Economic Psychology 30,* 263–276.

Fitzsimmons, S.R., Miska, C., and Stahl, G.K. (2011). Multicultural employees: Global business'untapped resource. *Organizational Dynamics 40*(3), 199–206.

Flak, L.S., and Dertz, W. (2005). Stakeholder theory and balanced scorecard to improve IS strategy development in public sector. In *Proceedings of the 28th Seminar on Information Systems Research in Scandinavia.* Kristiansand, Norway. August 6–9, 2005.

Flyvbjerg, B. (2013). Quality control and due diligence in project management: Getting decisions right by taking the outside view. *International Journal of Project Management 31*(5). 760–774, http://arxiv.org/ftp/arxiv/papers/1302/1302.2544.pdf.

Folkman, S., Lazarus, R.S., Dunkel-Schetter, C., Delongis, A., Gruen, P., and Sarason, R.J. (1986). Dynamics of a stressful encounter: Cognitive appraisal, coping, and encounter outcomes. *Journal of Personality and Social Psychology 50*(5), 992–1003.

Foster, S.M., Davis, H.P., and Kisley, M.A. (2013). Brain responses to emotional images related to cognitive ability in older adults. *Psychology and Aging 28,* 179–190.

Fouseki, K. (2006). Conflicting discourses on the construction of the New Acropolis Museum: Past and present. *European Review of History: Revue européenne d'histoire 13*(4), 533–548.

Frankel, R., Whipple, J.S., and Frayer, D.J. (1996). Formal versus informal contracts: Achieving alliance success. *International Journal of Physical Distribution and Logistics Management 26*(3), 47–63.

Fredrikson, M., Annas, P., Fischer, H., and Wik, G. (1996). Gender and age differences in the prevalence of specific fears and phobias. *Behavior Research and Therapy 34*(1), 33–39.

Freeman, R.E. (1984). *Strategic Management: A Stakeholder Approach.* Boston, MA: Pitman.

Frijda, N.H. (1986). *The Emotions.* Cambridge, UK: Cambridge University.

Friston, K.J., Holmes, A.P., Poline, J.B., Grasby, B.J., Willims, C.R., Frackowiak, R.J., and Turner, R. (1995). Analysis of *f*MRI time-series revisited. *NeuroImage 2,* 45–53.

Gagnon-Bartsch, J., Jacob, L., and Speed, T. (2013). Removing unwanted variation from high dimensional data with negative controls. Technical report, 820. Berkeley: Department of Statistics, University of California.

Galbraith, J.K. (1987). *Economics in Perspective: A Critical History.* Boston, MA: Houghton Mifflin.

Galbraith, J.K. (1992). *Brève histoire de l'euphorie financière.* Paris: Éditions du Seuil.

Galbraith, J.K. (1993). *La République des satisfaits: La culture du contentement aux États-Unis.* Paris: Éditions du Seuil.

Galbraith, J.K. (1995). *Voyage dans le temps économique: Témoignage de première main.* Paris: Éditions du Seuil.

Ganesan, S. (1994). Determinants of long-term orientation in buyer-seller relationships. *Journal of Marketing 58*(2), 1–19.

Garbarino, E., and Johnson, M.S. (1999). The different roles of satisfaction, trust, and commitment in customer relationships. *Journal of Marketing 63*(2), 70–87.

Garbarino, E., and Slonim, R. (2003). Interrelationships and distinct effects of internal reference prices on perceived expensiveness and demand. *Psychology et Marketing 20*(3), 227–238.

Garel, G., Giard, V., and Midler, C. (2005). *Management des projets et gestion des ressources humaines.* http://www.lamsade.dauphine.fr/~giard/2001-05.pdf.

Gayraud, J.F. (2011). *La grande fraude. Crime, subprimes et crises financières.* Paris: Odile Jacob.

Genus, A. (1997). Unstructuring incompetence: Problems of contracting, trust and the development of the channel tunnel. *Technology Analysis and Strategic Management 9*(4), 419–436.

Gherardi, F. (2006). Fighting behavior in hermit crabs: The combined effect of resource-holding potential and resource value in Pagurus longicarpus. *Behavioral Ecology and Sociobiology 59*, 500–510.

Gidel, T., and Zonghero, W.W. (2006). *Management de projet 1.* Paris: Ed. Lavoisier.

Girard, André. Former General Manager, Alcan. Interviewed in October 2015.

Gjerstad, S., and Smith, V.L. (2009). Monetary policy, credit extension, and housing bubbles: 2008 and 1929. *Critical Review: A Journal of Politics and Society 21*(2–3), 269–300.

Glass, S.J., and Newman, J.P. (2009). Emotion processing in the criminal psychopath: The role of attention in emotion-facilitated memory. *Journal of Abnormal Psychology 118*(1), 229–234.

Globerman, S., and Nielsen, B.B. (2007). Equity versus non-equity international strategic alliances involving Danish firms: An empirical investigation of the relative importance of partner and host country determinants. *Journal of International Management 13*(4), 449–471.

Gneezy, U. (2005). Deception: The role of consequences. *American Economic Review 95*(1), 384–394.

Goldberg, I., Harel, M., and Malach R. (2006). When the brain loses its self: Prefrontal inactivation during sensorimotor processing. *Neuron 50*(2), 329–339.

Goldberg, L.R. (1990). An alternative "description of personality": The big-five factor structure. *Journal of Personality and Social Psychology 59*, 1216–1229.

Goldfried, M.R., and Davison, G.C. (1994). *Clinical Behavior Therapy.* New York: Wiley, 352 p.

Gond, J.P., and Mercier, S. (2005). *Les théories des parties prenantes: une synthèse critique de la littérature.* Toulouse, France: Université des sciences sociales Toulouse (Social Sciences University of Toulouse): LIRHE (Interdisciplinary Laboratory for Research on Human Resources and Employment).

Gould, F.E., and Joyce, N.E. (2000). *Construction Project Management.* NJ: Prentice Hall.

Graham, J.H. (1996). Machiavellian project managers: Do they perform better? *International Journal of Project Management 14*(2), 67–74.

Graham, R.J., and Englund, R.L. (2003). *Creating an Environment for Successful Projects,* 2nd edn. San Francisco, CA: Jossey-Bass.

Grant, R.M. (1991). The Resource-based theory of competitive advantage: Implication for strategy formulation. *California Management Review 33*, 114–135.

Gray, J.A., and McNaughton, N. (2000). *The Neuropsychology of Anxiety.* New York: Oxford Medical Publications.

Grayson, K. (2007). Friendship versus business in marketing relationships. *Journal of Marketing 71*, 121–139.

Gregg, T.R., and Siegel, A. (2001). Brain structures and neurotransmitters regulating aggression in cats: Implications for human aggression. *Progress in Neuro-Psychopharmacology and Biological Psychiatry 25*, 91–140.

Grönroos, C. (1994). From marketing mix to relationship marketing: Towards a paradigm shift in marketing. *Management Decision 32*(2), 4–20.

Gross, J.J. (ed). (2013). *Handbook of Emotion Regulation..* New York: Guilford Press.

Guardian. (2013). Jellyfish clog pipes of Swedish nuclear reactor forcing plant shutdown. 1 October. http://www.theguardian.com/world/2013/oct/01/jellyfish-clog-swedish-nuclear-reactor-shutdown. Accessed Sept. 2, 2015.

Gundlach, G., and Cannon, J. (2010). "Trust but verify"? The performance implications of verification strategies in trusting relationships. *Journal of the Academy of Marketing Science 38*(4), 399–417.

Gurviez, P., and Korchia, M. (2002). Proposition d'une échelle de mesure multidimensionnelle de la confiance dans la marque. *Recherche et Applications en Marketing 17*(3), 41–61.

Guyana—Agricultural insurance component: Pre-feasibility study report. (2010). The World Bank. Report no: 75652. http://documents.worldbank.org/curated/en/2010/05/17394049/guyana-agricultural-insurance-component-pre-feasibility-study-report.

Haji-Kazemi, S., Andersen, B., and Klakegg, O.J. (2015). Barriers against effective responses to early warning signs in projects. *International Journal of Project Management 33*, 1068–1083.

Halman, J., and Braks, B.F.M. (1999). Project alliancing in the offshore industry. *International Journal of Project Management 17*(2), 71–76.

Hammer, M., Bennett, M.J., and Wiseman, R. (2003). Measuring intercultural sensitivity: The intercultural development inventory. *International Journal of Intercultural Relations 27*(4), 421–443.

Hampson, K.D., and Kwok, T. (1997). Strategic alliances in building construction: A tender evaluation tool for the public sector. *Journal of Construction Procurement 3*(1), 28–41.

Han, Y., Shaikh, M.B., and Siegel, A. (1996). Medial amygdaloid suppression of hostile attack behavior in the cat: II. Role of GABAergic pathway from the medial amygdala to the lateral hypothalamus. *Brain Research 716*(1), 72–83.

Hare, R.D. (1991). *The Hare Psychopathy Checklist: Revised.* Toronto: MultiPealth Systems.

Harris, G. (1993) The post-capitalist executive: An interview with Peter F. Drucker, *Harvard Business Review 71*(3), 114–122.

Harrison, F., and Lock, D. (2004). *Advanced Project Management: A Structured Approach*, 4th edn. Aldershot, UK: Gower.

Hauck, A.J., Walker, D.H.T., Hampson, K.D., and Peters, R.J. (2004). Project alliancing at National Museum of Australia: Collaborative process. *Journal of Construction Engineering Management 130*(1), 143–152.

Hedden, T., and Park, D.C. (2003). Contributions of source and inhibitory mechanisms to age-related retroactive interference in verbal working memory. *Journal of Experimental Psychology 132*, 93–112.

Hinds, A.L., Woody, E.Z., Drandic, A., Schmidt, L.A., Van Ameringen, M., Coroneos, M., and Szechtman, H. (2010). The psychology of potential threat: Properties of the security motivation system. *Biological Psychology 85*, 331–337.

Hobday, M. (2000). The project-based organization: An ideal form for managing complex products and systems? *Research Policy 29*(7/8), 871–894.

Hodgins, B.W, and Benidickson, J. (1978). Resource Management Conflict in the Temagami Forest, 1898 to 1914. (Editors: Cook, Terry and Lacelle, Claudette). *Historical Papers 13*(1), 148–175.

Hoff, H., Beneventi, H., Galta, K., and Wik, G. (2009). Evidence of deviant emotional processing in psychopathy: An *f*MRI case study. *International Journal of Neuroscience 119*, 857–878.

Hooge, S., and Dalmasso, C. (2015). Breakthrough R&D stakeholders: The challenges of legitimacy in highly uncertain projects. *Project Management Journal 46*(6), 54–73.

Houser, D., Schunk, D., and Winter, J. (2010). Distinguishing trust from risk: An anatomy of the investment game. *Journal of Economic Behavior and Organization 74*, 72–81.

Howell, J.M., Neufeld, D.J., and Avolio, B.J. (2005). Examining the relationship of leadership and physical distance with business unit performance. *Leadership Quarterly 16*(2), 273–285.

Hugo, V. (2001). *Quatrevingt Treize*. Paris: Le Livre de Poche.

Idrissou, L., van Paassen, A., Aarts, N., Vodouhè, S., and Leeuwis, C. (2013). Trust and hidden conflict in participatory natural resources management: The case of the Pendjari national park (PNP) in Benin. *Forest Policy and Economics 27*, 65–74.

Isaacowitz, D.M., Allard, E.S., Murphy, N.A., and Schlangel, M. (2009). The time course of age-related preferences toward positive and negative stimuli. *Journal of Gerontology: Psychological Sciences 64B*(2), 188–192.

Jaafari, A. (1990). Management know-how for project feasibility studies. *International Journal of Project Management 8*(3), 167–172.

Jap, S.D., Manolis, C., and Weitz, B.A. (1999). Relationship quality and buyer-seller interactions in channels of distribution. *Journal of Business Research 46*, 303–313.

Jarvenpaa, S.L., and Leidner, D.E. (1998). Communication and Trust in Global Virtual Teams. *Journal of ComputerMediated Communication 3*(4), 1–10.

Jarvis, C.B., MacKenzie, S.B., and Podsakoff, P.M. (2003). A critical review of construct indicators and measurement model misspecification in marketing and consumer research. *Journal of Consumer Research 30*(2), 199–218.

Jepsen, A.L., and Eskerod, P. (2009). Stakeholder analysis in projects: Challenges in using current guidelines in the real world. *International Journal of Project Management 27*(4), 335–343.

Jodl, H.G. (2012). Tunnel projects require cooperation/Tunnelprojekte brauchen Kooperation. *Geomechanics and Tunnelling 5*(6), 702–707.

Johnson, M.K., Rustichini, A., and MacDonald, A.W., III (2009). Suspicious personality predicts behavior on a social decisionmaking task. *Personality and Individual Differences, 47*, 3035.

Johnsson, J.I., Rydeborg, A., and Sundström, L.F. (2004). Predation risk and the territory value of cover: An experimental study. *Journal of Behavioral and Sociobiology Studies 56*, 388–392.

Jones, G.R., and George, J.M. (1998). The experience and evolution of trust: Implications for cooperation and teamwork. *Academy of Management Review 23*(3), 531–546.

Joslin, R., and Müller, R. (2015). Relationships between a project management methodology and project success in different project governance contexts. *International Journal of Project Management 33*, 1377–1392.

Jung, J.Y., and Wang, Y.J. (2006). Relationship between total quality management (TQM) and continuous improvement of international project management (CIIPM). *Technovation 26*(5–6), 716–722.

Juran, J. (n.d.) http://www.juran.com/.

Kadefors, A. (2005). Fairness in interorganizational project relations: Norms and strategies. *Construction Management and Economics 23*(8), 871–878.

Kahneman, D., and Tversky, A. (1979). Prospect theory: An analysis of decision under risk. *Econometrica 47*(2), 263–292.

Kamihigashi, T. (2008). The spirit of capitalism, stock market bubbles and output fluctuations. *International Journal of Economic Theory 4*(1), 3–28.

Kandel, E.R., Schwartz, J.H., and Jessell, T.M. (2000). *Principles of Neural Science*, 4th edn. New York: McGraw Hill.

Kasparov. G. (2006). *Échec et mat!* Paris: L'Archipel.

Kaye, K. (1996). When the family business is a sickness. *Family Business Review 9*(4), 347–368.

Kealey, D.J., Protheroe, D.R., MacDonald, D., and Vulpe, T. (2005). Re-examining the role of training in contributing to international project success: A literature review and an outline of a new model training program. *International Journal of Intercultural Relations 29*(3), 289–316.

Keil, M., Tan, B.C.Y., Wie, K.K., Saarinen, T., Tuunainen, V., and Wassenaar, A. (2000). A cross-cultural study on escalation of commitment behavior in software projects. *MIS Quarterly 24*(2), 299–325.

Kelly, P.M., and Adger, W.N. (2000). Theory and practice in assessing vulnerability to climate change and facilitating adaptation. http://foehn.colorado.edu/nome/HARC/Readings/Kelly.pdf.

Keltner, D., and Gross, J.J. (1999). Functional accounts of emotions. In *Handbook of Cognition and Emotion*. Edited by T. Dalgleish and M. Power, 467–480. New York: Wiley. http://www.personal.kent.edu/~dfresco/CBT_Readings/keltner_&_gross.pdf.

Kerzner, H. (1995). *Project Management: A System Approach to Planning, Scheduling, and Controlling*. New York: Van Nostrand Reinhold.

Keynes, J.M. (2006). *The General Theory of Employment, Interest and Money*. UK: Atlantic Publishers and Distributors (P) Limited. http://cas.umkc.edu/economics/people/facultypages/kregel/courses/econ645/winter2011/generaltheory.pdf.

Khang, D.B., and Moe, T.L. (2008). Success criteria and factors for international development projects: A life-cycle-based framework. *Project Management Journal 39*(1), 72–84.

Khodakarami, V., and Abdi, A. (2014). Project cost risk analysis: A Bayesian networks approach for modeling dependencies between cost items. *International Journal of Project Management 32*, 1233–1245.

Kindt, M., and Brosschot, J.F. (1997). Phobia-related cognitive bias for pictorial and linguistic stimuli. *Journal of Abnormal Psychology 106*(4), 644–648.

Kloppenborg, T., and Opfer, W. (2002). The current state of project management research: Trends, interpretations, and predictions. *Project Management Journal 33*(2), 5–18.

Klorman, R., Weerts, T.C., Hastings, J.E., Melaied, B.G., and Lang, L.J. (1974). Psychometric description of some specific-fear questionnaires. *Behavior Therapy 5*, 401–409.

Koutstaal, W. (2003). Older adults encode: But do not always use—Perceptual details: Intentional versus unintentional effects of detail on memory judgments. *Psychological Science 14*, 189–193.

Kryla-Lighthall, N., and Mather, M. (2009). The role of cognitive control in older adults' emotional well-being. In *Handbook of Theories of Aging*, 2nd edn. Edited by V. Berngtson, D. Gans, N. Putney, and M. Silverstein, 323–344. New York: Springer Publishing.

Kumar, K., van Fenema, P.C., and Von Glinow, M.A. (2004). Intense collaboration in globally distributed teams: Evolving patterns of dependencies and coordination. *Erim Report Series Research In Management* (ERIM Report Series): Reference number ERS-2004-052-LIS, 1–37.

Kumaraswamy, M.M., and Anvuur, A. (2008). Selecting sustainable teams for PPP projects. *Building and Environment 43*(6), 999–1009.

Kunzmann, U., Kappes, C., and Wrosch, C. (2014). Emotional aging: A discrete emotions perspective. *Frontiers in Psychology 5*, 380.

La Banque Mondiale (n.d.). Croissance du PIB (% annuel). http://donnees.banquemondiale.org/indicateur/NY.GDP.MKTP.KD.ZG.

Lam, K., Liu, T., and Wong, W.K. (2010). A pseudo-Bayesian model in financial decision making with implications to market volatility, under- and overreaction. *European Journal of Operational Research 203*(1), 166–175.

Lambert-Pandraud, R., Laurent, G., and Lapersonne, E. (2005). Repeat purchasing of new automobiles by older consumers: Empirical evidence and interpretations. *Journal of Marketing 69*(April), 97–113.

Lang, P.J., Bradley, M.M., and Cuthbert, B.N. (2008). International Affective Picture System (IAPS): Affective ratings of pictures and instructions manuel. Technical Report A-8, Gainesvilles, FL: University of Florida.

Lazarus, R.S. (1991a). *Emotion and Adaptation*. Oxford, UK: Oxford University Press.

Lazarus, R.S. (1991b). Progress on a cognitive-motivational-relational theory of emotion. *American Psychologist 46*(8), 819–834.

Le Boterf, G. (2005). *Construire les compétences individuelles et collectives*. Paris: Ed. d'Organisation.

Le Moigne, J.L. (1990). *La modélisation des systèmes complexes*. Paris: Dunod.

Leclerc, C.M., and Kensinger, E.A. (2011). Neural processing of emotional pictures and words: A comparison of young and older adults. *Developmental Neuropsychology 36*(4), 519–538.

Lee-Kelley, L., and Sankey, T. (2008). Global virtual teams for value creation and project success: A case study. *International Journal of Project Management 26*(1), 51–62.

Levenson, R.W. (1994). Human emotions: A functional view. In *The Nature of Emotion: Fundamental Questions*. Edited by P. Ekman and R.J. Davidson, 123–126. New York: Oxford University Press.

Lewicki, R.J., McAllister, D.J., and Bies, R.J. (1998). Trust and distrust: New relationships and realities. *Academy of Management Review 23*(3), 438–458.

Lewicki, R.J., Weiss, S.E., and Lewin, D. (1992). Models of conflict, negotiation and third party intervention: A review and synthesis. *Journal of Organizational Behavior 13*, 209–252.

Lewis, M.D. (2005). Bridging emotion theory and neurobiology through dynamic systems modeling. *Behavioral and Brain Sciences 28*, 169–245.

Lines, B.C., Sullivan, K.T., Smithwick, J.B., and Mischung, J. (2015). Overcoming resistance to change in engineering and construction: Change management factors for owner organizations. *International Journal of Project Management 33*, 1170–1179.

Liu, J., Meng, F., and Fellows, R. (2015). An exploratory study of understanding project risk management from the perspective of national culture. *International Journal of Project Management 33*, 564–575.

Liu, J.Y.C., and Yuliani, A.R. (2016). Differences between clients' and vendors' perceptions of IT outsourcing risks: Project partnering as the mitigation approach. *Project Management Journal 47*(1), 45–55.

Lj, M., Petrović, T.D.Č., Mihić, M.M., Obradović, V.L., and Bushuyev, S.D. (2015). Project success analysis framework: A knowledge-based approach in project management. *International Journal of Project Management 33*, 772–783.

Lösel, F., and Schmucker, M. (2004). Psychopathy, risk taking, and attention: A differentiated test of the somatic marker hypothesis. *Journal of Abnormal Psychology 113*(4), 522–529.

Loufrani-Fadida, S. (2008). Management des compétences et organisation par projets: une mise en évidence des leviers de gestion conjointe. *La Revue des Sciences de Gestion 231–232*, 73.

Lovallo, D., and Kahneman, D. (2003). Delusions of success: How optimism undermines executives' decisions. *Harvard Business Review 7*, 56–63.

Low, W.W., Abdul-Rahman, H., and Zakaria, N. (2015). The impact of organizational culture on international bidding decisions: Malaysia context. *International Journal of Project Management 33*, 917–931.

Lupien, S.J., McEwen, B.S., Gunnar, M.R., and Heim, C. (2009). Effects of stress throughout the lifespan on the brain, behavior and cognition: An overview. *Nature Reviews 10*, 434–445.

Lusthaus, C., Adrien, M., Anderson, G., Carden, F., and Montalvan, G.P. (2002). *Organizational Assessment: A Framework for Improving Performance*. Ottawa, CA: International Development Research Centre.

Lux, T. (1995). Herd behavior, bubbles and crashes. *Economic Journal 105*(431), 881–896.

Mackenzie, W. (2011). *Large Capital Projects Benchmarking*. Scotland (Edinburgh): Wood Mackenzie Research.

Mackey, M.C., and Glass. L. (1977). Oscillation and chaos in physiological control systems. *Science 197*, 287–289.

MacLaren, V.A., Best, L.A., and Bigney, E.E. (2010). Aggression-hostility predicts direction of defensive responses to human threat scenarios. *Personality and Individual Differences 49*, 142–147.

Magdol, L., Moffitt, T.E., Caspi, A., and Silva, P.A. (1998). Developmental antecedents of partner abuse: A prospective–longitudinal study. *Journal of Abnormal Psychology 107*(3), 375–389.

Mallalieu, L., and Nakamoto, K. (2008). Understanding the role of consumer motivation and salesperson behavior in inducing positive cognitive and emotional responses during a sales encounter. *Journal of Marketing Theory and Practice 16*(3), 183–197.

Manikutty, S. (2000). Family business groups in India: A resource-based view of the emerging trends. *Family Business Review XIII*(4), 279–292.

Mantel, S.J., Meredith, J.R., Shafer, S.M., and Sutton, M.M. (2011). *Project Management in Practice*. Hoboken, NJ: Wiley.

Marks, I.M., and Gelder, M.G. (1966). Different ages of onset in varieties of phobia. *American Journal of Psychiatry 123*(2), 218–221.

Martin, J.W. (2009). *Measuring and Improving Performance: Information Technology Applications in Lean Systems.* Hoboken, NJ: CRC Press.

Martin-Alcazar, F., Romero-Fernandez, P.M., and Sanchez-Gardey, G. (2011). Transforming human resource management systems to cope with diversity. *Journal of Business Ethics 107*(4), 511–531.

Marzouk, M., Amer, O., and El-Said, M. (2013). Feasibility study of industrial projects using SIMOS' procedure. *Journal of Civil Engineering and Management 19*(1), 59–68.

Mather, M. (2012). The emotion paradox in the aging brain. *Annals of New York Academic Sciences 1251*, 33–49.

Mather, M., and Carstensen, L.L. (2005). Aging and motivated cognition: The positivity effect in attention and memory. *Trends in Cognitive Sciences 9*, 496–502.

Mather, M., and Knight, M. (2005). Goal-directed memory: The role of cognitive control in older adults' emotional memory. *Psychology and Aging 20*, 554–570.

May, R.M. (1995). Necessity and change: Deterministic chaos in ecology and evolution. *Bulletin of the American Mathematical Society 32*(3), 291–308.

Mayer, R.C., Davis, J.H., and Schoorman, F.D. (1995). An integrative model of organizational trust. *Academy of Management Review 20*(3), 709–734.

McCullough, M.E., Brandon, A., Orsulak, P., and Akers, L. (2007). Rumination, fear, and cortisol: An in vivo study of interpersonal transgressions. *Health Psychology 26*(1), 126–132.

McFarland, R.G., Challagalla, G.N., and Shervani, T.A. (2006). Influence tactics for effective adaptive selling. *Journal of Marketing 70*, 103–117.

Meloy, J.R. (1997). Predatory violence during mass murder. *Journal of Forensic Sciences 42*, 326–329.

Melville. H. (1851). *Moby-Dick.* London: Harper and Brothers, Publishers.

Mentzer, J.T., Dewitt, W., Keebler, J.S., Min, S., Nix, N.W., Smith, C.D., and Zacharia, Z.G. (2001). Defining supply chain management. *Journal of Business Logistics 22*(2), 1–25.

Meredith, J.R., and Mantel, S.J. (2009). *Project Management: A Managerial Approach,* 7th edn. New York: Wiley.

Mesly, O. (2010). Voyage au cœur de la prédation entre vendeurs et acheteurs: une nouvelle théorie en vente et marketing. Doctoral thesis. Université de Sherbrooke. Bibliothèque nationale, 505 p.

Mesly, O. (2011). *Une façon différente de faire de la recherche en vente et marketing.* Presses de l'Université du Québec. Québec: Presses de l'Université du Québec, 202 p.

Mesly, O. (2013). Detecting financial predators ahead of time? A two-group longitudinal study. *Applied Financial Economics 23*(16), 1325–1336.

Mesly, O. (2014). The core of predation: The predatory core—Finding the neurobiological center of financial predators and preys. *Journal of Behavioral Finance 15*, 214–225.

Mesly, O. (2015a). *Creating Models in Psychological Research.* Cham: Springer Psychology, 125 p.

Mesly, O. (2015b). *Faisabilité de projets: Aspects oubliés de l'analyse –.* Montréal: Presses internationales Polytechnique, 200 p.

Mesly, O. (2015c) The role of physical distance in six interpersonal core competencies in international construction projects. *International Journal of Project Management 33* (2015):1425–1437.

Meyerson, D., Weick, K., and Kramer, R.M. (1996). Swift trust and temporary groups. In Research. Edited by R.M. Kramer and T.R. Tyler. Thousand Oaks, CA: Sage, 166–195.

Mikulincer, M. (1998). Attachment working models and the sense of trust: An exploration of interaction goals and affect regulation. *Journal of Personality and Social Psychology 74*(5), 1209–1224.

Mikulincer, M., Birnbaum, G., Woddis, D., and Nachmias, O. (2000). Stress and accessibility of proximity-related thoughts: Exploring the normative and intraindividual components of attachment theory. *Journal of Personality and Social Psychology 78*, 509–523.

Mikulincer, M., Florian, V., Birnbaum, G., and Malishkevich, S. (2002). The death-anxiety buffering function of close relationships: Exploring the effects of separation reminders on death-thought accessibility. *Personality and Social Psychology Bulletin 28*, 287–299.

Mikulincer, M., and Shaver, P.R. (2007). *Attachment in Adulthood–Structure, Dynamics, and Change.* New York: Guilford press.

Miles, M.B., and Huberman, A.M. (2003). *Analyse des données qualitatives,* 2nd edn. Bruxelles: De Boeck and Larcier.

Miller, R., and Lessard, D. (2000). The strategic management of large engineering projects. USA, *MIT,* 75–78.

Milosevic, D.Z. (2002). Selecting a culturally responsive project management strategy. *Technovation 22,* 493–508.

Ministry of Agriculture and Fisheries. (2011). Jamaica:Weather insurance for the coffee sector feasibility study. The World Bank, Report no: 75653. http://documents.worldbank.org/curated/en/2011/10/17394053/jamaica-weather-insurance-coffee-sector-feasibility-study.

Mobbs, D., Marchant, J.L., Hassabis, D., Saymour, B., Tan, G., Gray. M., Petrovic, P., Dolan, R.J., and Frith, C.D. (2009). From threat to fear: The neural organization of defensive fear systems in humans. *Journal of Neuroscience 29*(39), 12236–12243.

Monk, C.S., Nelson, E.E., McClure, E.B., Mogg, K., Bradley, B.P., Leibenluft, E., and Pine, D.S. (2006). Ventrolateral prefrontal cortex activation and attentional bias in response to angry faces in adolescents with generalized anxiety disorder. *American Journal of Psychiatry 163*(6), 1091–1097.

Montoya, E.R., Terburg, D., Bos, P.A. and van Honk, J. (2012). Testosterone, cortisol and serotonin as key regulators of social aggression: A review and theoretical perspective. *Motivation and Emotion 36,* 65–73.

Morgan, A.B., and Lilienfeld, S.O. (2000). A meta-analytic review of the relation between antisocial behavior and neuropsychological measures of executive function. *Clinical Psychology Review 20*(1), 113–136.

Morgan, R.M., and Hunt, S.D. (1994). The commitment-trust theory of relationship marketing. *Journal of Marketing 58*(3), 20–38.

Morris, P.W.G. (1988). Initiating major projects: The unperceived role of project management. *Proceedings of the 9th Internet World Congress on Project Management, Sept. 4th to 9th. 2,* 801–813.

Morris, P.W.G. (1989). Initiating major projects: The unperceived role of project management. *Project Management 7*(3), 180–185.

Morris, P.W.G., and Hough, G.H. (1987). *The Anatomy of Major Projects*. London, UK: Wiley.

Morris, P.W.G., and Pinto, J.K. (2004). *The Wiley Guide to Managing Projects*. Hoboken, NJ: Wiley.

Moschis, G.P., Mosteller, J., and Fatt, C.K. (2011). Research frontiers on older consumers' vulnerability. *Journal of Consumer Affairs 45*(3), 467–491.

Muchembeld, R. (2008). *Une histoire de la violence*. Paris: L'Univers historique.

Mulcahy, R. (2013). *Rita Mulcahy's PMP Exam Prep*, 8th edn. Minnetonka, MI: RMC Publications.

Musali, P. (1998). Issues, challenges and prospects of collaborative management of protected areas: A case of introducing people's participation in the management of mt. Elgon national park. *East African Geographical Review 20*(2), 80–86.

Napier, N.P., Keil, M., and Tan, F.B. (2009). IT project managers' construction of successful project management practice: A repertory grid investigation. *Information Systems Journal 19*(3), 225–282.

Neuman, L.W. (1994). *Social Research Methods, Qualitative and Quantitative Approaches*. Boston, MA: Allyn and Bacon, A Division of Simon and Schuster.

Neuvel, J.M.M., Jan de Boer, D., and Rodenhuis, W.K.F. (2015). Managing vulnerability: The implementation of vulnerability reduction measures. *Journal of Risk Research 18*(2), 182–198.

Newcombe, R. (2000). The anatomy of two projects: A comparative analysis approach. *International Journal of Project Management 18*(3), 189–199.

Ng, S.T., Wong, Y.M.W., and Wong, J.M.W. (2012). Factors influencing the success of PPP at feasibility stage A tripartite comparison study in Hong Kong. *Habitat International 36*, 423–432.

Nicholson, C.Y., Compeau, L.D., and Sethi, R. (2001). The role of interpersonal liking in building trust in long-term channel relationships. *Academy of Marketing Science 29*(1), 3–15.

Nicolini, D. (2002). In search of "project chemistry." *Construction Management and Economics 20*(2), 67–177.

Nienaber, A.M., Hofeditz, M., and Romeike, P.D. (2015). Vulnerability and trust in leader-follower relationships. *Personnel Review 44*(4), 567–591.

Nili, U., Goldberg, H., Weizman, A., and Dudai, Y. (2010). Fear thou not: Activity of frontal and temporal circuits in moments of real-life courage. *Neuron 66*(June), 949–962.

Norman, W.T. (1963). Toward an adequate taxonomy of personality attributes: Replicated factor structure in peer nomination personality ratings. *Journal of Abnormal and Social Psychology 66*, 574–583.

Nunnally, J.C. Jr. (1970). *Introduction to Psychological Measurement*. New York: McGraw-Hill.

Nunnally, J.C., and Bernstein, I.H. (1994). *Psychometric Theory* (3rd ed.). New York: McGraw-Hill.

Oakley, Graham. V.P. Construction. AF Thériault. Interviewed March 2nd, 2016.

Ochieng, E.G., and Price, A.D. (2009). Framework for managing multicultural project teams. *Engineering, Construction and Architectural Management 16*(6), 527–543.

Odean, T. (1998). Do investors trade too much? *American Economic Review 89*, 1279–1298.

Ogunlana, S., Zafaar, S., Silas, Y., and Olomolaiye, P. (2002). Factors and procedures used in matching project managers to construction projects in Bangkok. *International Journal of Project Management 20*(5), 385–400.

Ohlendorf, A. (2001). Conflict resolution in project management. *Information Systems Analysis MSIS 488*. http://www.umsl.edu/~sauterv/analysis/488_f01_papers/Ohlendorf.htm.

Okunoye, A., Frolick, M., and Crable, E. (2008). Stakeholder influence and ERP implementation in higher education. *Journal of Information Technology Case and Application Research 10*(3), 9–38.

Oliver, R.L. (1980). A cognitive model of the antecedents and consequences of satisfaction decisions. *Journal of Marketing Research (pre-1986) 17*(4), 460–469.

Olivier, L., and Payette, J.F. (2010). *Argumenter son mémoire ou sa thèse*. Québec: Les Presses de l'Université du Québec.

Olsen, R. (2012). Trust: The underappreciated investment risk attribute. *Journal of Behavioral Finance 13*, 308–313.

OPM3 (Organizational Project Management Maturity Model) (2008, 2013). Project Management Institute (PMI). *Organizational project management maturity model (OPM3®)*. Newtown Square, PA: Author.

O'Shaugnessy, W. (1992). *La faisabilité de projet*. Trois-Rivières, Canada: Les Éditions SMG.

Öst, L.G. (1987). Age of onset in different phobias. *Journal of Abnormal Psychology 96*(3), 223–229.

Owusu, R.A., and Welch, C. (2007). The buying network in international project business: A comparative case study of development projects. *Industrial Marketing Management 36*(2), 147–157.

Ozorhon, B., Arditi, D., Dikmen, I., and Birgonul, M.T. (2007). Effect of host country and project conditions in international construction joint ventures. *International Journal of Project Management 25*(8), 799–806.

Palmatier, R.W., Dant, R.P., Grewal, D., and Evans, K.R. (2006). Factors influencing the effectiveness of relationship marketing: A meta-analysis. *Journal of Marketing 70*, 136–153.

Palmer, R., Lindgreen, A., and Vanhamme, J. (2005). Relationship marketing: Schools of thought and future research directions. *Marketing Intelligence and Planning 23*(2/3), 313–330.

Paradi-Guilford, C. (2013). Feasibility study: Microwork for the Palestinian territories. Report no: acs3685. The World Bank. http://documents.worldbank.org/curated/en/2013/02/17732233/feasibility-study-microwork-palestinian-territories.

Parkhe, A. (1993). "Messy" research, methodological predispositions, and theory development in international joint ventures. *Academy of Management Review 18*(2), 227–268.

Patrick, C.J., Cuthbert, B.N., and Lang, P.J. (1994). Emotion in the criminal psychopath: Fear image processing. *Journal of Abnormal Psychology 103* (3), 523–534.

Pearson, A.W., Carr, J.C., and Shaw, J.C. (2008). Toward a theory of familiness: A social capital perspective. *Entrepreneurship Theory and Practice 32*(6), 949–969.

Pelletier, K.R. (1993). Between mind and body: Stress, emotions, and health. In *Mind/Body Medicine*. Edited by D. Goleman and J. Gurin, 19–38. Yonkers, NY: Consumer Reports Books.

Pels, J., Coviello, N.E., and Brodie, R.J. (2000). Integrating transactional and relational marketing exchange: A pluralistic perspective. *Journal of Marketing Theory and Practice 8*(3), 11–20.

Peteraf, M.A. (1993). The cornerstones of competitive advantage: A resource-based view. *Strategic Management Journal 14*, 179–191.

Petrick, J.A. (2011). Sustainable stakeholder capitalism: A moral vision of responsible global financial risk management. *Journal of Business Ethics 99*, 93–109.

Pheng, L.S., and Leong, C.H.Y. (2001). Asian management style versus Western management theories: A Singapore case study in construction project management. *Journal of Managerial Psychology 16*(2), 127–141.

Phillips, J.J., Bothell, T.W., and Snead, G.L. (2002). *The Project Management Scorecard: Improving Human Performance*. Boston, MA: Butterworth-Heinemann.

Phillips, M., Drevets, W.C., Rauch, S.L., and Lane, R. (2003). Neurobiology of emotion perception: The neural basis of normal emotion perception. *Biological Psychiatry 54*(5), 504–514.

Pinto, J.E., Henry, E., Robinson, T.R., and Stowe, J.D.S. (2010). *Equity Asset Valuation*. Hoboken, NJ: Wiley.

Pinto, J.K. (2014). Project management, governance, and the normalization of deviance. *International Journal of Project Management 32*(3), 376–387.

Pinto, J.K., and Mantel, S.J. Jr. (1990). The causes of project failure. *IEEE Transactions on Engineering Management 37*(4), 269.

Pinto, J.K., and Slevin, D.P. (1987). Critical factors in successful project implementation. *IEEE Transactions of Engineering Management 1*, 22–27.

Pinto, J.K., and Slevin, D.P. (1988). Project success: Definitions and measurement techniques. *Project Management Journal 19*(1), 67–72.

Plouffe, C.R., Hulland, J., and Wachner, T. (2009). Customer-directed selling behaviors and performance: A comparison of existing perspectives. *Journal of the Academy of Marketing Science 37*, 422–439.

Plutchik, R. (1980). *Emotion: A Psychoevolutionary Synthesis*. New York: Harper and Row.

Plutchik, R., and Platman, S.B. (1975). Personality connotations of psychiatric diagnoses. *Journal of Nervous and Mental Disease 165*, 418–422.

PM Network. (June 2015). Vol. 29, nbr. 6. Newton Square, PA: PMI, p. 17.

PM Network. (July 2015), Vol. 29, nbr 7. Newton Square, PA: PMI, p. 21.

PM Network. (January 2016). Vol. 30, nbr 1, Newton Square, PA: PMI, p. 9.

PMBOK 3. (2004). *A Guide to the Project Management Body of Knowledge* (PMBOK® Guide), 3rd edn. Newtown Square, PA: Project Management Institute.

PMBOK 4. (2008). *A Guide to the Project Management Body of Knowledge* (PMBOK® Guide), 4th edn. Newtown Square, PA: Project Management Institute.

PMBOK 5. (2013). *A Guide to the Project Management Body of Knowledge* (PMBOK® Guide), 5th edn. Newtown Square, PA: Project Management Institute.

PMBOK 5. (2013). *Guide du corpus des connaissances en management de projet* (Guide PMBOK®), 5th edn. Newtown Square, PA: Project Management Institute.

PMI (Project Management Institute). (2002). *Project Manager Competency Development (PMCD) Framework*. Newton Square, PA: Project Management Institute.

PMI Standard for Portfolio Management (PMI 2013).

PMI Today. (July 2015), Newton Square, PA: PMI, p. 3.

Prahalad, C.K., and Hamel, G. (1990). The core competence of the corporation. *Harvard Business Review 68*(3), 79–91.

Purves, D., Augustine, G.J., Fitzpatrick, D., Hall, W.C., LaManta, A.S., McNamara, J.O., and Williams, S.M. (eds). (2004). *Neuroscience*, 3rd edn. Sunderland, MA: Sinaver Associates.

Qureshi, S.M., and Kang, C.W. (2015). Analysing the organizational factors of project complexity using structural equation modelling. *International Journal of Project Management 33*, 165–176.

Rahman, M.M., and Kumaraswamy, M.M. (2005). Assembling integrated project teams for joint risk management. *Construction Management Economy 23*(4), 365–375.

Raichle, M.E. (2011). The restless brain. *Brain Connectivity 1*(1), 3–12.

Ravald, A., and Grönroos, C. (1996). The value concept and relationship marketing. *European Journal of Marketing 30*(2), 19–30.

Raz, N., Rodrigue, K.M., Head, D., Kennedy, K.M., and Acker, J.D. (2004). Differential aging of the medial temporal lobe: A study of a five-year change. *Neurology 10*, 433–438.

Reid, D.A., Pullins, E.B., Plank, R.E., and Buehrer, R.E. (2004). Measuring buyers' perceptions of conflict in business-to-business sales interactions. *Journal of Business and Industrial Marketing 19*(4), 236–249.

Reidy, D.E., Shelley-Tremblay, J.F., and Lilienfeld, S.O. (2011). Psychopathy, reactive aggression, and precarious proclamations: A review of behavioral, cognitive, and biological research. *Aggression and Violent Behavior 16*, 512–524.

Reimann, M., Schilke, O., Weber, B., Neuhaus, C., and Zaichkowsky, J. (2011). Functional magnetic resonance imaging in consumer research: A review and application. *Psychology and Marketing 28*(6), 608–637.

Riedl, R., and Javor, A. (2012). The biology of trust: Integrating evidence from genetics, endocrinology, and functional brain imaging. *Journal of Neuroscience, Psychology, and Economics 5*(2), 63–91.

Rose, T., and Manley, K. (2010). Motivational misalignment on an iconic infrastructure project. *Building Research and Information 38*(2), 144–156.

Rosebloom, P.H., Nanda, S.A., Bakshi, V.P., Trentani, A., Newman, S.M., and Kalin, N.H. (2007). Predator threat induces behavioral inhibition, pituitary-adrenal activation and changes in amygdala crf-binding protein gene expression. *Psychoneuroendocrinology, 32*, 44–55.

Roult, R., and Lefebvre, S. (2012). Le parc olympique de Montréal : un espace repensé selon les règles de la régénération urbaine. *Observatoire québécois du Loisir 9*(16), 1–4.

Rousseau, D.M., Sitkin, S.B., Burt, R.S., and Camerer, C. (1998). Not so different after all: A cross-discipline view of trust. *Academy of Management Review 23*(3), 393–404.

Rozin, P., and Royzman, E.B. (2001). Negativity bias, negativity dominance, and contagion. *Personality and Social Psychology Review 5*, 296–320.

Russell, R.F., and Stone, A.G. (2002). A review of servant leadership attributes: Developing a practical model. *Leadership & Organization Development Journal 23*(3), 145–157.

Ryan, G.W., and Bernard, H.R. (1994). Data management and analysis methods. In *Handbook of Qualitative Research*. Edited by N.K. Denzin and Y.S. Lincoln. London: Sage Publications.

Safari, M.S., Haghparast, A., and Semnanian, S. (2009). Effect of lidocaine administration at the nucleus locus coeruleus level on lateral hypothalamus induced antinociception in the rat. *Pharmacology, Biochemistry and Behavior 92*(4), 629–634.

Sama, L.M., and Shoaf, V. (2005). Reconciling rules and principles: An ethics-based approach to corporate governance. *Journal of Business Ethics 58*, 177–185.

Sartorius, K., Merino, A., and Carmichael, T. (2011). Human resource management and cultural diversity: A case study in Mozambique. *International Journal of Human Resource Management 22*(9), 1963–1985.

Sauser, B.J., Reilly, R.R., and Shenhar, A.J. (2009). Why projects fail? How contingency theory can provide new insights: A comparative analysis of NASA's Mars Climate Orbiter loss. *International Journal of Project Management 27*, 665–679.

Sawyer, A.G., and Peter, J.P. (1983). The significance of statistical significance tests in marketing research. *Journal of Marketing Research 20*(2), 122–133.

Scheibe, S., and Carstensen, L.L. (2010). Emotional aging: Recent findings and future trends. *Journals of Gerontology 10*, series b., 1–10.

Scheinkman, J.A., and Xiong, W. (2003). Overconfidence and speculative bubbles. *Journal of Political Economy 111*(6), 1183–1219.

Schneider, C. (2008). Fences and competition in patent races. *International Journal of Industrial Organization 26*(6), 1348–1364.

Schultz, J.T., Moodie, M., Mavoa, H., Utter, J., Snowdon, W., McCabe M.P., and Swinburn, B.A. (2011). Experiences and challenges in implementing complex community-based research project: The Pacific Obesity Prevention in Communities project. *International Association for the Study of Obesity 12*(2), 12–19.

Schweser, K. 2015. *CFA Exam Prep. Level 1, Vol. 4: Corporate Finance, Portfolio Management, and Equity Investments*. USA: Kaplan, Inc.

Senaratne, S., and Udawatta, N. (2013). Managing intragroup conflicts in construction project teams: Case studies in Sri Lanka. *Architectural Engineering and Design Management 9*(3), 158–175.

Seymen, O.A. (2006). The cultural diversity phenomenon in organisations and different approaches for effective cultural diversity management: A literary review. *Cross Cultural Management: An International Journal 13*(4), 296–315.

Seymour, B., Singer, T., and Dolan, R. (2007). The neurobiology of punishment. *Nature reviews. Neuroscience 8*(4), 300–311.

Shapiro, R. (2010). *The Trauma Treatment Handbook: Protocols across the Spectrum*. New York: W.W. Norton.

Sharpe, W.F. (1970). *Portfolio Theory and Capital Markets*. New York: McGraw-Hill.

Shen, L.Y., Tam, V.W.Y., Tam, L., and Ji, Y.B. (2010). Project feasibility study: The key to successful implementation of sustainable and socially responsible construction management practice. *Journal of Cleaner Production 18*, 254–259.

Shenhar, A.J., and Dvir, D. (2007). *Reinventing Project Management: The Diamond Approach to Successful Growth and Innovation*. Boston, MA: Harvard Business Press.

Shiller, R.J. (2005). *Irrational Exuberance*. New York: Crown Publishing Group, a division of Random House.

Shore, B. (2008). Systematic biases and culture in project failures. *Project Management Journal 39*(4), 5–16.

Siegel, A., and Sapru, H.N. (2011). *Essential Neuroscience*, 2nd edn. Baltimore, MD: Wolters Kluwer, Lippincott; Williams and Wilkins.

Sizemon, R. (2008). *The Human Aspect of Project Management*. Boston, MA: Addison-Wesley.

Skulmoski, G.J., and Hartman, F.T. (2010). Information systems project manager soft competencies: A project-phase investigation. *Project Management Journal* 41(1), 61–80.

Smith, J.B. (1998). Buyer-seller relationships: Similarity, relationship management, and quality. *Psychology et Marketing* 15(1), 3–21.

Smith, V.L., Suchanek, G.L., and Williams, A.W. (1988). Bubbles, crashes, and endogenous expectations in experimental spot asset markets. *Econometrica* 56(5), 1119–1151.

Soila-Wadman, M., and Köping, A.S. (2009). Aesthetic relations in place of the lone hero in arts leadership: Examples from film making and orchestral performance. *International Journal of Arts Management* 12(1), 31–43.

Soros, G. (2008). *The New Paradigm for Financial Markets: The Credit Crisis of 2008 and What it Means*. New York: PublicAffairs.

Sporns, O. (2011). *Networks of the Brain*. Cambridge, UK: MIT Press.

Squire, L.R., Bloom, F.E., McConnell, S.K., Roberts, J.L., Spitzer, N.C. Zigmond, M.J. (Eds). (2003). *Fundamental Neuroscience*, 2nd edn. Amsterdam; Boston: Academic Press (Elsevier Science).

Stahl, G.K., Mäkelä, K., Zander, L., Maznevski, ML. (2010). A look at the bright side of multicultural team diversity. *Scandinavian Journal of Management* 26(4), 439–447.

Stanley, E., Fawcett, J.A., Ogden, G., Magnan, M., and Cooper, M.B. (2006). Organizational commitment and governance for supply chain success. *International Journal of Physical Distribution and Logistics Management* 36(1), 22–35.

Starcke, K., and Brand, M. (2012). Decision making under stress: A selective review. *Neuroscience and Biobehavioral Reviews* 36, 1228–1248.

The Stationary Office (TSO). (2009). *Managing Successful Projects with Prince 2*. Belfast, Ireland: TSO.

Steffey, R.W., and Anantatmula, V.S. (2011). International projects proposal analysis: Risk assessment using radial maps. *Project Management Journal* 42(3), 62–74.

Stenstrom, E., and Saad, G. (2011). Testosterone, financial risk-taking, and pathological gambling. *Journal of Neuroscience, Psychology, and Economics* 4(4), 254–266.

Sterman, J.D. (2000). *Business Dynamics: Systems Thinking and Modeling for a Complex World*. Boston, MA: McGraw-Hill.

Stough, S., Eom, S., and Buckenmyer, J. (2000). Virtual teaming: A strategy for moving your organization into the new millennium. *Industrial Management and Data Systems* 100(8), 370–378.

Sudbury, L., and Simcock, P. (2009). A multivariate segmentation model of senior consumers. *Journal of Consumer Marketing* 26, 251–262.

Svensson, G. (2001). Perceived trust towards suppliers and customers in supply chains of the Swedish automotive industry. *Journal of Physical Distribution and Logistics Management* 31(9), 647–662.

Tailby, S., Richardson, M., Stewart, P., Danford, A., and Upchurch, M. (2004). Partnership at work and worker participation: An NHS case study. *Industrial Relations Journal* 35, 403–418.

Tardif, P.M. President VisiSyst. Interviewed April 2016.

Terburg, D., Morgan, B., and van Honk, J. (2009). The testosterone–cortisol ratio: A hormonal marker for proneness to social aggression. *International Journal of Law and Psychiatry* 32(4), 216–223.

Tett, R.P., Steele, J.R., and Beauregard, R.S. (2003). Broad and narrow measures on both sides of the personality-job performance relationship. *Journal of Organizational Behavior 24*(3), 335–356.

Thakurta, R., and Suresh, P. (2012). Impact of HRM policies on quality assurance under requirement volatility. *International Journal of Quality and Reliability Management 29*(2), 194–216,

Therrien, G. Former ACDI Manager. Interviewed Nov. 23, 2015.

Thomas, K.W., and Kilmann, R.H. (1975). The social desirability variable in organizational research: An alternative explanation for reported findings. *Academy of Management Journal 18*(4), 741–752.

Thompson, P.J., and Sanders, S.R. (1998). Partnering continuum. *Journal of Management Engineering 14*(5), 73–78.

Todorov, A., and Engell, A.D. (2008). The role of the amygdala in implicit evaluation of emotionally neutral faces. *Scan 3*, 303–312.

Tomaszczyk, J.C., and Fernandes, M.A. (2013). A positivity effect in older adults' memorability judgments of pictures. *Experimental Aging Research: An International Journal Devoted to the Scientific Study of the Aging Process 39*(3), 254–274.

Toor, S., and Ogunlana, S.O. (2010). Beyond the "iron triangle": Stakeholder perception of key performance indicators (KPIs) for large-scale public sector development projects. *International Journal of Project Management 28*(3), 228–236.

Tuckman, B. (1965). Developmental sequence in small groups. *Psychological Bulletin 63*(6), 384–399.

Tun, P.A., Wingfield, A.R., Merri, J., and Blanchard, L. (1998). Response latencies for false memories: Gist-based processes in normal aging. *Psychology and Aging 13*, 230–241.

Turner, B.L., Matson, P., McCarthy, J.J., Corell, R.W., Christensen, L., Eckley, N., Hovelsrud-Broda, G., et al. (2003). Illustrating the coupled human-environment system for vulnerability analysis: Three case studies. *Proceedings of the National Academy of Sciences of the United States of America 100*(14), 8080–8085.

Turner, R.J. (2004). Five necessary conditions for project success. *International Journal of Project Management 22*(5), 349–350.

Tyler, G.P., and Newcombe, P.A. (2006). Relationship between work performance and personality traits in hong kong organizational settings. *International Journal of Selection and Assessment 14*(1), 37–50.

Ueda, S., Isizuya-Oka, A., Nishimura, A., and Takeuchi, Y. (1999). Hypothalamic aggression area under serotonergic control in mouse-killing behavior of rats. *International Journal of Neuropsychopharmacology 2*, 255–261.

U.S. Department of State. (2001). Interview on Meet the Press. Secretary Colin L. Powell. http://2001-2009.state.gov/secretary/former/powell/remarks/2001/5012.htm. Accessed March 25, 2016.

Van Bruggen, G.H., Lilien, G.L., and Kacker, M. (2002). Informants in organizational marketing research: Why use multiple informants and how to aggregate responses. *Journal of Marketing Research XXXIX*, 469–478.

Van Goozen, S.H.M., Matthys, W., Cohen-Kettenis, P.T., Buitelaar, J.K., and Van Engeland, H. (2000). Hypothalamic-pituitary-adrenal axis and autonomic nervous system activity in disruptive children and matched controls. *Journal of the American Academy of Child and Adolescent Psychiatry 39*(11), 1438–1445.

Van Marrewijk, A. (2007). Managing project culture: The case of Environ megaproject. *International Journal of Project Management 25*(3), 290–299.

Van't Wout, M., and Sanfey, A.G. (2008). Friend of foe: The effect of implicit trust-worthiness judgments in social decision-making. *Cognition 108,* 796–803.

Vartanian, O., Mandel, D.R., and Duncan, M. (2011). Money or life: Behavioral and neural context effects on choice under uncertainty. *Journal of Neuroscience, Psychology, and Economics 4*(1), 25–36.

Verbeke, W.J.M.I., Riedtdjik, W.J.R., van den Berg, W.E., Dietvorst, R.C., Worm, L., and Bagozzi, R.P. (2011). The making of the Machiavellian brain: A structural MRI analysis. *Journal of Neuroscience, Psychology, and Economics 4*(4), 205–216.

Verburg, R.M., Bosch-Sijtsema, P., and Vartiainen, M. (2013). Getting it done: Critical success factors for project managers in virtual work settings. *International Journal of Project Management 31*(1), 68–79.

Vickland, S., and Nieuwenhujis, I. (2005). Critical success factors for modernizing public financial management information systems in Bosnia and Hergzegovina. *Public Administration and Development 25*(2), 95–103.

Vidal, L.A., and Marle, F. (2012). A systems thinking approach for project vulnerability management. *Kybernetes 41*(1/2), 206–228.

Vlachos, P.A., Theotokis, A., Pramatari, K., and Vrechopoulos, A. (2010). Consumer-retailer emotional attachment; Some antecedents and the moderating role of attachment anxiety. *European Journal of Marketing 44*(9/10), 1478–1499.

vom Brocke, J., and Lippe, S. (2015). Managing collaborative research projects: A synthesis of project management literature and directives for future research. *International Journal of Project Management 33,* 1022–1039.

Waldman, D.A., Atwater, L.E., and Davidson, R.A. (2004). The role of individualism and the five-factor model in the prediction of performance in a leaderless group discussion. *Journal of Personality 72*(1), 1–28.

Wang, E., Chou, H.W., and Jiang, J. (2005). The impacts of charismatic leadership style on team cohesiveness and overall performance during ERP implementation. *International Journal of Project Management 23*(3), 173–180.

Wang, J., and Yang, C.Y. (2012). Flexibility planning for managing Rand projects under risk. *International Journal of Production Economics 135,* 823–831.

Wang, M., Keller, C., and Siegrist, M. (2011). The less you know, the more you are afraid of: A survey on risk perceptions of investment products. *Journal of Behavioral Finance 12,* 9–19.

Weidenbaum, M. (1996). The Chinese family business enterprise. *California Management Review 38*(4), 141–156.

Weierich, M.R., Kensinger, E.A., Munnell, A.H., Sass, S.A., Dickerson, B.C., Wright, C.I., and Barrett, L.F. (2011). Older and wiser? An affective science perspective on age-related challenges in financial decision making. *Social Cognitive and Affective Neuroscience 6*(2), 195–206.

Wessinger, K-H. (2011). Identifying powerful project stakeholders using workflow, communication and friendship social networks. MBA dissertation, University of Pretoria, Pretoria. http://upetd.up.ac.za/thesis/available/etd-08112012-192621/. F/12/4/806/zw.

Wieseke, J., Lee, N., Broderick, A.J., Dawson, J.F., and Van Dick, R. (2008). Multilevel analyses in marketing research: Differentiating analytical outcomes. *Journal of Marketing Theory and Practice 16*(4), 321–340.

Williamson, O.E. (1975). *Markets and Hierarchies: Analysis and Anti-Trust Implications.* New York: Free Press.

Wishniewski, J., Windmann, S., Juckel, G., and Brüne, M. (2009). Rules of social exchange: Game theory, individual differences and psychopathology. *Neuroscience and Biobehavioral Reviews 33*, 305–313.

Wood, G., and McDermott, P. (2001). Building on trust: A co-operative approach to construction procurement. *Journal of Construction Procurement 7*(2), 4–14.

Wood, J.A., Boles, J.S., and Babin, B.J. (2008). The formation of buyer's trust of the seller in an initial sales encounter. *Journal of Marketing Theory and Practice 16*(1), 27–39.

Wood, V.F., and Bell, P.A. (2008). Predicting interpersonal conflict resolution styles from personality characteristics. *Personality and Individual Differences 45*(2), 126–131.

Woody, E.Z., and Szechtman, H. (2011). Adaptation to potential threat: The evolution, neurobiology, and psychopathology of the security motivation system. *Neuroscience and Biobehavioral Reviews 35*, 1019–1033.

World Bank. (2010). Urban accessibility/mobility index feasibility stage report. Report no: 69933. http://documents.worldbank.org/curated/en/2010/06/16377387/urban-accessibility-mobility-index-feasibility-stage-report.

www.standishgroup.com/.

www.unisdr.org/we/inform/terminology.

www.worldbank.org/projects.

Xia, W., and Lee, G. (2005). Complexity of information systems development projects: conceptualization and measurement development. *Journal of Management Information System 22*(1), 45–83.

Yang, L.R., Huang, C.-F., and Wua, K.-S. (2011). The association among project manager's leadership style, teamwork and project success. *International Journal of Project Management 29*, 258–267.

Yang, Y., and Raine, A. (2009). Prefrontal structural and functional brain imaging findings in antisocial, violent, and psychopathic individuals: A meta-analysis. *Psychiatry Research: Neuroimaging 174*(2), 81–88.

Yeo, K.T., and Ren, Y. (2009). Risk management capability maturity model for complex product systems (CoPS) projects. *Systems Engineering 12*(4), 275–294.

Youker, R. (1992). Managing the international project environment. *International Journal of Project Management 10*(4), 219–226.

Zebrowitz, L.A., Franklin, R.G., Hillman, S., and Boc, H.M. (2013). Older and younger adults' first impressions from faces: Similar in agreement but different in positivity. *Psychology and Aging 28*(1), 202–212.

Zeng, S.X., Xie, X.M., Tam, C.M., and Sun, P.M. (2009). Identifying cultural difference in R&D project for performance improvement: A field study. *Journal of Business Economics and Management 10*(1), 61–70.

Zimbardo, P. (2007). *The Lucifer Effect: Understanding How Good People Turn Evil.* New York: Random House.

Zwikael, O., and Smyrk, J. (2015). Project governance: Balancing control and trust in dealing with risk. *International Journal of Project Management 33*, 852–862.

Index